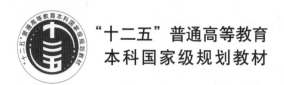

飞行器系统概论
（第2版）

Introduction to Flight Vehicle Systems
(Second Edition)

唐胜景 郭 杰 李 响 徐 瑞 王晓芳 编著

国防工业出版社

·北京·

内 容 简 介

本书系统地介绍了飞行器系统的基本概念和基本原理,主要包括空间环境、力学环境、飞行原理、系统组成、推进系统、外形与结构、制导与控制系统、地面设备与发射方式、有效载荷等。同时,对飞行器设计与研究的基本阶段、研究方法、典型试验及其发展等进行了介绍。本书再版修改、补充了较新的相关内容。

本书可作为高等学校有关专业教材,也可供从事飞行器设计、研究、试验和管理的相关人员使用参考。

The fundamental concepts and principles of flight vehicle systems are systematically introduced in this book, which includes atmospheric environment, mechanics circumstances, flight principles, system compositions, propulsion system, aerodynamic configuration and structures, guidance and control system, ground equipment and launch modes, payloads etc. As well, the basic concepts of flight vehicle design, research and development, research methods, typical experiments and so on are given. The new edition of this book is modified and supplemented with relatively new relevant contents.

The book can be used as the textbook for the related specialties in higher universities, as well as the reference for the relevant personnel engaged in flight vehicle design, research, test and management.

图书在版编目(CIP)数据

飞行器系统概论 = Introduction to Flight Vehicle Systems:英文/唐胜景等编著. —2 版. —北京:国防工业出版社,2021.7
ISBN 978-7-118-12467-5

Ⅰ.①飞… Ⅱ.①唐… Ⅲ.①飞行器–概论–英文
Ⅳ.①V47

中国版本图书馆 CIP 数据核字(2022)第 146802 号

※

国防工业出版社出版发行
(北京市海淀区紫竹院南路23号 邮政编码100048)
北京富博印刷有限公司印刷
新华书店经售

*

开本 787×1092 1/16 印张 25¾ 字数 598 千字
2021 年 7 月第 2 版第 1 次印刷 印数 1—3000 册 定价 108.00 元

(本书如有印装错误,我社负责调换)

国防书店:(010)88540777 书店传真:(010)88540776
发行业务:(010)88540717 发行传真:(010)88540762

第 2 版前言

本书自 2012 年 1 月出版以来,受到了领域内众多专家学者、教师和学生的广泛关注。2013 年本书被评为北京高等教育精品教材,2014 年获批为第二批"十二五"普通高等教育本科国家级规划教材。

本书第 1 版出版时,北京航空航天大学黄海教授、北京理工大学刘莉教授等均对本书提出了中肯的建议。出版后,得到了中国航天科工集团二院、中国工程院于本水院士,国家级教学名师、北京理工大学梅凤翔教授的指导;在使用过程中,我们还得到了同行领域专家学者的指点与帮助,得到了本书读者的关注与支持。特别是,九年多的教学实践过程中,编著者与读者对本书的认知也在不断地提升。

近年来,随着科学技术的快速发展,航空宇航领域也在飞速向前,特别是载人航天、深空探测、火星探测、空间站、火箭回收、人工智能、协同技术等的发展,推动了航空宇航领域科学技术的发展和人类对飞行器系统及其空间领域的认知不断深入。

对此,编著者结合教学科研实践过程中的体会与想法、同行专家学者的指导,以及读者使用过程中的建议,同时结合飞速发展的航空宇航领域的科学技术,着力进行本书的再版编写工作。

此次再版,我们主要在前一版的基础上进行了适量增补,强化飞行器系统知识的系统性、完整性,同时紧跟航空宇航领域科学技术的发展前沿,进一步突出飞行器系统的新理念与新发展,使原书内容更加完善。

本书的再版由唐胜景、郭杰主编;唐胜景、郭杰、李响、徐瑞、王晓芳共同完成。作为编著者,我们着力于教材的修改与完善、教材水平的不断提升。不足之处,诚请批评指正。本书再版过程中,部分研究生做出了重要贡献,对此,我们深表谢意。同时,再版参阅了大量国内外相关文献,对于作者们的辛勤劳动,我们表示诚挚的感谢。

本书再版得到了北京理工大学教务部和宇航学院的大力支持,各级领导、同事与朋友的关注与支持,以及国防工业出版社编辑的指导与帮助。对此,我们一并表示衷心的感谢!

编著者
2021 年 5 月

Preface to the Second Edition

Since the publication of the first edition in January 2012, extensive attention has been received from numerous experts, scholars, teachers and students in the field of aerospace science and technology. This book was rated as one of the Beijing top-quality teaching materials for higher education in 2013 and afterwards enrolled in the second batch of national-level planning textbooks for general higher undergraduate education in China's Twelfth Five-Year Plan in 2014.

When the first edition was created, Prof. Huang Hai from Beihang University and Prof. Liu Li from Beijing Institute of Technology (BIT) all put forward great suggestions for the publication of this book. After initial release, Mr. Yu Benshui, the academician of Chinese Academy of Engineering (ACE) working at the Second Research Institute of China Aerospace Science & Industry Corporation (CASIC), and Prof. Mei Fengxiang (1938—2020), the national level outstanding teacher, working at BIT, had proposed valuable evaluations and comments to the contents. Together with the feedbacks and support from experts, scholars and other readers of this book in their professional fields, and particularly the experience in the course of more than nine-year teaching practices, the comprehension of the author and the learner of this book is continuously improving.

At the same time, with the innovation of scientific technology in recent years, the field of aeronautical and astronautical engineering is also changing rapidly, especially the breakthroughs of manned spaceflight, deep space and Mars exploration, space station construction, rocket recovery, AI, and collaborative technology have promoted the development of aeronautical and astronautical science and technology, as well as human cognition of flight vehicle systems and the space.

Based on this background, with the consideration of the experience and ideas developed in teaching and research practices, the guidance from experts and scholars, and suggestions of the students when using it as the reference book, the authors decide to compile and reprint the second edition of this book.

In this reprint, we have mainly carried on the right amount supplement on the basis of the previous edition, strengthened the systematism and integrity of flight vehicle system knowledge, kept up with the latest development frontiers of aerospace science and technology, further highlighted the new concepts and innovations of flight vehicle system, and so as to make the content on the basis of the first edition more perfect.

Tang Shengjing and Guo Jie act as the editor-in-chief of the new edition. And Tang Shengjing, Guo Jie, Li Xiang, Xu Rui, Wang Xiaofang compile together. As authors, we are focusing on the successive improvement of the teaching material and always pursuing the best quality in the professional field. It is very pleased for us to receive readers' comments and corrections. We are grateful to some of our graduate students for their important contributions to the second edition of this book. Meanwhile, we have referred to a large number of relevant literatures and monographs during the process of the new edition. We would like to express our thanks to the people whose name are listed and not listed in the references for their hard work!

The reprint of this book has acquired wide support from the Department of Academic Affairs of BIT, the School of Aerospace Engineering of BIT, and the editors from National Defense Industry Press, as well as the close attention of many friends. The authors would like to express our heartfelt thanks hereon!

The authors
May, 2021

第 1 版前言

自 1903 年莱特兄弟的"飞行者"首次实现人类有动力连续飞行以来，航空宇航科学与技术成为 20 世纪以来人类取得卓越成果的领域之一。航空器、航天器、火箭与导弹及其他飞行器得以迅速发展。从微小型无人机到大型运输机，从火箭、导弹到运载火箭和远程洲际弹道导弹，从人造卫星到载人空间站和深空探测器，从大气层内到大气层外飞行以及星际航行，人类远古时代的诸多飞天梦想变成了今天的现实。这让我们联想起航空宇航领域著名科学家罗伯特·H. 戈达特 1904 年高中毕业时的一句话，"很难说有什么是不可能的！昨天的梦想是今天的希望和明天的现实。"如今，我们更应牢记他的名言，展开想象的翅膀，不断开拓创新，朝着更高、更快、更远的目标迈进。

基于多年来在飞行器系统与设计及其相关领域研究与教学实践的总结，综合了航空宇航科学与技术领域的国内外最新研究成果和发展动态，我们编写了本书，在内容安排上力求既注意反映本学科的基础理论、新技术和新发展，又注重理论与实践的结合。

本书内容以导弹为主，兼顾飞机与航天器，强化飞行器系统概念，从飞行器系统的角度组织教材的内容结构，并贯穿于本书的各个章节。其次，系统地介绍了飞行器的基本原理，突出基本概念、原理和方法。第三，关注国内外前沿发展动态，内容涵盖有关航空宇航科学与技术发展的新概念和新技术，如高超声速飞行器研究进展、多学科优化设计理念等。

本书可为航空航天类专业的本科生、研究生及相关人员学习飞行器系统基本概念和原理、理解飞行奠定基础；并有助于航空航天类专业双语教学的开展；同时可供航空航天技术爱好者、专业领域的科技人员参考。

全书由唐胜景主编，唐胜景、郭杰、李响、徐瑞、王晓芳共同完成。由于专业知识所限，书中一定存在不足之处，恳请读者批评指正。本书完成过程中，部分研究生做出了贡献，对此，我们表示诚挚的谢意。本书编写时参阅了大量国内外相关文献，对于作者们的辛勤劳动，我们深表感谢。

本书出版得到了北京理工大学教务处和国防工业出版社的大力支持，在此表示衷心感谢！

编著者

Preface to the First Edition

Since the Wright Brothers' Flyer realized the human continuous powered flight for the first time in 1903, aeronautical and astronautical science and technology have become one of the outstanding achievements of the humankind development field since the 20th century. Aircraft, spacecraft, rockets, missiles and other flying vehicles have been developed rapidly. Research fields range from micro UAVs to large transports, from rockets and missiles to carrier rockets and long-range intercontinental ballistic missiles, from artificial satellites to the manned space stations and deep space probes, from in atmosphere to extraatmospheric flight and interstellar travel. Many humankind dreams of flight in ancient times have become a reality today. This reminds us of a quote from Robert H. Goddard, a famous scientist in aerospace, said when he graduated from the high school in 1904, "It is difficult to say what is impossible, for the dream of yesterday is the hope of today and the reality of tomorrow." Today, We, engaged in the aerospace industry, should remember his famous word, begin to unfurl our imagination wings, keep on exploring and innovating, and march towards the higher, faster and farther goals.

Based on the summary of research and teaching practice in flight vehicle systems and design and related fields over the years, and integrating the latest research results and developments of aerospace science and technology at home and abroad, authors have made the English version of Introduction to Flight Vehicle Systems. We strive not only to reflect the fundamental theory, new technologies and development, but also focus on the combination theory with practice in the organization of this book.

The main characteristics of the book are as follows. Firstly, the main content of this book is about missiles, in addition to airplanes and space vehicles. As well the system of flight vehicles is emphasized. That is, the structure of this book is organized from the perspective of flight vehicle system and the system concept is throughout all the chapters of the book. Secondly, the fundamental principles of flight vehicle motion are introduced systematically, the essential concepts and methods are highly emphasized. Finally, the trends and developing state of the discipline at home and aboard is also focused on, including the new concepts and technologies with regard to the development of aerospace technology, such as the research progress of hypersonic vehicles and the concept of multidisciplinary design optimization.

This book is suitable for aeronautical and astronautical undergraduates, graduates and relevant personnel to study fundamental concepts and principles of flight vehicle systems, and understand the flight. This book is also beneficial to bilingual teaching of the aeronautical and astronautical science and technology. It can be used as the reference book for enthusiasts, professional scientists and technicians in such field.

Tang Shengjing acts as the editor-in-chief of the book which is compiled by Tang Shengjing, Guo Jie, Li Xiang, Xu Rui, Wang Xiaofang. Due to our limited professional knowledge, there must be some shortcomings in the book. It is very pleased for us to receive readers' comments and corrections. During the whole process to finish this book, some graduate students have contributed their efforts for it. The authors would express sincere thanks for their work. Meanwhile, we have referred to a large number of relevant literature and monographs in the process of completing this book. The authors would like to thank the authors whose names are listed or not listed in the references for their hard work.

The publication of this book has been supported by the Office of Undergraduate Education, Beijing Institute of Technology and guided and helped by the National Defense Industry Press. We would express our sincere thanks to them!

The authors

Contents

Chapter 1　Brief introduction to flight vehicles ⋯ 1

　1.1　Basic concepts ⋯ 1
　1.2　Overview of flight vehicles ⋯ 1
　　　1.2.1　Rockets ⋯ 2
　　　1.2.2　Missiles ⋯ 8
　　　1.2.3　Air vehicles ⋯ 22
　　　1.2.4　Spacecrafts ⋯ 27
　　　1.2.5　Hypersonic flight vehicles ⋯ 34
　1.3　Development history of flight vehicles ⋯ 35
　　　1.3.1　Flight vehicle exploration ⋯ 35
　　　1.3.2　Typical examples of aeronautics & astronautics ⋯ 37
　　　1.3.3　Prominent figures in aerospace ⋯ 40
　Questions ⋯ 44
　Words and phrases ⋯ 44
　References ⋯ 45

Chapter 2　Flight principles of flight vehicles ⋯ 46

　2.1　Rocket motion ⋯ 46
　　　2.1.1　Rocket movement ⋯ 46
　　　2.1.2　Rocket motion equation ⋯ 47
　　　2.1.3　Multistage rocket theory ⋯ 47
　2.2　Flight environment ⋯ 50
　　　2.2.1　Earth atmosphere ⋯ 50
　　　2.2.2　Near space ⋯ 52
　　　2.2.3　Outer space ⋯ 53
　2.3　Fundamental principle of fluid flowing ⋯ 54
　　　2.3.1　Basic concepts & typical parameters ⋯ 54
　　　2.3.2　Fundamental principles of fluid motion ⋯ 62
　　　2.3.3　Applications of the principles ⋯ 65
　　　2.3.4　High-speed flow ⋯ 68
　2.4　Reference system ⋯ 78

  2.4.1 Reference frames and transformations ……………………………… 78
  2.4.2 European-American coordinate system and Soviet Union coordinate system ……… 85
2.5 Flight mechanics description ……………………………………………… 85
  2.5.1 Forces acted on flight vehicles ………………………………… 86
  2.5.2 Moments acted on flight vehicles ……………………………… 91
2.6 Flight vehicle motion equation …………………………………………… 103
  2.6.1 Basic assumptions ………………………………………………… 103
  2.6.2 Fundamental principles …………………………………………… 103
  2.6.3 Flight vehicle motion equation establishment ………………… 103
2.7 Flight performances ………………………………………………………… 113
  2.7.1 Flight mission profile …………………………………………… 113
  2.7.2 Flight envelope …………………………………………………… 114
  2.7.3 Typical flight parameters ………………………………………… 115
2.8 Flight stability & control ………………………………………………… 119
  2.8.1 Stability …………………………………………………………… 119
  2.8.2 Control ……………………………………………………………… 122
2.9 Space vehicles flight principles ………………………………………… 126
  2.9.1 Kepler laws ………………………………………………………… 126
  2.9.2 Orbit equation & universal velocities ………………………… 130
  2.9.3 Satellite orbits …………………………………………………… 132
  2.9.4 Lunar exploration orbit ………………………………………… 135
  2.9.5 Mars exploration orbit …………………………………………… 137
Questions ………………………………………………………………………………… 139
Words and phrases ……………………………………………………………………… 140
References ……………………………………………………………………………… 141

Chapter 3 Flight vehicle system composition …………………………… 142

3.1 Flight vehicle system ……………………………………………………… 142
3.2 Missile system ……………………………………………………………… 142
  3.2.1 Missile system composition ……………………………………… 142
  3.2.2 Typical examples of the missile system ……………………… 144
3.3 Rocket system ……………………………………………………………… 149
  3.3.1 Multiple rocket launcher system ………………………………… 149
  3.3.2 Carrier rocket system …………………………………………… 151
3.4 Air vehicle system ………………………………………………………… 153
  3.4.1 Airplane system …………………………………………………… 153
  3.4.2 UAV system ………………………………………………………… 158
3.5 Spacecraft system …………………………………………………………… 162
  3.5.1 Space engineering system ………………………………………… 162
  3.5.2 Typical examples of the spacecraft system …………………… 164

Questions ··· 175
Words and phrases ··· 176
References ··· 176

Chapter 4 Flight vehicle propulsion system ··· 177

4.1　Aeroengine system ··· 177
　　4.1.1　Aviation piston engines ··· 177
　　4.1.2　Turbojet engines & turbofan engines ··· 180
　　4.1.3　Turboprop engines & turboshaft engines ··· 185
　　4.1.4　Ram-engines ··· 187
4.2　Rocket engine system ··· 188
　　4.2.1　Main performance parameters ··· 188
　　4.2.2　Solid rocket propulsion system ··· 192
　　4.2.3　Liquid rocket propulsion system ··· 194
　　4.2.4　Combination engines ··· 201
4.3　Thrust vector control system ··· 202
　　4.3.1　Classification of thrust vector control system ··· 203
　　4.3.2　Performance and selection of thrust vector control system ··· 205
4.4　New-type propulsion system ··· 205
　　4.4.1　Solar energy propulsion system ··· 205
　　4.4.2　Nuclear energy propulsion system ··· 206
　　4.4.3　Electric propulsion system ··· 207
　　4.4.4　Micro propulsion system ··· 209
Questions ··· 211
Words and phrases ··· 211
References ··· 213

Chapter 5 Flight vehicle aerodynamic configuration ··· 214

5.1　Design requirements for configuration ··· 214
5.2　Flight vehicle configuration & parameters ··· 214
　　5.2.1　Airfoil ··· 215
　　5.2.2　Wing form & tail configuration ··· 216
　　5.2.3　Fuselage ··· 217
5.3　Typical flight vehicle configuration ··· 219
　　5.3.1　Typical missile configuration ··· 219
　　5.3.2　Typical aircraft configuration ··· 223
　　5.3.3　Control surfaces ··· 224
　　5.3.4　High-lift devices ··· 228
5.4　Configuration of hyper-velocity vehicles ··· 230
　　5.4.1　Lifting body configuration ··· 230

    5.4.2 Blended-wing-body ··· 231
    5.4.3 Waverider configuration ·· 232
  5.5 Flight vehicle configuration variation & development ································ 233
    5.5.1 Missile configuration variations ··· 233
    5.5.2 Aircraft configuration variations ·· 235
    5.5.3 Morphing flight vehicle configuration ····································· 243
 Questions ··· 247
 Words and phrases ··· 247
 References ·· 248

Chapter 6 Flight vehicle structures ··· 249

  6.1 Requirements of flight vehicle structure design ······································· 249
    6.1.1 Flight vehicle structure design philosophy ······························· 251
    6.1.2 Basic flight vehicle structure design requirements ····················· 252
    6.1.3 Basic flight vehicle structure design contents ···························· 253
    6.1.4 Flight vehicle structure design methods ·································· 253
  6.2 Wing ··· 254
    6.2.1 Function and loads ·· 254
    6.2.2 Main load-bearing members of a wing ···································· 256
    6.2.3 Load transfer analysis ·· 259
    6.2.4 Wing structure form ·· 260
    6.2.5 Connection ·· 265
  6.3 Body ··· 268
    6.3.1 Function and loads ·· 268
    6.3.2 Main load-bearing members of body ······································ 270
    6.3.3 Body structure form ·· 271
  6.4 Tail ··· 274
    6.4.1 Function and constitution of tail ·· 274
    6.4.2 Basic requirements of tail ··· 275
    6.4.3 Loads of tail ··· 276
  6.5 Flight vehicle laminated composite structures ··· 276
    6.5.1 Characteristics of composites ·· 277
    6.5.2 Application of composite materials to airframe structures ·········· 278
 Questions ··· 280
 Words and phrases ··· 280
 References ·· 282

Chapter 7 Flight vehicle guidance and control system ··································· 283

  7.1 Basic concepts of guidance and control system ······································· 283
    7.1.1 Function and composition ·· 283

  7.1.2 Classification of guidance system ·················· 284
  7.1.3 Basic design requirements ························ 285
 7.2 Guidance and control components ························ 286
  7.2.1 Sensitive components ····························· 286
  7.2.2 Seeker ··· 300
  7.2.3 Star sensor ·· 306
  7.2.4 Actuation ·· 309
 7.3 Guidance system ·· 311
  7.3.1 Autonomous guidance system ··················· 312
  7.3.2 Remote guidance system ························ 318
  7.3.3 Homing guidance system ························ 321
  7.3.4 Combination guidance system ··················· 322
 7.4 Guidance laws ·· 323
  7.4.1 Autonomous guidance law ······················· 323
  7.4.2 Remote guidance law ···························· 326
  7.4.3 Optimal guidance law ···························· 327
  7.4.4 Differential game guidance law ·················· 330
  7.4.5 Cooperative guidance law ······················· 331
 7.5 Control system ··· 333
  7.5.1 Control modes ···································· 333
  7.5.2 Skid-to-turn (STT) and bank-to-turn (BTT) ······ 336
  7.5.3 Autopilot ·· 337
 Questions ·· 341
 Words and phrases ··· 341
 References ··· 342

Chapter 8 Ground equipment and launch ·················· 344

 8.1 Carrier rocket & space vehicle launch ················· 344
  8.1.1 Space launch base ······························· 344
  8.1.2 Space vehicle return & reentry ················· 349
 8.2 Missile launch ·· 350
  8.2.1 Missile launch modes ···························· 350
  8.2.2 Missile launch equipment ························ 352
 8.3 Airport and ground equipment ························· 355
  8.3.1 Airport ·· 355
  8.3.2 Ground equipment ······························· 357
  8.3.3 Weather observation ····························· 358
 Questions ·· 359
 Words and phrases ··· 359
 References ··· 360

Chapter 9 Payloads 361

9.1 Rocket payloads 361
9.2 Missile payloads 363
    9.2.1 Warhead composition 363
    9.2.2 Typical warheads 364
    9.2.3 Fuses 367
9.3 Air vehicle payloads 372
9.4 Space vehicle payloads 373
Questions 376
Words and phrases 377
References 377

Chapter 10 Flight vehicle system research and development 378

10.1 Flight vehicles design concepts 378
    10.1.1 The R&D process 378
    10.1.2 System analysis 379
10.2 Basic stages for R&D 379
    10.2.1 Idea proposition 380
    10.2.2 Proof of tactical and technical requirements 380
    10.2.3 Preliminary design stage 380
    10.2.4 Technical design stage 381
    10.2.5 Trial-manufacture and test stage 381
    10.2.6 Finality of design stage 381
    10.2.7 Manufacture and utilization stage 381
10.3 Basic design requirements 381
    10.3.1 Design requirements for flight vehicle systems 381
    10.3.2 Quality management for flight vehicle system R&D 382
10.4 Design methods 383
10.5 Typical tests 386
    10.5.1 Wind tunnel test 386
    10.5.2 Simulation tests 389
    10.5.3 Flight test 390
    10.5.4 Environmental test 391
    10.5.5 System performance test 391
    10.5.6 Reliability test 392
10.6 Summary 392
Questions 392
Words and phrases 392
References 393

Chapter 1　Brief introduction to flight vehicles

Generally, the man-made machines that fly near the earth or among the planets are called flight vehicles. It is a usual term. In the 20th century, the development of flight vehicles was one of the most effective activities and the great achievements of the humankind. In this chapter, the basic concepts of flight vehicles, the characteristics of different types of flight vehicles, the development history and current status of flight vehicles and related technologies will be introduced.

1.1　Basic concepts

Generally speaking, two terms are often mentioned in aerospace field, which are aeronautics and astronautics. The brief expressions are as follows.

Aeronautics, usually called aviation, means the research, design, development and production of flight vehicles flying in the aerosphere. While this term originally referred solely to the science of operating the aircraft, it has since been expanded to include technology, business and other aspects related to aircraft.

Astronautics implies the research, development, design and construction of space vehicles which fly into the space. Astronautics is the branch of engineering that deals with machines designed to work outside of earth's atmosphere, whether manned or unmanned. In other words, it is the science and technology of space flight.

However, a spacecraft has to fly across the aerosphere when it is launched or returned. Particularly, for a space shuttle, though its main activities occur out of the aerospace, the courses of takeoff and landing are similar to a plane's. So it is accompanied with the characteristics of aviation and spaceflight. It is difficult to clearly distinguish the aeronautics and astronautics on technology, and the aerospace not only means how to develop the aerospace vehicles, but also contains the technology which is involved during the activities.

Just as we know, aerospace science is the high technology. As well it is the development symbol of science and technology in a country. That is to say, it is the synthesized technology, including many fields such as mechanical engineering, material science, electronic science, control science and so on. The development of flight vehicles will bring about the development and applications in different fields, such as military, civil, scientific research etc. In the 21st century, man will have more activities to explore space and utilize space resources.

1.2　Overview of flight vehicles

Generally speaking, flight vehicles can be divided into three categories, as shown in Figure 1.1.

The first is air vehicles which fly in the air. The second is called space vehicles which fly in the space. Rockets and missiles belong to the third category, and they can fly both in the air and space.

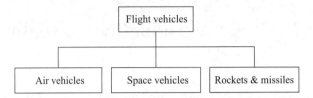

Figure 1.1 Classification of flight vehicles

1.2.1 Rockets

A rocket is the flight vehicle whose propulsion is supplied by the rocket engine. The rocket engine with propellant does not depend on the air or other actuating mediums to produce thrust, working by the principle of acting and reacting force which pushes rocket forward simply by throwing their exhaust backward extremely fast.

The basic compositions of a rocket include propulsion system, payloads and body of the rocket. Propulsion system is the power supply of the rocket. The body loads all the components of the rocket together. Payloads are the main objects carried by the rocket. Payloads of the armament rocket are the warheads, while those of the carrier rocket are a variety of space vehicles. Different kinds of instruments are the payloads of rocket for scientific research.

Some main parts of compositions are same for rockets and missiles. Actually, there are relatively large differences between them. Rockets always have no guidance system, especially in early times, no controlled rockets existed. By means of guidance and control system, missiles, which have been developed since the World War II on the basis of rockets, can hit the targets with high precision.

Sometimes, the terms of rockets and missiles are confused. For example, in Russia, the concepts are same for rockets and missiles. Russian strategically rocket army is in fact strategic missile army. While in the United States, they are different concepts for missiles and rockets. At the same time, with the development of aerospace technologies, the difference between rockets and missiles are getting smaller. For example, some rockets have simplified guidance control system as weapons, but they are still classified as rockets instead of missiles.

1.2.1.1 Classification

Rockets can be divided into chemical rocket, nuclear rocket and electrical rocket according to the different supplied power for engines. Considering the applications, the rockets can be used as fireworks, weapon, launch vehicles for artificial satellites, human spaceflight and space exploration etc., i.e. rocket can be divided into rocket weapon, space sounding rocket and carrier rocket.

1) Rocket weapon

The rocket weapons are used as weapons, the aim of which is to cause damage or harm (either physical or mental) to living beings, artificial structures or systems. Additionally, rocket weapons can also be called rocket armament. Rockets for military and recreational uses date back to at least the 13th century in China.

As mentioned above, a rocket is a self-propelled, unguided weapon system powered by the rocket engine. However, the distinction can become somewhat blurred, especially where a weapon begins as an unguided rocket and is then fitted with a guidance system. For example, some types of rocket are developed with terminal guidance are still called rockets instead of missiles. Moreover, the GMLRS (Guided Multiple Launch Rocket System) is still referred to as rocket artillery, despite employing guided munitions. Rocket artillery is a kind of artillery equipped with rocket launchers instead of conventional guns or mortars.

The rocket weapons'payloads, which are multifarious warheads such as blast warhead, fragmenting warhead, shaped-charge warhead, chemical warhead etc., are the symbolic dissimilarity comparing with other rocket classifications. Figure 1.2 and Figure 1.3 respectively show the uncontrolled rocket and the MLRS (Multiple Launch Rocket System).

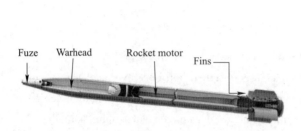

Figure 1.2　Uncontrolled rocket　　　　　　Figure 1.3　Rocket bomb launcher

2) Space sounding rocket

Asounding rocket, sometimes called a research rocket, is an instrument-carrying rocket designed to take measurements and perform scientific experiments during its sub-orbital flight. Sounding rockets take their name from the nautical term "to sound" which means to take measurements of the water's depth by a weight line. Sounding in the rocket context is equivalent to taking a measurement.

A common sounding rocket is basically divided into two parts: the solid fueled rocket motor and the payload. The payload is the section which carries the instruments to conduct the experiment and send the data back to ground station on earth.

The sounding rockets are used to carry instruments from 50km to 1500km above the surface of the Earth, and the altitude generally between weather balloons and satellites. The average flight time is less than 30min, usually between 5min to 20min. Sounding rockets are commonly used for research such as aeronomy, ultraviolet and X-ray astronomy, and microgravity. Sounding rockets are low cost and the payload can be developed as quickly as six months. These rockets allow scientists to conduct investigations at specified times and altitudes. The experiments provide a variety of information on the upper atmosphere, about the Sun, stars, galaxies and other planets. Figure 1.4 shows a typical mission profile of the sounding rocket.

3) Carrier rocket

A carrier rocket (or called launch vehicle) is a rocket used to carry a payload from the Earth's surface into outer space. A carrier rocket system includes the launch vehicle, the launch pad and other infrastructure. Carrier rocket has evolved from the missile and the payloads of carrier rockets

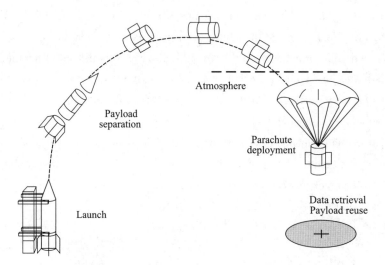

Figure 1.4 Typical mission profile of a sounding rocket

usually include artificial satellite, manned spaceship, space station and space probe.

As expendable launch vehicles, carrier rockets are designed for one-time use. They usually separate from their payload, and may break up during atmospheric reentry. Reusable launch vehicles, on the other hand, are designed to be recovered intact and used again for subsequent launches. For orbital spaceflights, the space shuttle was once the only launch vehicle with an orbiter which has been used for multiple flights. But now the new generation of repeatable launch vehicles represented by SpaceX's Falcon 9 is developing and growing, and has become an important trend in the development of future launch vehicles. In addition, non-rocket space launch alternatives are at the planning stage.

Carrier rockets are often characterized by the amount of mass they can lift into orbit. For example, a Proton rocket has a launch capacity of 22000kg into low earth orbit. Carrier rockets are also characterized by the number of stages they employ. Rockets with as many as five stages have been successfully launched, and there have been designs for several single-stage-to-orbit vehicles. Additionally, launch vehicles are very often supplied with boosters, which supply high thrust early on in the flight, and normally in parallel with other engines on the vehicle. Boosters allow the remaining engines to be smaller, which reduces the burnout mass of later stages, and thus allows for larger payloads.

The primary carrier rocket families in the word include Long March series, Delta series, Soyuz series, Ariane series and Falcon series.

1.2.1.2 Typical rockets

1) Chang Zheng series

Chang Zheng rocket (also called Long March) is the rocket family of expendable launch systems operated by the People's Republic of China. Development and design falls under the auspices of the China Academy of Launch Vehicle Technology. In English, the rockets are abbreviated as LM or CZ.

The Long March is China's primary expendable launch system family, as shown in Figure 1.5.

The Shenzhou spacecrafts and Chang'e 1 lunar orbiter are also launched by the Long March rocket. The maximum payload for LEO (Low earth orbit) is 12000kg (CZ-3B), the maximum payload for GTO (Geostationary transfer orbit) is 5500kg (CZ-3B/E). The development of CZ-2F began in 1992, which is a man-rated version of CZ-2E. Its first launch was in November 19, 1999, and accomplished a successful mission that carrying the first spacecraft of China-Shenzhou 1 into the space. The CZ-5 is a large cryogenic liquid bundled carrier rocket that can launch various spacecraft such as low-Earth orbit satellites, geosynchronous transfer orbit satellites, sun-synchronous orbit satellites, space stations, lunar probes, and Mars probes. On July 23, 2020, the Tianwen-1 was put into the scheduled Earth-Mars transfer orbit. On November 24, 2020, the Chang'e 5 probe was sent into the Earth-Moon transfer orbit. On April 29, 2021, the core module of Chinese space station Tianhe was sent into the scheduled orbit. The CZ-7 launch vehicle is a new type of liquid fuel medium-sized launch vehicle with a "two-and-a-half-stage" configuration. The CZ-7 carrier rocket was successfully launched for the first time from China's Wenchang Space Launch Site on June 25, 2016. On March 12, 2021, China successfully completed the launch of the improvement of CZ-7 Yao-2 carrier rocket at the Wenchang Space Launch Site.

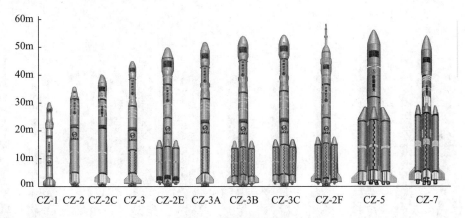

Figure 1.5　CZ series

2) Delta series

Delta, shown in Figure 1.6, is a versatile family of expendable launch systems that has provided space launch capability in the United States since 1960. There have been more than 300 Delta rockets launched, with a 95% success rate. Two Delta launch systems—Delta II and Delta IV—are in active use. Delta rockets are currently manufactured and launched by the United Launch Alliance. Currently development of the Delta rocket series is focused on the Delta IV Heavy, which uses three common booster cores to lift higher masses to orbit and escape velocity.

3) Soyuz series

Vostok (Russian word, translated as "East"), of which the total number of launched rockets are the most in the world, was a family of rockets designed for the human spaceflight program, as shown in Figure 1.7. This famous family of rockets launched the first artificial satellite ("Sputnik"), the first Venus probe, the first Mars probe, the first unmanned expendable freighter spacecraft, and the first manned spacecraft in human history. It was a subset of the R-7 family of rockets. The Soyuz

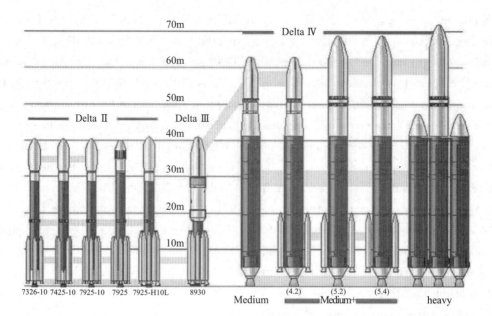

Figure 1.6 Delta series

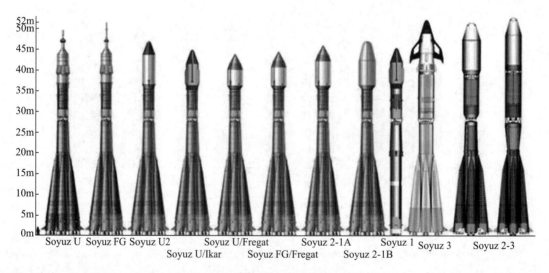

Figure 1.7 Soyuz series

rocket family is the subset of the Vostok rocket family, mainly charged in launching the Soyuz manned spacecraft and the progress resupply spacecraft.

4) Ariane series

Ariane is a series of a European civilian expendable launch vehicles developed by European Space Agency (ESA) for space launching. The name comes from the French spelling of the mythological character Ariadne; the word is also used in French to describe some types of humming bird. The several versions of the launcher include: Ariane 1 ~ Ariane 5. Ariane 4 and Ariane 5 are in active service. The maximum payloads of them for LEO are from 9.4t to 25t, while for GTO are from 4.2t to 7.5t. The maximum payload of the recent version of Ariane 5 for GTO has supposedly increased to 13t ~ 15t. Figure 1.8 shows the Ariane series.

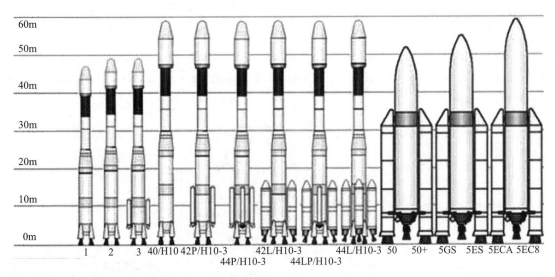

Figure 1.8 Ariane series

5) Falcon series

The successive models of the Falcon Rocket family mainly include Falcon 1, Falcon 5, Falcon 9 and Falcon Heavy, as shown in Figure 1.9. The Falcon 9 rocket is a partially reusable two-stage mid-lift launch vehicle designed and manufactured by SpaceX in the United States. Both the first and second stages are powered by SpaceX Merlin engines, using cryogenic liquid oxygen and rocket-grade kerosene (RP-1) as propellants. Falcon 9 completed its first launch on June 4, 2010, and completed its first recovery on December 21, 2015.

The "Falcon Heavy" rocket, which was successfully launched on February 6, 2018, is the most powerful carrier rocket currently in service. Its low-Earth orbit carrying capacity is 63.8t, and its geosynchronous orbit carrying capacity is 26.7t.

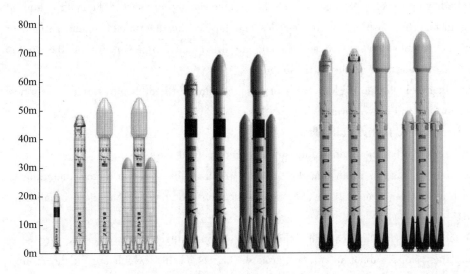

Figure 1.9 Falcon series

1.2.2 Missiles

Usually, a missile contains a warhead, pushed by its own propulsion system and guided by the guidance and control system, which is used to attack the target. Figure 1.10 shows the most common classification of missiles.

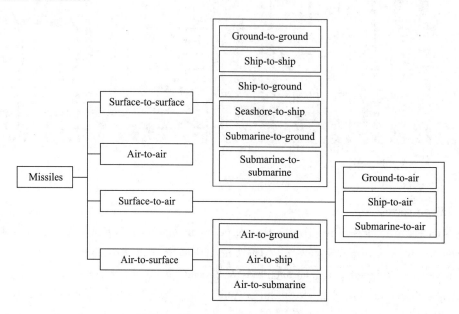

Figure 1.10 Classification of missiles

1.2.2.1 Classification

To define the missile categories, there are different methods which are as follows.

1) Surface & air combination missiles

The following is one of them, which is defined according to the launching and attacking instances. In fact, the positions of the missiles and targets are determined on the surface or in the air. Then, the combinations of the surface and air are made. Surface means on the ground, ship, seashore and submarine. And the air is simple.

According to Figure 1.10, four types of missiles can be divided into, surface-to-surface, air-to-air, surface-to-air and air-to-surface.

2) Strategic & tactical missiles

Furthermore, it can be divided into strategic and tactical missiles.

The strategic missiles mean that the missiles would attack strategic targets including important bases, such as nuclear weapon bases, key communications equipment, headquarters, military airports and so on.

The tactical missiles indicate that the missiles would attack some objective targets, such as, tanks, airplanes, ships, submarines, radar, armored targets, castles, which are on the ground, in the air, at sea and so on.

3) Ballistic & cruise missiles

Based on the trajectory mode, two special types of missiles can be exclusively classified which are ballistic missiles and cruise missiles.

A ballistic missile is a missile that follows a sub-orbital ballistic flight path with the objective of delivering one or more warheads to a predetermined target. The missile is only guided during the relatively brief initial powered phase of flight and its course is subsequently governed by the laws of orbital mechanics and ballistics. A cruise missile is a guided missile designed to deliver a large warhead over long distances with high accuracy. Modern cruise missiles can travel at supersonic or high subsonic speeds, are self-navigating, and can fly on a non-ballistic, extremely low altitude trajectory.

Figure 1.11 shows the flying trajectories of a ballistic missile and a cruise missile. The lower one is the trajectory of a cruise missile and the higher one is the trajectory of a ballistic missile.

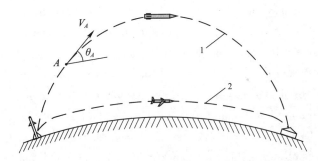

Figure 1.11 Trajectories of a ballistic missile and a cruise missile
1—A ballistic missile;2—A cruise missile.

1.2.2.2 Typical missiles
1) Surface-to-surface missile

A surface-to-surface missile is a guided projectile launched from a hand-held, vehicle mounted, trailer mounted, fixed installation or from a silo, a ship, or even from a submarine. Surface-to-surface missiles are often powered by a rocket engine or sometimes fired by an explosive charge, since the launching platform is typically stationary or moving slowly. They usually have fins or wings for lift and stability, although hyper-velocity or long-range missiles may utilize body lift or fly a ballistic trajectory. The Fieseler Fi 103 (also known as the "V1") was the first surface-to-surface missile.

(1) Ground-to-ground missile

A ground-to-ground missile (GGM) is a missile designed to be launched from the ground or the sea and strike targets on land or at sea. They may be fired from hand-held or vehicle mounted devices, from fixed installations, or from a ship. They are often powered by a rocket engine or sometimes fired by an explosive charge, since the launching platform is typically stationary or moving slowly. They usually have fins and/or wings for lift and stability, although hyper-velocity or short-range missiles may use body lift or fly a ballistic trajectory. The V1 flying bomb was the first operational surface-to-surface missile.

Ⅰ. V2 ballistic missile

For a ballistic missile, rocket engine is adopted, which only works at the beginning of the trajectory. We could only control the missile at this short segment of trajectory. After a few seconds of vertical launching, the missile turns as required procedure. The engine won't stop working until the missile turns to certain angle and its velocity reaches required value, then the warhead separates from the fuselage. This short segment of trajectory is called as powered trajectory. Hereafter, without power and control, warhead and fuselage do inertia flight just like trajectile during the rest of long trajectory, which is called as coasting path.

The word "trajectory" roots in Greek firstly, original intention of it means "slinging". So the missile whose trajectory possesses the character of slinging is named as ballistic missile. In recent time, to increase the accuracy, warhead will be guided after it returns into the aerosphere. This means it is not only a trajectile any more. Ballistic missile has lost its inherent meaning, though we still call it the original name.

Early ballistic missiles all use liquid propellant (for example liquid oxygen and alcohol) which is filled into the missile before launching. Thus, devices to stock and transport propellant must be required. It results in that not only missile ground equipment is bulky and complex but also reaction time is very long. So for the ballistic missiles developed in the late period of 1950s, "prepackage" liquid propellant which could be stocked and solid propellant are adopted. Such missiles are able to be launched any time. Missiles using solid propellant not only are convenient to work but also possess simple structure, preparation time for launching is relatively short as well. In recent time, warhead of ballistic missile has been developed. Main warhead can fall into a great many guided sub-warheads. This is not only favorable for penetration and existence but also increase killing probability. There are a variety of modes to launch ballistic missiles, besides launching from underground well missiles could be launched from submarine, plane and vehicles as well.

Figure 1.12 shows V2 ballistic missile composed of liquid rocket engine with the range of 320km developed by Germany in 1942.

II. Russia "dagger" SS-19

SS-19 is the third-generation intercontinental ballistic missile of USSR. It is designed to replace SS-11. SS-19 is two-stage missile which could stock liquid propellant. Its length, diameter and range are all approximately equivalent to those of SS-11. Throw weight is increased by a factor 4, six guided multiple warheads (each equals to 55 trillion TNT) are carried in it. Inertia guidance system and third generation electron devices are all adopted in it. Its precision could reach 400m.

There are three types of SS-19. SS-19 I carries four or six guided multiple warheads. Single warhead is adopted by SS-19 II developed from SS-19 I. It has high precision and large power. High precision terminal guidance system and guided multiple warheads are taken by SS-19 III.

SS-19 is launched from underground well. Now, over 360 SS-19 have been equipped since 1975. They could be utilized to attack hard targets such as underground wells.

III. America battle-axe cruise missile

There is a pair of big missile wings for the cruise missile whose shape looks like a plane. Aerojet engines are utilized to provide power and work during the whole flight. These missiles are weak in maneuverability, so they are only suitable to attack targets which are fixed or moving slowly. Trajectory has its own characteristics. Let's take a ground-launched cruise missile for an example. Relying on the solid roll booster, the missile takes off from a launcher, it climbs firstly, then turns to level flight. When approaching the target it turns into a dive. Because flight altitude is relative higher

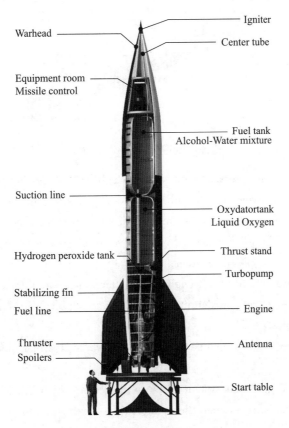

Figure 1.12 V2 ballistic missile

and airspeed is slow, cruise missiles developed in 1950s are easily detected and held up. In 1970s, these missiles are developed to be super miniaturization. Inertial guidance system and map-matching guidance system are adopted to increase the guidance accuracy. The missile could fly into the target area at very low altitude, thus it won't be easily detected by the radar. Project of maneuver variable trajectory enhances its ability of penetration. Now the cruise missile is able to execute the mission of strategic bombardment.

"Battle-axe" is developed by General Dynamics Corporation of USA. It is one of the most advanced cruise missiles. It is subsonic and could be used in any weather. The missile was designed since 1972 and equipped in the army in 1983. There are three types of "battle-axe" cruise missiles: ground-launched cruise missile (GLCM), air-launched cruise missile (ALCM) and sea-launched cruise missile (BGM-109). Figure 1.13 shows the ground-launched type.

No matter which kind the missile is, its configuration, weight, and roll booster are all the same. Warhead, engine and guidance system are the main differences among three kinds of missiles. Their launching ways are different as well. Inertia guidance system, terrain matching guidance system or satellite navigation system could all be adopted in the missile. Its range is between 450km and 2500km, and the velocity could reach to 800km/h.

(2) Anti-ship missile

The missiles are designed to attack to the naval ships. The range covers from dozens of

Figure 1.13 Battle-axe (GLCM)

kilometers to hundreds of kilometers. The missiles can be divided into ship-to-ship missile, seashore-to-ship missile, ship to submarine missile, submarine-to-ship missile, submarine-to-submarine missile and air to ship missile. For the naval ships are weak in maneuverability and their sizes and volumes are relatively large, anti-ship missiles are not required to fly like surface-to-air missiles. There are no too strict requirements are proposed for them.

Including aerojet engine and rocket engine, no matter which one of them is adopted to be as the main engine, solid rocket engine is definitely used as its roll booster. Actually, most of anti-ship missiles are cruise missiles.

The warheads of the missiles include shaped-charge warhead, armor-piercing warhead and blast warhead.

Usually, combination guidance systems such as the combination of automatic guidance system and automatic homing guidance system, the combination of remote guidance system and automatic homing guidance system are adopted in these missiles.

Ⅰ. Russia "Granit"

P-700 (SS-N-19) Granit is a Russian naval anti-ship cruise missile (Figure 1.14). The P-700 has a distinctive annular air intake in the nose. Maximum speed is believed to be between Mach 1.6 to Mach 2.5. Its range is estimated to reach 550km to 625km. The guidance system is mixed-mode, with inertial guidance, terminal active radar homing guidance and also anti-radar homing. Mid-course correction is probable.

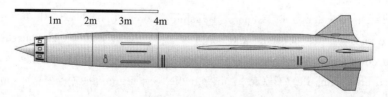

Figure 1.14 P-700 (SS-N-19) anti-ship missile

Ⅱ. China YJ-12

YJ-12 is a Chinese supersonic anti-ship cruise missile (Figure 1.15). It is the first self-developed supersonic anti-ship missile equipped by China.

The YJ-12 anti-ship missile is mainly launched by airborne aircraft, and can also be loaded onto large surface ships. It has the characteristics of high speed, high maneuverability, large range and low-altitude penetration. It is the core strength of Chinese anti-ship missiles and plays an important

Figure 1.15 YJ-12 anti-ship missile

role in defense against aircraft carriers.

(3) Anti-tank missile

The missiles are specially designed to attack the armored targets, especially the tanks. The range covers from dozens of meters to 4 ~ 5km or farther. Usually the missile has two solid rocket engines. The thrust of the takeoff engine is thousands of Newton, and the working hour is less than a second. Although the thrust of the main engine is smaller, its working time lasts to dozens of seconds. Laser guidance system, radar command guidance system and image guidance system and so on are utilized in the missiles.

Figure1.16 shows a typical anti-missile system. As shown in Figure 1.17, for the early antitank missile, shooter uses the telescope to aim the target and observe the missile at the same time. The control command is transmitted to the missile through the lead wire to guide the missile to fly in the line of sight (connection of aiming point and target). Guaranteeing that aiming point, missile and target are all in the same line, thus the missile will hit the target in the end. This method is called three-point guidance method. There is a 2 ~ 3km lead wire on the missile. Because the shooter uses the telescope to aim the target and observe the missile at the same time, velocity of the missile has been limited only around 100 ~ 150m/s.

For the second-generation antitank missile, shooter only needs to aim the target, moving deviation between the missile and target is detected by the infrared goniometer, and the lead command is formed automatically. In this way, the first generation missile is developed from visual tracking to aid tracking. Its velocity has been increased up to 250m/s. The missile is launched from the launching tube which could turn around random, so it could fire the target directly even the tank is only 20m away.

Laser-guided missile is the typical third generation antitank missile. There are two schemes for

Figure 1.16   HJ-73 anti-tank missile system

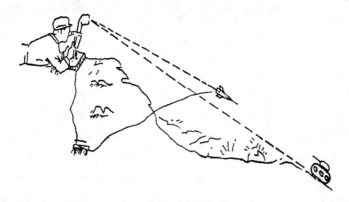

Figure 1.17   Sketch of missile operations

this kind of missile. One scheme is trying to make the missile fly in the laser beam which points to the object directly, the other scheme is letting the receiver on the missile accept the reflected laser from the target, thus the missile will be guided. Omitting lead wire is the obvious merit of laser-guided antitank missile. In this way, velocity will be increased. In addition, imaging guided missiles (thermal imaging guidance and millimeter wave guidance) have also been designed. Such missiles could track targets automatically. Shooter does not need to care about them any more after launching. Such missiles could work at any time. They won't be limited by night and poor weather conditions.

Ⅰ. America Javelin

The FGM-148 Javelin is an American man-portable fire-and-forget anti-tank missile (Figure 1.18). It uses an automatic infrared guidance that allows the user to seek cover immediately after launch. The Javelin's warhead is capable of defeating modern tanks by attacking them from above where their armor is thinnest (known as top-attack).

Ⅱ. China HJ-12

The HJ-12 is a man-portable, fire-and-forget infrared homing anti-tank missile of China

Figure 1.18 Javelin anti-tank missile

(Figure 1.19). The HJ-12 is a fire-and-forget system utilizing Lock-On before Launch (LOBL) and is capable of being fired within buildings and bunkers due to its soft launch system. Once launched, it will home autonomously onto its target, allowing the operator to immediately take cover or reload to engage another target.

Figure 1.19 HJ-12 anti-tank missile

2) Air-to-air missile

The missiles are launched from the airplanes to attack the air targets. They are the main weapons of fighter and bomber.

As the missile is carried by the plane, high velocity and altitude have been got, so the roll booster is cancelled. This is in favor of decreasing the size and weight. Because warhead could not be over size, guidance system is required to have high degree of precision. Remote guidance system could not be used, because the equipment is complex and its guidance precision will fall in company with the increase of range. Infrared homing guidance system and radar homing guidance system are adopted in the missiles.

The missile also has four wings and four rudders, so its flight direction can be changed rapidly. When the target enters into the attacking area, the pilot can get the signal and the missile will be launched as quickly as possible.

Stern-chase attack is applied by the early air-to-air missile. Due to the increasing of the target's velocity, technology of omnidirectional attack has been developed for the missile. To hit the target with high maneuverability, short range air-to-air missile whose least range is only a few hundred meters has been designed. To attack the target far away from the missile, long range missile which is able to fly at any altitude has also been developed. The ability of attacking target at low altitude is especially enhanced.

Air-to-air missiles are broadly divided into two groups. The first consists of missiles designed to engage opposing aircraft at ranges of less than approximately 20miles (32km), these are known as short-range or "within visual range" missiles and are sometimes called "dogfight" missiles because they emphasize agility rather than range. These usually use infrared guidance, and are hence also called heat-seeking missiles. The second group consists of medium—or long-range missiles, which both fall under the category of beyond visual range missiles. Beyond visual range missiles tend to rely upon some sort of radar guidance, in which there are many forms, and modern ones also using inertial guidance or "mid-course updates" technique.

Ⅰ. America "Sidewinder" AIM-9S

This type of missile is a derivative of the type AIM-9M. Subsequently, the missile has been equipped in the air force and produced till now. Main improvements are introduced as follows. For the homing head, its sensitivity and ability of resisting infrared jamming are all improved. Solid rocket engine generates less smoke. Besides these, weight of warhead is increased as well. Structure and performance of AIM-9S are basically the same as those of AIM-9M. One of those types is shown in Figure 1.20.

Figure 1.20 "Sidewinder" AIM-9X air-to-air missile

AIM-9M has a normal configuration measurement in length, diameter and wingspan, of which the values are respectively 2.87m, 0.127m and 0.64m. Its weight of fuselage is 86kg. Its warhead can generate 10.15kg fragment after explosion. AIM-9M adopts one solid rocket engine as its power system, and its range is 0.5~17.7km. Its flight velocity and altitude are respectively Mach 2.5 and

25km that reflect its good flight performance, and which can also be testified by high maneuver overload value of 35. AIM-9M has an active laser fuse and adopts passive infrared homing guidance system.

II. Russia "Archer" R73

The "Archer" R73 is a short-range air-to-air missile developed by Vympel NPO of Russia that entered service in 1984 (Figure 1.21). The R73 is an infrared homing (heat-seeking) missile with a sensitive, cryogenic cooled seeker with a substantial "off-boresight" capability: the seeker can "see" targets up to 40° off the missile's centerline. It can be targeted by a helmet-mounted sight (HMS) allowing pilots to designate targets by looking at them. Minimum engagement range is about 300 meters, with maximum aerodynamic range of nearly 30km at altitude.

Figure 1.21　R73 air-to-air missile

III. China PL-10

The PL-10 is a short-range, infrared-homing air-to-air missile developed by China (Figure 1.22). The missile is fitted with a multi-element IIR seeker capable of +/− 90 degree off-boresight angles. The missile seeker can be slaved to a Helmet Mounted Display (HMD), allowing the pilot to track a target beyond the aircraft's radar scan envelope using the missile's high off-boresight capability, achieved by the pilot turning his head towards the target to lock-on, better known as "look and shoot". Flight is controlled by a thrust-vector controlled solid rocket motor and free-moving type control wings on the missile's tail. The central portion of the missile has long, thin strakes, which help maintain missile maneuverability in the terminal homing stage after the rocket motor stops firing.

3) Surface-to-air missile

This kind of missile is launched from the land or sea surface to attack targets in the air. They belong to antiaircraft weapon, so they are called antiaircraft missile as well. Owing to the targets including airplanes, cruise missiles and ballistic missiles, this kind of missile could be divided into ground (ship) -to-air missile and antiballistic missile.

In the past, both long range bomber and cruise missile fly at high altitude to penetrate. Owing to the enhancing of radar's searching ability, low altitude penetration has to be adopted. Thus low

Figure 1.22 PL-10 air-to-air missile

altitude and minimum altitude ground (ship)-to-air missiles have been designed. Minimum altitude ground-to-air missile is a kind of mini-type field air defense weapon. It is carried by one soldier, launched from the shooter's shoulder to attack invasive hostile aircrafts flying at minimum altitude. Homing guidance system is adopted in this missile so that it has high accuracy. However, owing to its size and relatively smaller weight, sometimes it cannot destroy the target even it hit the object directly.

In order to make surface-to-air missile take-off rapidly and obtain high airspeed, solid rocket engine is used as its roll booster which will shed automatically after running out. Then main engine will work to maintain its velocity. Generally, liquid or solid rocket engine and ramjet engine are introduced to be as the main engine. In recent years, combination of solid rocket engine and ramjet engine are developed which will simplify the missile's structure and decrease its size and weight.

Generally, ship-to-air missile or ground-to-air missile has four wings and four rudders. The wings are placed in " + " or " × " form, so do the rudders. Under the control of guidance system, the missile can change its flight direction rapidly to chase the target.

Antiballistic missile is used to attack the ballistic missile. Because the target's velocity is very high, it requires that the missile has maneuverability, velocity and guidance precision.

Solid booster that provides large thrust is equipped in antiballistic missile. One kind of antiballistic missile uses a three-stage solid rocket engine as its power system. The first stage is its accelerator, the second stage is the main engine, and the third stage will push the missile onto the target accurately. Missile is guided by the radio command sent from launch site. To intercept the object effectively, antiballistic missile system is composed of antiballistic missile, early warning system, identification system, tracking system and interception system.

I. America Patriot Advanced Capability-3

The PAC-3 (Patriot Advanced Capability) is an upgrade of the Patriot air defense missile system (Figure 1.23). A new MIM-104F missile was introduced. It is also referred as the PAC-3 missile. It is a lot smaller than previous Patriot missiles. Also the MIM-104F missile is more maneuverable. It destroys targets by ramming them. Yet still the missile has a small High Explosive Fragmentation (HE-FRAG) warhead to enhance the kill probability. The MIM-104F missile has a range of about 40km and can reach targets at an altitude of up to 20km. The Patriot PAC-3 is a part of three-layer missile defense network. The first line of defense comes from AEGIS missiles, designed to knock-out ballistic missiles in space. If that fails, ballistic missiles are intercepted by THAAD antiballistic missile system, just as they reenter the atmosphere. The final layer of defense is the Patriot PAC-3 missile.

Figure 1.23 PAC-3 air defense missile

II. Russia "Growler" S-400

The S-400 is an anti-aircraft weapon system developed in the 1990s by Russia's Almaz Central Design Bureau as an upgrade of the S-300 family (Figure 1.24). It has been in service with the Russian Armed Forces since 2007.

Figure 1.24 S-400 air defense missile

Ⅲ. China HQ-9

The HQ-9 is a medium, long-range, active radar homing surface-to-air missile (Figure 1.25). The HQ-9 is a two-stage missile. The thrust vector control (TVC) of HQ-9 is the most obvious visual identification. TVC of HQ-9 is exposed and thus can be observed from the side. The HQ-9's guidance system is composed of inertial guidance plus mid-course uplink and active radar terminal guidance systems.

Figure 1.25　HQ-9 air defense missile

4) Air-to-surface missile

The missile is launched from the plane (bomber, fighter or strafer) or helicopter to attack target on the land, sea or under the water. Air-to-surface missiles include air-to-ground missiles, air-to-ship missiles and air-to-submarine missiles. Typical guidance for air-to-surface missiles includes laser guidance, infrared guidance, and television tracking guidance or GPS guidance. The type of guidance depends on the type of target. Ships, for example, may be detected via passive or active radar, while this would not work very well against land targets which typically do not contain such a large mass of metal surrounded by empty space.

Sub-categories of air-to-surface missiles include air-launched antitank missiles, air-launched cruise missiles, air-launched antiship missiles and antiradiation missiles. Guided aerial bomb and aerial torpedo both belong to air-to-surface precision guided munitions. Without engine, aerial bomb is guided by laser, GPS, or others, so it has high precision. No matter on or under the water, aerial torpedo is specially designed to attack ships or submarines.

One of the major advantages of air-to-surface missiles over other weapons available for aircraft to use to attack ground targets is the standoff distance they provide. This allows them to launch the weapons outside the most intense air defenses around the target site. Most air-to-surface missiles are fire-and-forget in order to take most advantage of the standoff distance—they allow the launching platform to turn away after launch. Some missiles have enough range to be launched over the horizon. These missiles (typically either cruise or anti-ship missiles) need to be able to find and home in on the target autonomously.

Typically, the higher and faster the launching aircraft is flying, the further away the missile's target can be. For long range missiles, this difference can be small, but short range missiles (like the AGM-65 Maverick) often dramatically increase in range when launched at altitude.

There have been examples of air-launched ballistic missiles, but they are rare. Sometimes air-to-surface missiles are divided into the categories of tactical and strategic. Typically this indicates conventional explosive or small nuclear warhead (tactical) and large nuclear warhead (strategic).

Ⅰ. America AGM-65 Maverick air-to-ground missile

The AGM-65 Maverick is an air-to-ground missile (AGM) designed for close air support (Figure 1.26). It is the most widely produced precision-guided missile in the Western world, and is effective against a wide range of tactical targets, including armor, air defenses, ships, ground transportation and fuel storage facilities. The AGM-65 measures more than 2.4m in length and 30cm in diameter.

Figure 1.26 AGM-65 Maverick air-to-ground missile

Ⅱ. France "flying fish" AM39

This missile was equipped in army in 1980. In 1982, during the war between Argentina and England, Argentina used a "flying fish" AM39 (Figure 1.27) to destroy the chaser of "Sheffield". In the war between Iraq and Iran, 12 ships of Iran were destroyed by this kind of missiles. The configuration of AM39 reaches 4.69m in length and 0.35m in diameter. The weight of its fuselage is 625kg. A combined guidance system, which consists of inertia guidance and active radar homing guidance, is adopted to upgrade its performance. The attack range of AM39 is 70km.

Figure 1.27 "Flying fish" AM39 air-to-ship missile

1.2.3 Air vehicles

1.2.3.1 Classification

There are several viewpoints for the classification of air vehicles, such as the viewpoint of heavier or lighter than air, the viewpoint of operator (manned or unmanned), etc. In each viewpoint, there are also sub-viewpoints for more detailed classifications. In this book, we will discuss the classification of air vehicles by classifying them into two broad categories, i.e. the air vehicles heavier than air and lighter than air, shown in Figure 1.28.

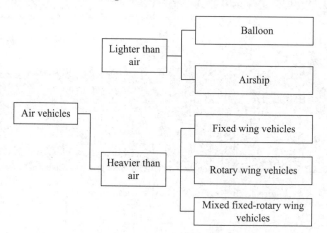

Figure 1.28 Classification of air vehicles

1) Lighter than air

Balloons and airships are lighter-than-air (LTA) air vehicles which use buoyant force generated from the air around it as its lifting force. While the principles of aerodynamics do have some application to balloons and airships, LTA crafts operate principally as a result of aerostatic principles relating to the pressure, temperature and volume of gases. The distinction between a balloon and an airship is that an airship has some means of controlling both its forward motion and steering itself, while balloons are carried along with the wind. In other words, a balloon is an unpowered aerostat, or

LTA craft, whilean airship is a powered LTA craft able to maneuver against the wind.

2) Heavier than air

Heavier-than-air (HTA) vehicles use aerodynamic force generated by moving wing relative to the wind as its lifting force. A heavier-than-air vehicle is usually characterized by a long span object called wings. These so-called wings are designed with a certain cross-section shape called airfoil, causing wings to generate aerodynamic lift force if it moves relative to the wind.

In general, heavier-than-air vehicles can be classified again based on the relative movement between wing and fuselage. The classification consists of fixed-wing airplane, rotary-wing airplane, and mixed-fixed-rotary-wing airplane.

1.2.3.2 Typical air vehicles

1) Balloon

There are only two practical methods of producing a buoyant balloon. If the air inside a suitably large and lightweight envelope is heated to a high temperature, the gas expands and a sufficient amount of fluid (air is a fluid) may be forced out of the interior so that its weight decreases and the total weight of the craft becomes less than the amount of fluid (or air) displaced. The other means of achieving buoyant flight is to fill the envelope with a gas that is sufficiently lighter than air. Figure 1.29 and Figure 1.30 show the outline of balloon and airship respectively.

Figure 1.29   Balloons

Figure 1.30   Airship

2) Airship

Vehicles of commerce and weapons of war, Zeppelin airships, shown in Figure 1.31, offered the first practical means of transporting very heavy loads over very long distances through the air. Airships are traditionally divided into three classes: rigid, semi-rigid and non-rigid. Non-rigid airships, or pressure airships, depend on the internal pressure of the gas in the envelope to maintain their shape. Semi-rigid airships are pressure airships with a rigid keel structure.

3) Fixed wing airplane

Fixed-wing airplane is an airplane with one or more pairs of wings commonly attached symmetrically to the left and right side of its fuselage permanently, thus not allowing wing any relative movement to the fuselage. In order to generate its lift, fixed-wing airplane generates wing's relative movement to the wind by moving the whole airplane forward. As the consequence, a fixed-wing airplane must keep moving forward with a minimum velocity called stall speed so that the airplane's

Figure 1.31 The famous Zeppelin airship

wing can generate enough lift to counteract aircraft's weight and fly. There are several standards to class fixed wing airplanes, for example, usage, types of engines, engines number, flying speed and so on.

I. Airbus A380

The A380 is a double-decker passenger jet that is the Airbus's largest airliner ever built. Lengthwise, it would nearly stretch from goal line to goal line of a football field while its wing tips would hang well beyond the sidelines. Three full decks will run along the entire length of the plane. Upper and main decks will serve as passenger areas, and will be connected by a grand staircase near the front of the plane and by another smaller staircase at the back. Although the lower deck will be reserved primarily for cargo, it could be outfitted for special passenger uses such as sleeper cabins, business centers or even child care service. In a one-class configuration, the A380 could accommodate as many as 840 passengers. The more likely three-class configuration will still offer an unprecedented 555 passenger seats. Either way, the A380 would offer 30% ~ 50% more seating than its direct competition, the Boeing 747-400.

II. F-22 Raptor

The Lockheed Martin F-22 Raptor is a fifth-generation, single-seat, twin-engine, all-weather stealth tactical fighter aircraft. The result of the USAF's Advanced Tactical Fighter (ATF) program, the aircraft was designed primarily as an air superiority fighter, but also has ground attack, electronic warfare, and signal intelligence capabilities. F-22 is also one of the most expensive fighters, costing billions of dollars. Due to the high cost, the production of F-22 has been terminated and the last F-22 was delivered to the USAF in 2012. Figure 1.32 shows civil airbus A380 and military fighter F-22.

4) Rotary wing airplane

Rotary wing airplane is an airplane with a pair of wings and its wing axis commonly attached to the upper section of its fuselage, thus allowing the wing to rotate around its axes and move relative to the fuselage. These rotating wings are commonly called rotor. By rotating its rotor, rotary wing airplane is capable of generating wing's relative movement to the wind without the needs to move the whole

Figure 1.32　Airbus A380 and fighter F-22

airplane. Consequently, rotary wing airplane is capable of vertical take-off and hovering in the air.

Helicopter and rotorcraft all belong to rotary wing airplanes, as shown in Figure 1.33 and Figure 1.34 respectively. The great difference between them is that helicopter's propeller is connected with engines but rotorcraft not, in other words, there is no or just a little torque on rotorcraft's propeller.

Figure 1.33　Helicopter　　　　　　　　Figure 1.34　Rotorcraft

5) Tiltrotor aircraft

Mixed fixed-rotary airplane is an airplane with certain configuration that allows it to act as fixed-wing airplane or rotary wing airplane under certain condition. Among this kind of airplane, MV-22 is a famous one, shown in Figure 1.35.

Figure 1.35　MV-22

6) UAV

An unmanned aerial vehicle (UAV), commonly known as a drone, is an aircraft without a human pilot on board. The most significant feature of drones is unmanned, so its shape and structure are more variable. Figure 1.36 shows the representatives of military and civilian drones.

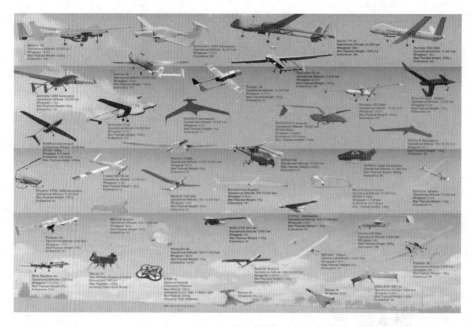

Figure1.36 UAVs

As seen in Figure 1.37, there are many kinds of aircraft, for example, the transport aircraft, the military airplanes, the gliders, the training airplanes, and so on. Different aerodynamic configurations are for different kinds of the airplanes, such as, normal configurations, single wing and dual-wing configurations, and so on.

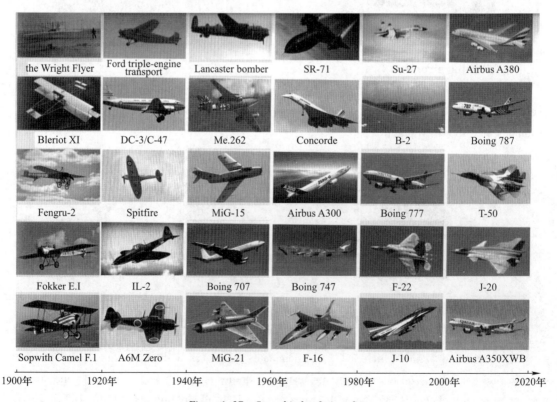

Figure 1.37 Some kinds of aircraft

1.2.4 Spacecrafts

1.2.4.1 Classification

Spacecrafts (or Space vehicles) are vehicles or robotic machines designed for space exploration or other purposes, which enter into a closed orbit around the Earth or around other celestial bodies. Based on whether the spacecraft carries human on board or not, it is divided into two categories as shown in Figure 1.38. Unmanned space vehicles without mankind in it always use telemetry and telecontrol technology to carry out communication and control. Satellites have been used widely in our daily life, such as satellite TV, communications, navigation, weather forecast, agriculture and fisheries investigation. And space probe can be used to explore other celestial bodies, e.g., the moon, the Mars, and others. Manned space vehicles are used for human spaceflight which carry people on board as crew or passengers. Manned spaceships, space stations and space shuttles all belong to this category.

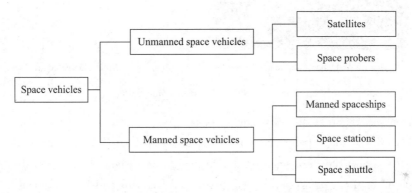

Figure 1.38 Categories of space vehicles

Artificial satellite is the most important unmanned space vehicle. The Sputnik 1, launched by the Soviet Union in 1957, is the first artificial satellite. Since then, thousands of satellites have been sent into orbit around the Earth. About ten countries have the capabilities to launch satellite and more than 50 nations can make their own artificial satellites. A few hundred satellites are currently operational, while thousands of nonoperative satellites and satellite fragments fly around the Earth freely as space debris. A few space probes have been launched up, fly to the other planetary, orbit as artificial satellites, such as around the Moon, Mercury, Venus, Mars, Jupiter, Saturn, etc.

Astronauts need manned spaceship to operate in outer space and then return to the earth. Space station is always been launched in parts and assembled in orbit, which was used for human to work in space for a long time. It also can be used as a mother spacecraft that other spacecraft can dock to it. A space station is another manned spacecraft, which need other vehicles to transport people and supplies to and from the station.

1.2.4.2 Typical space vehicles

1) Satellite

Artificial satellites are unmanned robotic spacecraft which are mostly sent to the Earth orbit for various purposes. Human uses them to observe specific area on the ground, transfer video, obtain navigation information, precast the weather, investigate natural resource, and research space environment.

I. Global Positioning System

The Global Positioning System (GPS), shown in Figure 1.39, is a space-based global navigation satellite system that can give reliable location and time information at anywhere on or near the Earth. Anyone can use a GPS receiver to access the satellite freely and determine your latitude and longitude.

 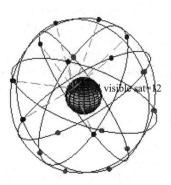

(a) Navigation satellite of GPS　　　　　　　　(b) Orbits and visible satellites

Figure 1.39　Global Positioning System

II. BeiDou Navigation Satellite System

TheBeiDou Navigation Satellite System (BDS) is a global satellite navigation system developed by China, and it is also the third mature satellite navigation system after GPS and GLONASS. The BDS is composed of three parts: space segment, ground segment and user segment. It can provide high-precision, high-reliability positioning, navigation, and timing services for all kinds of users around the world, all-weather and all-day, and it has short message communication capabilities, with preliminary regional navigation, positioning and timing capabilities. The positioning accuracy is decimeter and centimeter level, the speed measurement accuracy is 0.2m/s, and the timing accuracy is 10 nanoseconds. On July 31st, 2020, the BDS-3 was formally commissioned, marking the completion of the "three-step" BDS development strategy. Figure 1.40 shows the satellites and orbits of BDS.

 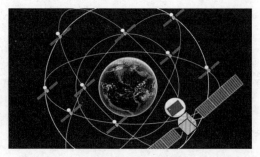

(a) Navigation satellite of BDS　　　　　　　　(b) Orbits and visible satellites

Figure 1.40　BeiDou Navigation Satellite System

2) Spaceship

Spaceship is a spacecraft that transports astronauts and cargo to space and returns safely. Spaceship can be divided into two types: one-time use and reusable. Using a launch vehicle, the spaceship is send into the earth's satellite orbit, after completing the mission, it then returns back into the atmosphere. In addition to the basic system equipment of ordinary artificial satellites, the spaceship also has a life support system, a reentry system for returning to the earth, and a recovery and landing system.

I. Apollo

The Apollo spaceship composes of three parts designed to accomplish the American Apollo program's goal of landing astronauts on the Moon by the end of the 1960s and returning them safely to the Earth. The expendable (one-use) spacecraft consisted of a combined command and service module (CSM) and a lunar module (LM). Two additional components complemented the spacecraft stack for space vehicle assembly: a spacecraft - LM adapter (SLA) designed to shield the LM from the aerodynamic stress of launch and to connect the CSM to the Saturn launch vehicle; and a launch escape system (LES) to carry the crew in the command module safely away from the launch vehicle in the event of a launch emergency. The design was based on the lunar orbit rendezvous approach: two docked spacecraft were sent to the Moon and went into lunar orbit. While the LM separated and landed, the CSM remained in orbit. After the lunar excursion, the two craft rendezvoused and docked in lunar orbit, and the CSM returned the crew to the Earth. The command module was the only part of the space vehicle that returned with the crew to the Earth's surface. Figure 1.41 shows the LM and CSM of Apollo 11.

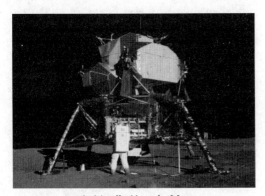

(a) Apollo 11 on the Moon (b) Return capsule

Figure 1.41 Apollo 11 spaceship

II. Dragon

Dragon, shown in Figure 1.42, is a reusable cargo spacecraft developed by SpaceX, an American private space transportation company. Dragon is launched into orbit by the company's Falcon 9 two-stage-to-orbit launch vehicle. Dragon is designed to deliver both cargo and people to orbiting destinations. It is the only spacecraft currently flying that is capable of returning significant amounts of cargo to the Earth. Currently Dragon carries cargo to space, but it was designed from the beginning to carry humans. The first demonstration flight under NASA's Commercial Crew Program

(a) Separation

(b) Docking

Figure 1.42 Dragon spaceship

launched on March 2, 2019. The Dragon spacecraft successfully docked with the space station. On March 3, 2019, becoming the first American spacecraft in history to autonomously dock with the International Space Station.

Ⅲ. Shenzhou

Shenzhou is the first manned space program of China. The first four unmanned test flights launched in 1999, 2001 and 2002. And the manned flight Shenzhou 5 was launched on October 15, 2003 with the CZ-2F from the Jiuquan Satellite Launch Center. Yang Liwei became the first astronaut of China sent to the Earth orbit. Shenzhou 5 enables the PRC as the third country which can send people into space independently.

Shenzhou manned spacecraft consists of orbital module, reentry capsule and service module, as shown in Figure 1.43. To flight independently the orbital module was also equipped with its own propulsion, power system, and control systems.

(a) Shenzhou 5 spaceship

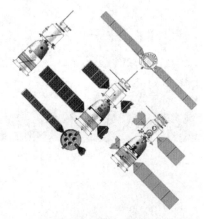
(b) Shenzhou 6 spaceship

Figure 1.43 Shenzhou manned spaceship

Shenzhou 6 was launched on October 12, 2005, which is the second human spaceflight of China. The spacecraft carried a crew of Fei Junlong and Nie Haisheng for five days in low Earth orbit. Shenzhou 7 was the third human spaceflight mission of the Chinese space program, which carried out with crew members Zhai Zhigang, Liu Boming and Jing Haipeng. On September 25, 2008,

Zhai Zhigang, wearing a Chinese-developed space suit, conducted a 22-minute spacewalk.

The Shenzhou 12, or Divine Vessel, was launched on a CZ-2F carrier rocket on June 17, 2021, from the Jiuquan Satellite Launch Center in northwest China's Gobi Desert. It sent the crew to the core module of the planned space station, called Tiangong or Heavenly Palace, which is still under construction in a low-Earth orbit.

3) Space station

A space station (also called an orbital station) is a manned satellite designed to remain in space (or most commonly low Earth orbit) for a long period of time. The biggest difference between space station and other manned spacecraft is that the space station doesn't include major propulsion and landing facilities, and all the astronauts and supplies needed must be launched or taken back by other vehicles.

The International Space Station (ISS), shown in Figure 1.44, is the most famous internationally-developed space station in orbit.

The on-orbit construction of ISS began in 1998 and had been become the largest space station ever constructed. It was expected to be finished in 2012 and serve in orbit until at least 2020. The ISS serves as a space science laboratory which crews can conduct experiments of biology, human biology, physics, astronomy and meteorology in the microgravity environment. The station also can be used for the testing of the spacecraft systems and key technologies that will be required for the future deep space missions. The ISS is operated by astronauts or scientists, and has been an uninterrupted human presence in space for the past 12 years. As of June 2011, the crew of expedition 28 was aboard.

The Chinese Space Station, shown in Figure 1.45, refers to a space station system planned by the People's Republic of China. The construction of a large-scale space station is the third step of China's manned spaceflight strategy. The ultimate goal is to independently build a permanent large-scale space station of 60t to 180t in low-Earth orbit. The Chinese space station will be completed around 2022, with a total of 12 missions planned. On April 29, 2021, the CZ-5B Yao-2 launch vehicle carried the space station Tianhe core module and launched at the Wenchang Space Launch Site in Hainan.

Figure 1.44　The International Space Station

Figure 1.45　Chinese Space Station program

4) Space shuttle

In order for people to be able to travel economically to space, reusable launch vehicles are needed. Space shuttle is usually used repeatedly as a transportation tool, which travels between the Earth and space just like an aircraft. Each space shuttle is made up of three main assemblies: the reusable orbiter vehicle (OV), the expendable external tank (ET), and the two reusable solid rocket boosters (SRB). Only the orbiter stays in orbit shortly after the tank and boosters are discarded. The vehicle is launched vertically, and the orbiter landing to the ground horizontally likes an airplane for reuse.

Ⅰ. America space shuttle

The U.S. Space Shuttle is the world's first reusable space vehicle that travels between the ground and outer space. It consists of an orbiter, an outer tank and a solid booster. Each orbiter can be reused one hundred times, each time a maximum of 29.5t of payload can be sent into a low-Earth orbit of 185km to 1110km, and 14.5t of payload can be brought back to the ground. The space shuttle has a total length of 56.14m and a height of 23.34m. The orbiter can enter a low-inclination orbit or a high-inclination orbit, and can perform rounds, docking, docking, crew and cargo transportation, space testing, satellite launch, maintenance, and recovery tasks.

The first orbiter, the Enterprise, was built in 1976 for approach and landing tests, but it has no orbital capabilities. Four fully operational orbiters were originally built: Columbia, Challenger, Discovery (Figure 1.46), and Atlantis. In 1991, the fifth operational satellite Endeavor was built to replace Challenger. It is one of the four space shuttles built by NASA and one of the space shuttles still in use.

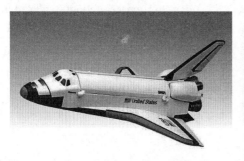

Figure 1.46 The space shuttle "Discovery"

Ⅱ. Soviet space shuttle

Buran was the first space shuttle produced as part of the Soviet Union's Buran program. In addition to describing the first operational Soviet space shuttle, "Buran" is also the title of the entire Soviet space shuttle project and its orbiter, known as the "Buran-class orbiter." The Buran-class orbiter uses the consumable Energia rocket, which is a type of super-heavy launch vehicle.

On November 15, 1988, the Soviet Space Shuttle Blizzard launched from the Baikonur Space Center for the first time and entered a circular orbit 250km above the ground 47min later. It orbited the earth twice, and after three hours in space, it returned safely as scheduled, landed accurately on a concrete runway 12km away from the launch point, and completed an unmanned test flight.

5) Space probe

A space probe is a robotic spacecraft that explores the outer space. A space probe may approach the Moon, travel through interplanetary space, flyby, orbit, or land on other planetary bodies, or enter interstellar space. The space agencies of the USSR/Russia, the United States, the European Union, Japan, China, India, and others have collectively launched probes to several planets and moons of the Solar System, as well as to a number of asteroids and comets.

Lunar probes are robotic space probes thatcan fly by, impact, orbit or land on the Moon for the purpose of lunar exploration.

The Chinese Lunar Exploration Program, also known as theChang'e Project after the Chinese moon goddess Chang'e, is an ongoing series of robotic Moon missions by the China National Space Administration (CNSA). The program incorporates lunar orbiters, landers, rovers and sample return spacecraft, launched using Long March rockets.

The first spacecraft of the program, theChang'e 1 lunar orbiter, was launched from Xichang Satellite Launch Center on October 24, 2007, having been delayed from the initial planned date of April 17-19, 2007. A second orbiter, Chang'e 2, was launched on October 1, 2010. Chang'e 3, which includes a lander and rover, was launched on December 1, 2013 and successfully soft-landed on the Moon on December 14, 2013. Chang'e 4, which includes a lander and rover, was launched on December 7, 2018 and landed on January 3, 2019 on the South Pole-Aitken Basin, on the far side of the Moon. On November 24, 2020, the CZ-5 carrier rocket successfully sent the Chang'e 5 probe into the scheduled orbit. On December 1, the Chang'e 5 landed in the pre-selected landing area on the front of the moon, and then completed a series of tasks such as sampling. On December 17, the Chang'e 5 returner carried lunar samples and landed on Earth. Figure 1.47 shows the mission profile of the Chang'e 5 program.

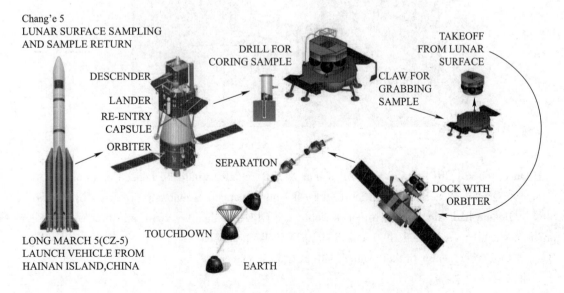

Figure1.47 Chang'e 5 program

1.2.5 Hypersonic flight vehicles

In aerodynamics, a hypersonic speed is defined as a highly supersonic speed. Since the 1970s, the term has generally been assumed to refer to speeds of Mach 5 and above. The hypersonic regime is a subset of the supersonic regime. Generally speaking, hypersonic flight vehicles are these vehicles which can fly at a hypersonic speed. The representative hypersonic flight vehicles are X-37 and X-51 flight test vehicles.

Ⅰ. X-37B

The X-37, also known as the Orbital Test Vehicle, is an unmanned vertical-takeoff, horizontal-landing space plane developed by the Boeing Company. The X-37 began as a NASA (National Aeronautics and Space Administration) project in 1999, and then was transferred to the U.S. Department of Defense in 2004. The latest X-37 project, started on November 17, 2006, is X-37B Orbital Test Vehicle (OTV), of which the actual program builds on industry and government investments by DARPA (Advanced Research Projects Agency), NASA and the Air Force of America. The Boeing Company is the prime contractor for the OTV program.

The X-37B has a typical Lifting body configuration, of which the length, the Height and the wingspan are 8.9m, 2.9m and 4.5m respectively. The size of the X-37B's payload bay is 2.1m × 1.2m and the loaded weight is up to 4,990kg. The electric power of orbital flight consists of Gallium arsenide solar cells with lithium-ion batteries. The X-37B can remain in Low Earth orbit for up to 270 days at a time and with an Orbital speed of 28,200km/h. Figure 1.48 shows the configuration of the X-37B.

Figure 1.48　X-37B

On October 8, 2014, NASA confirmed that X-37B vehicles would be housed at Kennedy Space Center in Orbiter Processing Facilities (OPF) 1 and 2, hangars previously occupied by the Space Shuttle. Boeing had said the space planes would use OPF-1 in January 2014, and the Air Force had previously said it was considering consolidating X-37B operations, housed at Vandenberg Air Force Base in California, nearer to their launch site at Cape Canaveral.

Ⅱ. X-51

The Boeing X-51 is an unmanned scramjet demonstration aircraft for hypersonic flight testing. The X-51 Waverider program is run as a cooperative effort of the United States Air Force,

DARPA, NASA, and Boeing Company. The X-51's General characteristics have rarely been revealed except an approximate length of 7.9m and a rough empty weight of 1,800kg. Figure 1.49 shows the configuration of the X-51.

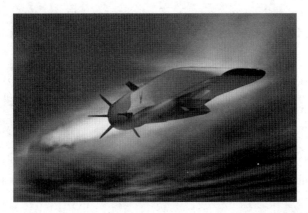

Figure 1.49 X-51

The X-51 completed its first powered flight successfully on May 26, 2010 by flying for over 200s and reaching a speed of Mach 5; however, it did not meet the planned 300s flight duration, The flight had the longest scramjet burn time of 140s.

The second test flight was initially scheduled on March 24, 2011, but was not conducted due to unfavorable test conditions. The flight took place on June 13, 2011. However, the flight over the Pacific Ocean ended early due to an inlet unstart event after being boosted to Mach 5 speed. The flight data from the test is being investigated.

The third test flight took place on August 14, 2012. The X-51 was to make a 300s experimental flight at speeds of Mach 5. After separating from its rocket booster, the craft lost control and crashed into the Pacific. The Air Force Research Laboratory (AFRL) determined the problem was the X-51's upper right aerodynamic fin unlocked during flight and became uncontrollable; all four fins are needed for aerodynamic control. The aircraft lost control before the scramjet engine could ignite.

On May 1, 2013, the X-51 performed its first fully successful flight test on its fourth test flight. The X-51 and booster detached from a B-52H and was powered to Mach 4.8 by the booster rocket. It then separated cleanly from the booster and ignited its own engine. The test aircraft then accelerated to Mach 5.1 and flew for 210s until running out of fuel and plunging into the Pacific Ocean off Point Mugu for over six minutes of total flight time; this test was the longest air-breathing hypersonic flight. Researchers collected telemetry data for 370s of flight. The test signified the completion of the program.

1.3 Development history of flight vehicles

1.3.1 Flight vehicle exploration

1.3.1.1 Rockets and missiles

In Chinese ancient time, a rocket meant an arrow with flame. A. C. 10, a powder rocket made by means of the principles of acting force and opposite acting force, in China, was the first all over the world.

Missiles were from the rockets since1936 in Germany. V1 and V2 were the first missiles in the world. Till 1948, Germany had about 48 types of missiles for different purposes. In World War II, USA and Russia got a lot of experts and information from Germany. The research for missiles was developed rapidly in those two countries.

In 1950's, researches on ballistic missile which already drawn much attention became more and more important to many countries in the world. There was a prediction that ballistic missile will ultimately supersede the long distance bomber. America and Soviet Union had developed many different types of medium range ballistic missile, long-range ballistic missile and ocean-spanning ballistic missile respectively. Due to the main technology level then, those long-range ballistic missiles and ocean-spanning ballistic missiles adopted step rocket as their power systems which were mainly liquid rockets.

The ballistic missile carrying nuclear warhead is considered as a kind of powerful offensive weapon with great destructiveness. Since 1960's, in order to improve the nuclear ballistic missile's flexibility, maneuverability, target accuracy and the ability to cope with the antimissile systeminterception, many advanced technologies were developed such as liquid propellant storage, solid propellant, multiple warheads, motive launch, silo launch, etc.

The winged missile, developed since 1950's, is another important kind of tactical missile which has high permissible overload and maneuverability including the air-to-air missiles and ground-to-air missiles. The development of winged missile passed through three phases. Air-to-air missiles in phase 1 mainly have close, medium altitude and chase-tail attack characteristics. Phase 2 developed medium range interception and all-weather attack abilities, while missiles in phase 3 evolved into two parts, one is long range interception type and another is close fight type. The development directions at present are that improving the guidance accuracy, multi-target tracking, omnidirectional launching, etc.

Cruise missile develops rapidly at the present time. Since the middle 1970's, because of the appearance of high efficiency mini-type turbofan engine and development of the mid-course guidance, final guidance and nuclear warhead miniaturization, numbers of cruise missiles, including air launching type , ground launching type and submarine launching type, have been developed with advantaged characteristics such as low-altitude penetration, ultra-low-altitude penetration, high target accuracy, etc.

1.3.1.2 Airplanes

Explorationfor airplanes since long time ago, man simulated the birds that fly in the sky in ancient times. The imitation is just as the bionics we called today. During B. C. 476-B. C. 221 (the Warring States Period), wooden magpie was made by Gongshu Ban who lived in B. C. 507-B. C. 444. In A. D. 25-220 (the Eastern Han Dynasty), wooden birds were developed by Zhang Heng who lived in A. D. 78-A. D. 139.

InItaly, Leonardo da Vinci (达·芬奇, 1452—1519), who observed the bird flight, wrote the article *Sul Volo degli Uccelli* in 1505; at the same time, he gave the drawings of flapping-wing air vehicle which used man's effort to shake down, though it was difficult to realize at that time. In

France, de Montgolfier manufactured the hot air ball that was a flight vehicle less than the air in 1783.

1.3.1.3 Space vehicles

Humans have always wondered about themysterious objects in the sky. Ancient astronomers observed moving stars and called these objects planets, meaning wanderers. They also found comets, and meteors or shooting stars apparently falling from the sky. With the development of aerospace technology in the 20th century, it has the capability to send machines and people into outer space.

Space vehicles offer human a better tool to explore the solar system and the rest of the universe, to understand the many objects and phenomena, and the resources for human benefit and attributes of the space environment.

The Cold War between the Soviet Union and the United States activated a space engineering and scientific competition. The "Space Race" began with the launch of Sputnik in 1959 and reached the climax in 1969, when the American astronaut Neil Armstrong land on the Moon. The Space exploration of human beings continues in various forms and for a range of purposes all the time.

Sputnik 1, the first artificial Earth satellite, was launched by the Soviet Union on October 4, 1957.

Yury Gagarin, the first human to go into space, was launched by the Soviet Union on April 12, 1961.

Apollo 11, the first lunar lander, was launched by the United States on July 20, 1969 and sent American astronauts Neil Armstrong and Edwin ("Buzz") to walk on the surface of the Moon. Between July 1969 and December 1972, 12 Americans set foot on the Moon in Apollo missions.

Several hundred men and women have spent more than 438 days in space. From the early 1970s, Soviet Union (Russian from December 1991) and U.S. had launched many space stations and space shuttles for varying periods of human space activity, for example, the Skylab station. With the development of the International Space Station (ISS) in the 1990s, the intent has been to have human living and working in space on a permanent basis. The first ISS crew took up long-duration residence on November 2, 2000.

By now, only the United States has sent automated spacecraft, then human-crewed expeditions, to explore the Moon. Unmanned machines have orbited and landed on Venus and Mars, explored the Sun's environment, comets, and asteroids, and made near fly by exploration for Mercury, Jupiter, Saturn, Uranus and Neptune.

1.3.2 Typical examples of aeronautics & astronautics

The development for flight vehicles hasa very fast pace since 1903, when the first man flight was made. A lot of dreams of mankind have been realized gradually.

1) The first flight of airplane

The Flyer I (or called the Wright Flyer), was made by the Wright brothers. It will definitely be remembered as the first powered man-made aircraft in Aeronautics for ever. The Flyer I flew four times on December 17, 1903 near the Kill Devil Hills, about four miles south of Kitty Hawk, North Carolina, U.S. The U.S. Smithsonian Institution describes the aircraft as "... the first powered, heavier-than-air machine to achieve controlled and sustained flight with a pilot aboard. "

The Flyer was based on the Wrights' experience testing gliders at Kitty Hawk between 1900 and 1902. Their last glider, the 1902 Glider, led directly to the design of the Flyer. The Flyer I has a wing span of 12.3m and a wing area of $47.4m^2$. And its total weight including the pilot is about 360kg.

The Wrights built the aircraft in 1903 using giant spruce wood as their construction material. The wings were designed with a camber. Since they could not find a suitable automobile engine for the task, they commissioned their employee Charlie Taylor to build a new design from scratch. A sprocket chain drive, borrowing from bicycle technology, powered the twin propellers, which were also made by hand. The Flyer was a canard biplane configuration. As with the gliders, the pilot flew lying on his stomach on the lower wing with his head toward the front of the craft in an effort to reduce drag. He steered by moving a cradle attached to his hips. The cradle pulled wires which warped the wings and turned the rudder simultaneously.

The Wright brothers built a movable track to help launch the Flyer. This downhill track would help the aircraft gain enough airspeed to fly. After two attempts to fly this machine, one of which resulted in a minor crash, Orville Wright took the Flyer for a 12s, sustained flight with an approximate range of 36.6m on December 17, 1903. This was the first successful, powered, piloted flight in history.

In 1904, the first flight lasting more than five minutes took place on November 9. The Flyer II was flown by Wilbur Wright. In 1908, passenger flight took a turn for the worse when the first fatal air crash occurred on September 17. Orville Wright was piloting the plane. Orville Wright survived the crash, but his passenger, Signal Corps Lieutenant Thomas Selfridge, did not. The Wright Brothers had been allowing passengers to fly with them since May 14, 1908.

2) The first man-made satellite

Sputnik-1, launched by Soviet Union, was the first artificial satellite to be put into Earth's orbit in the world. It was launched into an elliptical low Earth orbit on October 4, 1957, and was the first in a series of satellites collectively known as the Sputnik program. The Russian word "Sputnik" literally means "co-traveler", "traveling companion" or "satellite", and then becomes the synonym of the artificial satellite.

Sputnik-1 was launched during the International Geophysical Year, at the 5th Tyuratam range, in Kazakh SSR (now at the Baikonur Cosmodrome). The satellite travelled at 29000km/h with a total weight of 83kg, taking 96.2min to complete an orbit, and emitted radio signals at 20.005MHz and 40.002MHz which were monitored by amateur radio operators throughout the world. The signals continued for 22 days until the transmitter batteries ran out on October 26, 1957. Sputnik-1 burned up on January 4, 1958, as it fell from orbit upon reentering Earth's atmosphere, after travelling about 60 million kilometers and spending three months in orbit.

Apart from its value as a technological, Sputnik also helped to identify the upper atmospheric layer's density, through measuring the satellite's orbital changes. It also provided data on radio-signal distribution in the ionosphere. Pressurized nitrogen in the satellite's body provided the first opportunity for meteoroid detection. If a meteoroid penetrated the satellite's outer hull, it would be detected by the temperature data sent back to Earth.

Sputnik-1 also brought political impact to the world. The unanticipated announcement of Sputnik-1's success resulted in the Sputnik crisis in the United States and ignited the Space race within the Cold War. The launch ushered in new political, military, technological, and scientific developments. While the Sputnik launch was a single event, it marked the start of the space age.

3) The first man into the space

Yuri Alekseyevich Gagarin (March 9, 1934—March 27, 1968) was a Soviet pilot and cosmonaut. He was the first human being to journey into outer space, along with the Vostok spacecraft completed an orbit of the Earth on April 12, 1961.

As one of the 20 inductees from the 3400 Soviet air force pilots who were younger than 35 years old, Gagarin was sent to Moscow and began the training in the Soviet cosmonaut training center in March 1960. In the training course, Gagarin was behaving superiorly and became the first Soviet astronaut.

On April 12, 1961, Gagarin took the Vostok 1 spacecraft launched from the Baikonur launch site and set sail. The flight lasted 1 hour 48 minutes in a maximum height of 301 km orbit around the earth and Gagarin safely returned and landed, completed the world's first manned space flight, to fulfill the aspirations of mankind into space. He was driving the Vostok 1 spacecraft to become the world's first manned spacecraft into outer space. During the 108-minute flight, Gagarin was promoted to major from captain.

Gagarin became an international celebrity, and was awarded many medals and honors, including Hero of the Soviet Union, the nation's highest honor. Vostok 1 marked his only spaceflight, but he served as backup crew to the Soyuz 1 mission (which ended in a fatal crash). Gagarin later became deputy training director of the Cosmonaut Training Centre outside Moscow, which was later named after him. Gagarin died in 1968 when a MiG 15 training jet he was piloting crashed.

On April 7, 2011 United Nations General Assembly adopted a resolution declaring April 12 as the International Day of Human Space Flight.

4) The first footprint on the Moon

At July 16, 1969, a huge Saturn 5 rocket carrying the Apollo 11 spacecraft launched from Cape Kennedy of the United States, began a significant journey to achieve the first human landing on the Moon's space. The U.S. astronauts Neil Armstrong, Edwin Aldrin and Michael Collins droved the Apollo 11 spacecraft across a long distance of 380000 km, carrying the dreams of all mankind to set foot on the lunar surface. This goal was first accomplished on July 20, 1969 when astronauts Neil Armstrong and Buzz Aldrin landed, while Michael Collins remained in lunar orbit. This was indeed a little step, but the whole of mankind had taken a great step. They witnessed the realization of a great dream that flying onto the Moon and this step had crossed 5000 years of time. All of this is due to a great science and engineering project, the Apollo program.

The Apollo program was the United States spaceflight effort which landed the first human on the Moon. Conceived during the Eisenhower administration and conducted by the National Aeronautics and Space Administration (NASA), Apollo began in earnest after President John F. Kennedy's 1961 address to Congress declaring his belief in a national goal of "landing a man on the Moon" by the end of the decade in a competition with the Soviet Union for the superiorities in space.

The Apollo program ran from 1961 until 1975, and was America's third orbital human spaceflight program (following Mercury and Gemini). It used Apollo spacecraft and Saturn launch vehicles, which were also used for the Skylab program in 1973—1974. These subsequent programs are thus often considered part of the Apollo program. The Apollo 11 mission firstly achieved the goal of step to the moon, and then five subsequent Apollo missions also landed astronauts on the Moon, the last in December 1972. In these six Apollo spaceflights, 12 men walked on the Moon. These are the only times that humans have landed on another celestial body.

The program was successfully carried out despite two major setbacks. One is the 1967 Apollo 1 launch pad fire that killed three astronauts; the other is an oxygen tank rupture during the Apollo 13 flight in 1970 which disabled the Command Module. Using the Lunar Excursion Module as a "lifeboat", the three crewmen narrowly escaped with their lives, thanks to their skills and the efforts of flight controllers, project engineers, and backup crew members.

Apollo program set major milestones in human spaceflight. It stands alone in sending manned missions beyond low Earth orbit; Apollo 8 was the first manned spacecraft to orbit another celestial body, while Apollo 17 marked the last moonwalk and the last manned mission beyond low Earth orbit. The program spurred advances in many areas of technology incidental to rocketry and manned spaceflight, including avionics, telecommunications, and computers. Apollo also sparked interest in many fields of engineering and left many physical facilities and machines developed for the program as landmarks. Its command modules and other objects and artifacts are displayed throughout the world, notably in the Smithsonian's Air and Space Museums in Washington, D. C. and at NASA's centers in Florida, Texas and Alabama. The Apollo 13 Command Module is housed at the Kansas Cosmosphere and Space Center in Hutchinson, Kansas.

5) First space probe

Luna 1 was the first spacecraft to reach the vicinity of the Earth's Moon, and the first spacecraft to be placed in heliocentric orbit. Intended as an impactor, Luna 1 was launched as part of the Soviet Luna program in 1959, however due to an incorrectly timed upper stage burn during its launch, it missed the Moon, in the process becoming the first spacecraft to leave geocentric orbit.

6) First probe landed on the Moon

Luna 9 was the first probe to soft land on the Moon andtransmits pictures from the lunar surface on February 3, 1966. It was proven that a lunar lander would not sink into a thick layer of dust, as had been feared.

7) Earth's Farthest probe

Voyager 1 is a space probe launched by NASA on September 5, 1977. Voyager 1 is the first spacecraft to reach interstellar space. Having operated for 41 years, 8 months and 9 days as of May 14, 2019, the spacecraft still communicates with the Deep Space Network to receive routine commands and to transmit data to Earth.

1.3.3 Prominent figures in aerospace

1) Johannes Kepler

Johannes Kepler (December 27, 1571—November 15, 1630), shown in Figure 1.50, was a

German mathematician, astronomer and astrologer. Between 1609 and 1619, Kepler developed his three laws of planetary motion in Astronomia Nova and Harmonices Mundi using Tycho Brahe's data. Thanks in part to a telescope he received from Galileo, Kepler also advanced the science of optics. His achievements in astronomy and mathematics shaped our current understanding of the solar system. These works also provided one of the foundations for Isaac Newton's theory of universal gravitation.

Figure 1.50   Johannes Kepler

2) Isaac Newton

SirIsaac Newton PRS (December 25, 1642—March 20, 1727), shown in Figure 1.51, was an English physicist, mathematician, astronomer, natural philosopher, alchemist, and theologian.

Figure 1.51   Isaac Newton

His monograph Philosophiæ Naturalis Principia Mathematica, published in 1687, lays the foundations for most of classical mechanics. In this work, Newton described universal gravitation and the three laws of motion, which dominated the scientific view of the physical universe for the next three centuries. Newton showed that the motions of objects on Earth and of celestial bodies are governed by the same set of natural laws, by demonstrating the consistency between Kepler's laws of planetary motion and his theory of gravitation, thus removing the last doubts about heliocentrism and advancing the Scientific Revolution. The Principia is generally considered to be one of the most important scientific books ever written.

3) Konstantin Eduardovich Tsiolkovsky

KonstantinEduardovich Tsiolkovsky (September 17, 1857—September 19, 1935), shown in Figure 1.52, was an Imperial Russian and Soviet rocket scientist and pioneer of the astronautic theory. Along with the German Hermann Oberth and the American Robert H. Goddard, he is considered to be one of the founding fathers of rocketry and astronautics. During his lifetime he published over 88 works on space travel and related subjects. Among his works are designs for rockets with steering thrusters, multi-stage boosters, space stations, airlocks for exiting a spaceship into the vacuum of space, and closed cycle biological systems to provide food and oxygen for space colonies. Since 1896, Tsiolkovsky systematically studied the theory of motion of jet apparatus. Thoughts on the use of the rocket principle in the cosmos were expressed as early as 1883 by Tsiolkovsky. But a rigorous theory of jet propulsion described them in 1896. His Tsiolkovsky formula, published in 1903, established the accurate relationships between the mass of the rocket and its propellant, the speed of the gas at exit, and rocket speed.

Figure 1.52 Tsiolkovsky

4) The Wright brothers

The Wright brothers, Orville (August 19, 1871—January 30, 1948) and Wilbur (April 16, 1867—May 30, 1912), shown in Figure 1.53, were two Americans credited with inventing and building the world's first successful airplane and making the first controlled, powered and sustained heavier-than-air human flight, on December 17, 1903. In the two years afterward, the brothers developed their flying machine into the first practical fixed-wing aircraft. Although not the first to build and fly experimental aircraft, the Wright brothers were the first to invent aircraft controls that made fixed-wing powered flight possible.

Figure 1.53   The Wright brothers

5) Von Braun

Baron Weiner Magnus Maximilian Von Braun (March 23, 1912—June 16, 1977), shown in Figure 1.54, was born in East Prussiaville, Germany Sitz. He is the German rocket expert. One of the pioneers of the aerospace industry in the 20th century. He was the chief designer of the famous V2 rocket. After World War II, von Braun and his team moved to the United States. He served as the chief designer of NASA's space research and development project and presided over the design of the Apollo 4 launch vehicle Saturn 5. He served as NASA Marshall during the commander-in-chief of the Space Flight Center, the first human landing on the moon was achieved in July 1969.

Figure 1.54   Von Braun

6) Qian Xuesen

Qian Xuesen (December 11, 1911—October 31, 2009), shown in Figure 1.55, was born in Shanghai, graduated from Jiaotong University. He is the world-renowned scientist, aerodynamicist, founder of China's manned spaceflight, member of Chinese Academy of Sciences and Chinese Academy of Engineering. He is also known as the "Father of China's Aerospace", "Father of Chinese Missiles" and "Father of China's Automatic Control". Qian Xuesen was a professor at Massachusetts Institute of Technology and California Institute of Technology. Since Qian Xuesen returned to work in China, the launch of Chinese missiles and atomic bombs has advanced for at least 20 years.

Figure 1.55 Qian Xuesen

Questions

1. Try to describe the similarities and differences between aeronautics and astronautics.
2. Please describe the typical classification of missiles and give an example of each type.
3. What are the differences between ballistic and cruise missiles?

Words and phrases

aerial bomb 航空炸弹	armament 军备,武器
aerial torpedo 航空鱼雷	astronautics 太空航空学
aeronautics 航空学	aviation 飞行,航空,航空学,航空术
aerospace 航空航天	ballistic missile 弹道导弹
America "harpoon" anti-ship missile 美国"鱼叉"通用反舰导弹	blast warhead 爆破战斗部
	bulky 体积庞大
America "Siderwinder" AIM-9S air-to-air missile 美国"响尾蛇"(Sidewinder) AIM-9S 空空导弹	cabin 弹舱、机舱
	carrier rocket 运载火箭
	chaser 驱逐舰
America battle-axe (BGM-109) cruise missile 美国"战斧"(BGM-109) 通用巡航导弹	coasting path 被动段弹道
	cruise missile 巡航式导弹

dive 俯冲	洲际弹道导弹
early warning system 预警系统	shaped-charge warhead 聚能破甲战斗部
fixed wing airplane 固定翼飞机	Sheffield "谢菲尔德"号
fragmenting warhead 破片杀伤战斗部	slinging 抛射
France "flying fish" AM39 air to ship missile 法国"飞鱼"AM39 空舰导弹	sounding rocket 探空火箭
fuselage 弹身、机身	space shuttle 航天飞机
General Dynamics Corporation 通用动力公司	spaceflight 航天,宇宙飞行
identification system 目标识别系统	step rocket 多级火箭
infrared jamming 红外干扰	stern-chase attack 尾追目标进行攻击
intercept 拦截	strafer 强击机
jet engine 喷气发动机	strategic missile 战略导弹
Lockheed Martin Corporation 洛克希德·马丁公司	submarine 潜水艇
payload 有效载荷	tactical missile 战术导弹
penetration 突防	The Boeing Company 波音公司
powered trajectory 主动段弹道	three-stage solid rocket engine 三级固体火箭发动机
radio command 无线电指令	throw weight 投掷重量
ramjet engine 冲压式喷气发动机	torque 扭矩,转矩
roll booster 助推器	tracking system 跟踪系统
rotary wing airplane 旋翼飞机	trajectile 被抛射物
Russia "dagger" SS-19 [RS-18] international ballistic missile 俄罗斯"匕首"SS-19[RS-18]	turbojet engine 涡轮喷气发动机
	wooden magpie 木鹊

References

[1] 谢础,贾玉红. 航空航天技术概论[M]. 北京:北京航空航天大学出版社,2005.
[2] 宋笔锋,谷良贤. 航空航天技术概论[M]. 北京:国防工业出版社,2006.
[3] 何庆芝. 航空航天概论[M]. 北京:北京航空航天大学出版社,2001.
[4] 范剑峰,黄祖蔚. 载人飞船工程概论[M]. 北京:国防工业出版社,2000.
[5] 刘桐林. 世界导弹知识大全[M]. 北京:军事科学出版社,1998.
[6] 钱学森. 星际航行概论[M]. 北京:科学出版社,1963.
[7] 吴国兴. 空间站和航天飞机[M]. 北京:宇航出版社,2003.
[8] 赵育善,吴斌. 导弹引论[M]. 西安:西北工业大学出版社,2002.
[9] 文仲辉. 导弹系统分析与设计[M]. 北京:北京理工大学出版社,1989.
[10] 钱杏芳,林瑞雄,赵亚男. 导弹飞行力学[M]. 北京:北京理工大学出版社,2000.
[11] 王春利. 航空航天推进系统[M]. 北京:北京理工大学出版社,2004.
[12] 罗格. 世界航天器与运载火箭集锦[M]. 北京:宇航出版社,2000.
[13] 伍德科克 G. 空间站和空间平台[M]. 褚桂柏译. 北京:中国科学技术出版社,1990.
[14] 谭维炽,胡金刚. 航天器系统工程[M]. 北京:中国科学技术出版社,2009.

Chapter 2 Flight principles of flight vehicles

Flight is the process by which an object moves either through the atmosphere, or beyond it, by generating lift and propulsive thrust, or aero statically using buoyancy, or by simple ballistic movement.

Man has always wanted to fly. Since the first successful flight by the Wright brothers at Kitty Hawk, North Carolina, aircraft designers have spent much time and effort in developing that first crude flying machine into the modern aircraft of today.

A flight vehicle designer needs to master the flight principles of flight vehicles. This chapter will talk about the basic principles of flight vehicles, such as flight environment, forces and moments acted on a flight vehicle, reference frames, and transformations, flight vehicle motion equations, flight performance, stability, and control, etc.

2.1 Rocket motion

As is known, missiles are derived from rockets. Rockets obtain thrust from the rocket engine. Rocket vehicles are often constructed in tall thin shapes and taken off vertically. Rockets can be classified into single-stage rockets and multistage rockets. At present, there is a limitation in the technique of rockets that a single-stage rocket cannot reach the first universal velocity. As a result, multistage rockets are introduced for intercontinental flight and injection into orbit flight.

2.1.1 Rocket movement

According to Newton's third law, the acting forces and reacting forces are equal, but in opposite directions.

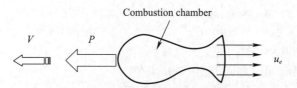

Figure 2.1 Schematic diagram of rocket movement Principle

As shown in Figure 2.1, an incendiary agent burns in the combustion chamber of a rocket, and the combustion gas flow with highspeed is emitted afterward outside, whose speed is u_e. On the other hand, the gas flow reacts on the rocket to form acting and reacting, which produces a thrust P, to push the rocket forward with the speed of V.

2.1.2 Rocket motion equation

At t time, the mass of a rocket is M, the speed of the rocket is V. At $t + dt$ time, the mass is divided into two parts, which are $M - \dfrac{dM}{dt}dt = M - \dot{m}dt$ and $\dot{m}dt$; $\dot{m}$ is the rate of mass flow variations. The speeds of those two parts are $V + \dfrac{dV}{dt}dt$ and $V + \dfrac{dV}{dt}dt + u_e$ respectively.

According to the principle of momentum, the change of momentum is equal to the impulse of external forces.

$$(M - \dot{m}dt)(V + dV) + \dot{m}dt(V + dV + u_e) - MV = \sum F_i dt \qquad (2-1)$$

Simplified,

$$M\frac{dV}{dt} = -\dot{m}u_e + \sum F_i \qquad (2-2)$$

where, $-\dot{m}u_e = \dfrac{dM}{dt}u_e$ is the rate of momentum of gas fluid, whose direction is opposite to u_e. It is called gas reacting force, which is the dynamic power to push the rocket forward. $\sum F_i$ is the total external forces.

Eq. (2-2) is the general equation of motion of a varied mass object.

Furthermore, if $\dfrac{dM}{dt} = 0$, then

$$M\frac{dV}{dt} = \sum F_i \qquad (2-3)$$

It is the equation of motion of a constant-mass object.

2.1.3 Multistage rocket theory

As a simplification, the gravity and air resistance are ignored, and only the rocket engine thrust is considered. In this case, the rocket's flight speed is seen as an ideal speed. The ideal speed is larger than the actual flight speed because the gravity and air resistance will decelerate the actual speed of the rocket.

If the gravity and air resistance are ignored, then

$$M\frac{dV}{dt} = P = -\frac{dM}{dt}u_e \qquad (2-4)$$

The equation can also be written as:

$$dV = -u_e \frac{dM}{M} \qquad (2-5)$$

Supposing that the initial values are, $V = V_0, M = M_0$, the equation of rocket's ideal speed can be written as follows:

$$V = V_0 + u_e \ln \frac{M_0}{M} \qquad (2-6)$$

When the rocket propellantis burnt out, the mass of the rocket is M_K, and the ideal speed reaches the maximum speed, i. e. :

$$V_k = V_0 + u_e \ln \frac{M_0}{M_K} \qquad (2-7)$$

Define the effective mass ratio, $N = \frac{M_0}{M_K} = \frac{M_0}{M_0 - M_P}$, where M_P is the mass of the propellant, and the reciprocal of N, $\mu = M_K/M_0$ is called rocket structure coefficient.

A rocket with the least quality of its structure and the most quality of the propellant will get the maximum ideal speed.

If $V_0 = 0$, then the Tsiolkovsky equation is obtained,

$$V_k = u_e \ln N \qquad (2-8)$$

According to the Tsiolkovsky equation, it is concluded that there are two ways to improve the ideal speed of the rockets. One is to increase the effective gas flow velocity u_e using a high-energy propellant. The other is to increase the effective mass ratio N, by reducing the ratio between rocket structural mass and propellant mass.

The relationship between N and V_k/u_e is shown in Figure 2.2.

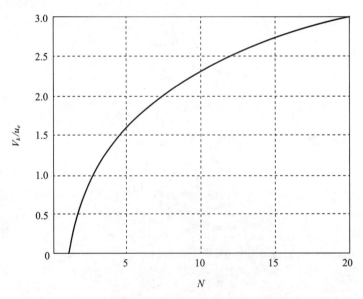

Figure 2.2 Tsiolkovsky equation

As shown in Table 2.1, even when the effective gas flow velocity u_e reaches 2800m/s, and the mass ratio is 10% (the total mass of the rocket is 100t, the mass of propellant is 90t, and the mass of structural is 10t), the maximum ideal speed of the rocket is only 6452m/s, which is less than the first cosmic speed 7900m/s. With the consideration of gravity and air resistance, the speed value will be even lower.

Table 2.1 Ideal rocket speed

$\mu = \dfrac{1}{N}$	Ideal rocket speed $V/$ (m/s)		
	$u_e = 2000$	$u_e = 2400$	$u_e = 2800$
0.9	211	253	295
0.7	715	856	1001
0.5	1385	1661	1940
0.3	2414	2895	3375
0.1	4617	5545	6452

It is clear that using a single-stage rocket for the current space flight is impossible. To solve this problem, Tsiolkovsky proposed the idea of designing a multistage rocket, which is a rocket that uses two or more stages, each of which contains its engines and propellant.

If there are n-stage rockets, the final speed is

$$V_k = V_1 + V_2 + \cdots + V_n = \sum_{i=1}^{n} V_i \quad (2-9)$$

where, V_i is the velocity increment, which is given by the ith-stage rocket, the expression is as follows,

$$V_i = u_{ef}^{(i)} \ln \frac{M_0^{(i)}}{M_K^{(i)}} \quad (2-10)$$

where, $u_e^{(i)}$ is the first i-stage rocket motor combustion gas flow velocity.

$M_0^{(i)}$ is the ignition quality of the i-stage rocket, $M_K^{(i)}$ is the quality after the rocket propellant combustion.

Multistage rocket has three combination modes: serial, parallel, and mix modes. A serial stage is mounted on top of another stage. A parallel stage is attached alongside another stage. In serial stage schemes, the first stage is set at the bottom and is usually the largest one, the second stage and subsequent upper stages are above it, usually decreasing in size. In parallel staging schemes, solid or liquid rocket boosters are used to assist with lift-off. These are sometimes referred to as stage 0. In the typical case, the first stage and booster engines are fired to propel the entire rocket upwards. When the boosters run out of fuel, they are detached from the rest of the rocket (usually with some kind of small explosive charge) and fall away. The first stage then burns up and falls off. This leaves a smaller rocket, with the second stage on the bottom, which is fired afterward. Known in rocketry circles as staging, this process is repeated until the final stage's motor burns to completion.

Once the fuel is ignited, the space and structure that contained it and the motors themselves are useless but only add weight to the vehicle and slow down the acceleration. The rocket lightens itself by dropping the stages which are no longer useful. The thrust of future stages can provide more acceleration than if the earlier stage were still attached, or a single, large rocket would be capable of. When a stage drops off, the rest of the rocket is still traveling near the speed that the whole assembly reached at burn-out time. This means that it needs less total fuel to reach a given velocity and altitude.

The advantage is that each stage can use a different type of rocket motor, each of which is tuned for its particular operating conditions. The lower stage motors are designed for use at atmospheric pressure, while the upper stages can use motors suited to near-vacuum conditions. Lower stages tend to require more structure than upper stages because they need to bear their mass plus that of the stages above them. Optimizing the structure of each stage decreases the mass of the total vehicle and provides further advantages.

On the downside, staging requires the vehicle to lift motors which are not being used until later, as well as making the entire rocket more complex and harder to build. Besides, each staging event is a significant point of failure during a launch, with the possibility of separation failure, ignition failure, and stage collision. To make the flight speed meet requirements, the series of failures should be minimized.

2.2 Flight environment

The flight environment includes both the earth environment and planetary environment where flight vehicles fly.

For the planetary environment, it is near-vacuum, the effects on the flight vehicles are from the sun, solar wind, and others. The main discussions are as follows. The air is around the Earth. Flight vehicles fly mostly in the atmosphere range. The flight is connected with the atmosphere, and the reaction between them is mutual. The forces act on the flight vehicles.

When flight vehicles move in the atmosphere, the flight environment is called the atmospheric environment. It is the only flight operating environment for aerostats, and important for missiles and spacecrafts. As for space environment, it refers to the deep space where spacecrafts fly in.

2.2.1 Earth atmosphere

As shown in Figure 2.3, according to the temperature change with the height away from the Earth, the atmospheric layer around it can be divided into 5 layers, which are troposphere, stratosphere, mesosphere, thermosphere and outer sphere.

Aerostats fly in the troposphere and stratosphere. The atmospheric layers have great impacts on aviation. The temperature, pressure, humidity, wind speed, and direction in different layers will affect flight performances.

2.2.1.1 Troposphere

Near the Earth, the air thickness varies with the season and latitude. In the equator, the height of the troposphere is about 17~18km, in the polar region, it is about 8~9km, in the mid-latitude region, it is about 10~12km.

In this layer, the air mass is about three-fourths of all air mass, and the air density is the largest at sea level. It changes dramatically, including wind speed, air density, temperature, pressure, humidity, and so on.

2.2.1.2 Stratosphere

It starts from the top of the troposphere to about 50~55km away from the sea level. There is no flowing air in the vertical direction, only in the horizontal direction. The mass of air in this layer is about one-fourth of the total air mass.

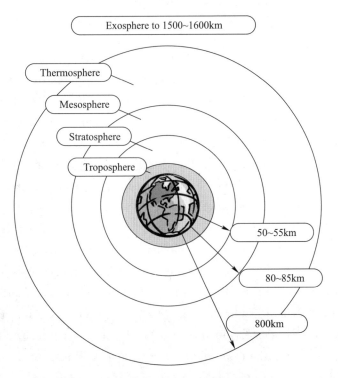

Figure 2.3 Atmospheric layers around the earth

Generally, airplanes fly in the troposphere and stratosphere layers. Airlines and transports are in the stratosphere and cross through the troposphere during the taking-off and landing processes.

2.2.1.3 Mesosphere

It is the portion of the atmosphere from the top of the stratosphere to about 80~85km above the earth's surface. It is characterized by temperatures that decrease from 10℃ to -90℃ with the increase of altitude. Only a little air in this layer, it is about one three-thousand.

It is too high above the Earth for jets and balloons to enter and too low for orbiting spacecraft.

2.2.1.4 Thermosphere

It is distributed from the top of the mesosphere starting to the hot top at an altitude of 500~1000km above sea level. The height of the thermosphere varies considerably due to variations in solar activity. The temperature of the thermosphere gradually increases with height and can be as high as 1500℃.

It is in this layer that many of the satellites orbiting the Earth are located. The International Space Station orbits in this layer, and its operating altitude is between 350km and 420km.

2.2.1.5 Exosphere

It extends from the exosphere (at the top of the thermosphere, at an altitude of about 700km) to about 10000km, where it merges with the solar wind. The exosphere lies too far above the Earth for any meteorological phenomena to occur. However, the Northern and Southern Lights sometimes appear in the lower part of the exosphere, where they overlap with the thermosphere. The exosphere contains many satellites orbiting the Earth.

2.2.2 Near space

Generally speaking, near space is the region of Earth's atmosphere that lies 20km to 100km above sea level, including the stratosphere, mesosphere, and part of the thermosphere. The area aroused widespread concern in the international area because of its significant value for exploitation and utilization. The near space has two distinct sides in the natural environment. On the one hand, the law of gravitation followed by traditional aircraft and Kepler's law of the universe followed by satellite systems are not applicable. In addition, due to extreme harsh conditions such as low temperature and low pressure, thin air and high radiation, this special airspace seems to "prohibit" human visits. On the other hand, due to the rapid development of aerospace science and technology, the near space is "attracting" explorers by virtue of its unique environmental characteristics such as few clouds, rain and lightning, stable air flow and almost constant temperature.

As shown in Figure 2.4, flight vehicles that fly in near space are generally air balloon, airships, UAVs, and hyper-supersonic vehicles. The near space flight vehicle is the craft that can stay in near space for a longtime, sustained flight or suborbital flight vehicle, or hypersonic cruise vehicle that fly in near space. Near space vehicles have the advantages of continuous coverage in the same area and close distance to the target. Therefore, they have unique advantages in regional intelligence collection, surveillance, reconnaissance, communication relay, navigation and electronic warfare. Near space vehicles can continuously monitor and observe key areas for a long time, which is helpful to accurately evaluate the battlefield; It can be used as an electronic jamming and countermeasure platform to implement electronic jamming and countermeasure against incoming aircraft, missiles and other targets, so as to make them deviate from the route or reduce the hit rate; it can be used as a wireless communication relay platform to provide over the horizon communication.

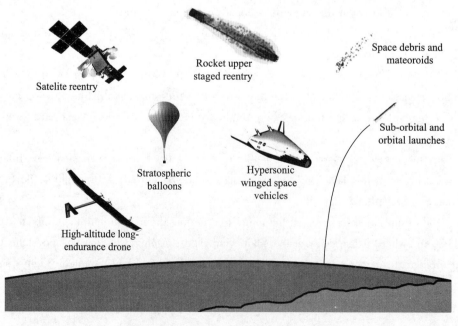

Figure 2.4   Flight vehicles in near space

2.2.3 Outer space

As shown in Figure 2.5, outer space is the vast space that exists beyond the Earth and between celestial bodies, which is formed by vacuum, electromagnetic radiation, the radiation of energetic particles, plasma, and so on.

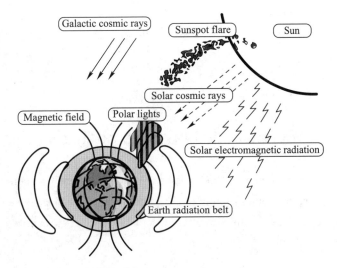

Figure 2.5　Space environment

2.2.3.1 Geospace

The geospace environment is composed of the upper atmosphere environment, ionosphere environment, and magnetic environment. The density and pressure of the upper atmosphere decrease as the exponential with the altitude increases. There is more information shown in Figure 2.6.

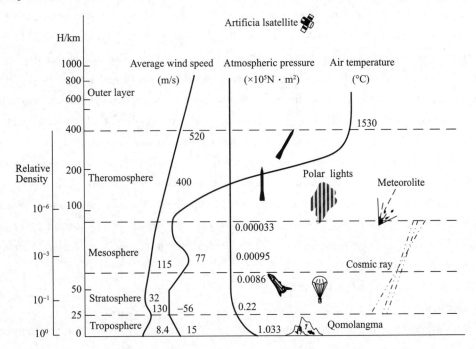

Figure 2.6　Geospace environment

2.2.3.2 Interplanetary space

Interplanetary space is an environment that is very similar to the vacuum environment, where contains electromagnetic radiation, energetic particle radiation, and stable plasma jet. Besides mainly impacted by solar activity, the environment is also affected by micrometeoroids and the cosmic rays from galactic cluster.

2.2.3.3 Interstellar space

Interstellar space is the physical space within a galaxy, beyond the influence of each star on the enclosing plasma.

2.2.3.4 Intergalactic space

Intergalactic space is the physical space between galaxies.

2.3 Fundamental principle of fluid flowing

In the above equation of motion of the rocket, the air resistance is ignored, so it is an ideal description. In order to obtain a more accurate description of the aircraft movement, it is necessary to clarify the interaction force between the aircraft and the air. This section mainly introduces the basic concepts and principles of fluid.

2.3.1 Basic concepts & typical parameters

2.3.1.1 Continuous media hypothesis & fluid micelle

The term "medium" is commonly used in fluid mechanics and aerodynamics to denote the fluids they deal with. Fluids include both liquids and gases, and in this course, it refers mainly to air. The gas is composed of a large number of molecules, each of which is in constant irregular thermal motion, colliding with each other from time to time. The distance traveled by a gas molecule from one collision to the next is called the free range. This distance is negligible compared to the size of the vehicle, and the extent to which the air changes significantly is generally of the same order of magnitude as the size of the vehicle. When the air is disturbed by the aircraft and moves, it does not move as a molecular unit, but must be a large number of air molecules moving together. The performance characteristics exhibited by air will not be the behavior of each molecule, but must exhibit the overall properties. In this premise, aerodynamics adopts the concept of "continuous medium", that is, the medium is seen as a continuous fluid, assuming that the space occupied by the medium is densely packed with this medium everywhere and there are no more voids.

In the analysis, we often take a piece of the very small size of the medium to see its motion or analyze the forces it is subjected to, such a piece of the very small size of the medium we call "micelle". The fluid micelle contains enough molecules, but it is small enough concerning the characteristic size of the object.

2.3.1.2 Basic parameters of fluids

1) **Density**

In the case of a continuous medium, the density at a point P inside the medium can be discussed. As shown in Figure 2.7, we draw a tiny space around point P, let the volume of this space be ΔV, where the mass of the enclosed medium is Δm, Then the average density of the medium in the space is

$$\bar{\rho} = \frac{\Delta m}{\Delta V} \qquad (2-11)$$

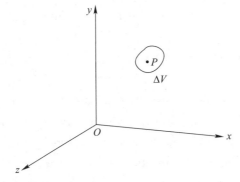

Figure 2.7 Fluid micelle

When $\Delta V \to 0$, $\Delta m/\Delta V$ is defined as the density of the medium at point P, i. e.

$$\rho_P = \lim_{\Delta V \to 0} \frac{\Delta m}{\Delta V} \qquad (2-12)$$

2) Pressure

In inviscid fluid, regardless of whether the fluid is at rest or flowing, the pressure at any point inside the fluid is isotropic, which means the pressure inside the ideal fluid does not vary depending on the orientation of the pressurized surface. Pressure is a function of spatial coordinates. If the flow parameter varies with time, the pressure is still a function of time.

3) Temperature

Temperature refers to the degree of heat and cold of an object, which essentially indicates the intensity of the thermal motion of molecules. For air, temperature indicates the average speed of irregular movement of air molecules.

In daily life, Celsius T_C (℃) and Fahrenheit T_F (℉) are used to be as the units of temperature. In Celsius, the freezing point of water is 0℃ and the boiling point of water is 100℃. In Fahrenheit, they are 32℉ and 212℉, respectively. The conversion formula for the two units is:

$$T_F = \frac{9}{5} T_C + 32 \qquad (2-13)$$

In theoretical calculations, absolute temperature is often used. The temperature at which the molecules stop doing irregular thermal motion, that is, the temperature at which the molecular motion speed is zero, is taken as the zero degree of the absolute temperature. If its scale is the same as Celsius, it is called Kelvin T_K (K). The conversion formula of Kelvin temperature and Celsius temperature is:

$$T_K = T_C + 273.15 \qquad (2-14)$$

2.3.1.3 Gas state equation

The parameters of atmosphere are mainly expressed by pressure, temperature and density.

Density, $\rho = m/v$, v refers to the volume. Temperature is T. Pressure is expressed as p.

The relation among them can be determined by gas state equation. That is,

$$p = \rho RT \qquad (2-15)$$

The state parameters of atmosphere vary with the altitude. They affect the aerodynamic forces and moments. The thrust generated by most aviation engines is also influenced by the state parameters.

2.3.1.4 Typical characteristics of gas

1) Compressibility

When pressure is applied to gas, the volume of the gas changes. The property that the volume or density of a gas with a certain mass change with pressure at a certain temperature is called compressibility. The compressibility of a gas is usually measured by its bulk modulus of elasticity, defined as the increase in pressure required to produce a unit change in relative volume that is $E = -\dfrac{\mathrm{d}p}{\dfrac{\mathrm{d}V}{V}}$. For a certain mass of gas, its volume is inversely proportional to its density $\dfrac{\mathrm{d}\rho}{\rho} = -\dfrac{\mathrm{d}V}{V}$,

thus the bulk modulus of elasticity of the gas can be written as $E = \rho \dfrac{\mathrm{d}p}{\mathrm{d}\rho}$.

At the usual pressure, the bulk modulus of elasticity of air is quite small, about one-twentieth of water. Therefore, the density of air can easily change with the pressure, that is, the air has compressibility. For specific flow problems, whether the compressibility of air should be considered should be based on whether the change in pressure resulting from the flow process has caused a significant change in density. In general, when the airflow rate is low, the change in density caused by the change in pressure is very small, and the compressibility of air can be disregarded.

2) Viscosity

The friction of two neighboring layers in fluid is called viscosity. It is the internal friction, whose expression is using viscosity coefficient μ.

Viscosity of atmosphere is the result of the random motion of the gas molecule. In order to explain it, it is supposed that the flowing atmosphere can be divided into some layers.

It can be observed from Figure 2.8, that velocities in the two layers are different. Due to random motion, the gas molecules in the lower layer whose velocity is larger can intrude the upper layer. It will accelerate the upper layer. Similarly, the gas molecules in the upper layer can also intrude the lower layer. It will decelerate the lower layer. In this way, internal friction is generated between all the neighboring layers.

3) Thermal conductivity

When gas has a temperature gradient along a certain direction, heat is transferred from a higher temperature to a lower temperature; this property is called the thermal conductivity of the gas. Experiments show that the heat transferred per unit time is proportional to the heat transfer area and proportional to the temperature gradient along the direction of heat flow, that is $q = -\lambda \dfrac{\partial T}{\partial n}$. λ is the thermal conductivity parameter. The thermal conductivity of the fluid varies with the fluid

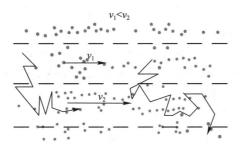

Figure 2.8 Viscosity

medium, and the value of the thermal conductivity of the same fluid medium varies slightly with temperature. Considering that the thermal conductivity of air is very small, the effect of heat transfer from the air on the flow characteristics can be ignored in engineering when the temperature gradient is not large.

2.3.1.5 Basic model of fluid

When studying a specific flow problem, taking into account all the physical properties of the fluid necessarily makes the problem very complex. For some specific problems, it is important to grasp some physical properties that play a dominant role and ignore some physical properties that take a secondary position, so that the problem can be simplified. According to the simplification of different cases of physical properties of real fluids, the following fluid models can be derived.

1) Ideal fluid

This is a model that does not take into account the viscosity of the gas. Because the viscosity coefficient of air is very small, in the actual flow process, only in the range of a very thin layer immediately above the surface of the object, the velocity of each layer of airflow varies greatly, and thus the gradient is large and the viscous force is large. In the region beyond this thin layer, the viscous force is small and the viscosity can usually be ignored because of the slow velocity change between the layers of airflow and the small velocity gradient.

2) Incompressible fluid & Compressible fluid

Incompressible fluid is a model that does not take into account the compressibility or elasticity of the gas. It can be considered to have a bulk elastic modulus of infinity or it has a fluid density equal to a constant. Liquids are very close to this case. When solving the flow law of an incompressible fluid, only the laws of mechanics need to be obeyed without considering thermodynamic relations, thus making the solution of the problem simple and easy. The compressible ideal fluid model is obtained by considering only the effect of the compressibility of the gas, but not the effect of the viscosity of the gas. In this case, the viscosity coefficient of the gas is considered to be equal to zero, while its bulk elastic modulus is not equal to zero.

3) Insulation fluid

This is a model that does not consider the thermal conductivity of the fluid, i.e., it treats the thermal conductivity of the fluid as zero. In general flow, when the temperature gradient is not too large, the heat transfer between gas micelles is minimal, and ignoring the heat transfer between gas micro clusters has little effect on the flow characteristics, so the effect of heat transfer can be disregarded.

2.3.1.6 Speed of sound

If an infinitesimal disturbance occurs in a fluid, the disturbance will propagate through the fluid at a well-defined velocity, which is called the sonic velocity or speed of sound. The speed of sound depends on the properties of the fluid,

$$a^2 = \left(\frac{\mathrm{d}p}{\mathrm{d}\rho}\right) \qquad (2-16)$$

The equation expresses the definition of the speed of sound, which can be further described by other fluid properties. When the bulk modulus K of the fluid is considered, it is derived from the definition of K that

$$\mathrm{d}p = -K\frac{\mathrm{d}v}{v} = K\left(\frac{\mathrm{d}p}{p}\right) \qquad (2-17)$$

And

$$a^2 = \frac{K}{\rho} \qquad (2-18)$$

Furthermore, since the process is adiabatic, the relationship is obtained as follows:

$$pv^k = \text{const.} \qquad (2-19)$$

or

$$p\left(\frac{1}{\rho}\right)^k = \text{const.} \qquad (2-20)$$

where, k is the specific heat ratio defined as $k = c_p/c_v$,

$$\frac{\mathrm{d}p}{\mathrm{d}\rho} = \frac{kp}{\rho} \qquad (2-21)$$

Subsequently, the equation can be written as

$$a^2 = \frac{kp}{\rho} = kRT \qquad (2-22)$$

For the air, giving the properties of air, it is obtained that

$$a = 20\sqrt{T} \qquad (2-23)$$

As is shown above, the speed of sound is related to the propagation medium. Proclaimed by the experiments, the more compressible the media is, the smaller the speed of sound is.

In water, $a = 1440 \text{m/s}$.

In the air, $a = 340.29 \text{m/s}$ (at sea level).

The speed of sound is not a constant in the air. Particularly, it changes with the air temperature and thus with altitude. The air temperature decreases with altitude, so does the speed of sound, but not so quickly.

2.3.1.7 Mach number

Mach number is a very important parameter to express the magnitude of the flight speed of a flight vehicle.

$$Ma = \frac{V}{a} \tag{2-24}$$

It is a dimensionless parameter. The bigger the Mach number is, the bigger the speed of the flight vehicle is. Here, a stands for the local speed of sound.

$0 \leqslant Ma \leqslant 0.3$ low subsonic

$0.3 < Ma \leqslant 0.85$ high subsonic

$0.85 < Ma \leqslant 1.3$ transonic

$1.3 < Ma \leqslant 5.0$ supersonic

$5.0 < Ma$ high supersonic

Low-speed flight is subsonic flight. High-speed flight can be broken into three basic categories: transonic, supersonic, and hypersonic. As their names imply, the categories are related to the speed of sound. As stated before, the Mach number is the airplane's speed in units of the speed of sound. At Mach 1 the plane is going exactly at the speed of sound. Subsonic refers to speeds below Mach 1. Transonic refers to speeds approaching Mach 1. Commercial transports, most military transports, bombers, and business jets fly at high subsonic speeds. Supersonic refers to speeds above Mach 1 and is usually left to fighters and interceptors for short bursts. Hypersonic refers to speeds of high Mach numbers. At present, the only hypersonic vehicle is the Space Shuttle during reentry.

2.3.1.8 Reynolds number

For truly accurate simulation of the full-scale flow, it is necessary to achieve geometric and dynamic similarity. The latter requires the relative magnitudes of the inertia and viscous forces associated with the moving fluid to be modeled correctly and the ratio of those forces is given by a dimensionless parameter known as Reynolds number (Re).

$$Re = \frac{\rho V d}{\mu} \tag{2-25}$$

where, ρ is the fluid (air) density, V is the relative wind speed, d is a characteristic dimension, and μ is the viscosity of the fluid.

2.3.1.9 Standard atmosphere

Why there is the definition of the standard atmosphere?

As we know, the atmosphere around the Earth varies with the season, time, height, and geographical position.

In order to design and research, analyze, perform experiments, and compare, the standard atmospheric state parameters are to be the basis for flight vehicle design.

The contract about the standard atmosphere was founded by the international aviation association.

The revised average values of the atmospheric state at the middle latitude in the northern hemisphere are to be the standard atmosphere. The model of atmospheric parameters was founded, which was called Standard Atmosphere Table. It was defined by the COSPAR (Committee on the space research).

CIRA stands for COSPAR International Reference Atmosphere. The national standard in the former USSR is CA. In 1975, the standard atmosphere was established in China.

Actually, below 30km, the standard atmospheric parameters are identical in different countries. Next is the modeling of the standard atmosphere.

At the sea level,

Height $H = 0$m

Temperature $T = 15°C = 288.15$K

Pressure $P_0 = 101325$Pa $= 101325$N/m^2

Density $\rho_0 = 1.225$kg/m^3

Temperature varies with different altitudes below 32km.

$$T(K) = \begin{cases} 288.15 - 0.0065H & 0 \leq H < 11000\text{m} \\ 216.65 & 11000\text{m} \leq H \leq 20000\text{m} \\ 216.65 + 0.001(H - 20000) & 20000\text{m} < H \leq 32000\text{m} \end{cases} \quad (2-26)$$

According to Figure 2.9, the variations of pressure and density are discussed as follows.

The differential equation is $dp = -gdm = -g\rho dh$.

And gas state equation is

$$p(H) = \rho(H) RT(H)$$

where, specific gas constant $R = 287.053$ J/(kg · K).

Pressure and density change with different altitudes in the Troposphere, $0 \leq H \leq 11000$m.

$$\begin{cases} \dfrac{p_H}{p_a} = \left(\dfrac{T_H}{T_a}\right)^{5.25588} \\ \dfrac{\rho_H}{\rho_a} = \left(\dfrac{T_H}{T_a}\right)^{4.25588} \end{cases} \quad (2-27)$$

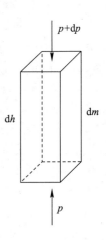

Figure 2.9 Mass Model

Here, p_a, ρ_a and T_a are the pressure, density and temperature at the sea level respectively.

At Stratosphere, $11000\text{m} \leq H \leq 20000\text{m}$, pressure and density change with different altitudes are as follows:

$$\frac{p_H}{p_{11}} = \frac{\rho_H}{\rho_{11}} = e^{-\frac{H-11000}{6341.62}} \quad (2-28)$$

where, p_{11} and ρ_{11} are the pressure and density in corresponding to the altitude $H = 11000$m.

At Stratosphere, $20000\text{m} \leq H \leq 32000\text{m}$, pressure and density change with different altitudes are given as follows:

$$\begin{cases} \dfrac{p_H}{p_{20}} = \left(\dfrac{T_H}{216.65}\right)^{-34.1632} \\ \dfrac{\rho_H}{\rho_{20}} = \left(\dfrac{T_H}{216.65}\right)^{-35.1632} \end{cases} \quad (2-29)$$

where, p_{20} and ρ_{20} are the pressure and density in corresponding to the altitude $H = 20$km.

The Simplified table of the 1976 U.S. Standard Atmosphere is given by Table 2.2.

Table 2.2 1976 U. S. Standard Atmosphere

H/km	T/K	p/ (N/m^2)	ρ/ (kg/m^3)	a/ (m/s)	$\mu \times 10^5$/ (Pa·s)
−2	301.2	127800	1.4780	347.9	1.851
0	288.1	101300	1.2250	340.3	1.789
2	275.2	79500	1.0070	332.5	1.726
4	262.2	61660	8.1930×10^{-1}	324.6	1.661
6	249.2	47220	6.6010×10^{-1}	316.5	1.595
8	236.2	35650	5.2580×10^{-1}	308.1	1.527
10	223.3	26500	4.1350×10^{-1}	299.5	1.458
12	216.6	19400	3.1190×10^{-1}	295.1	1.422
14	216.6	14170	2.2790×10^{-1}	295.1	1.422
16	216.6	10350	1.6650×10^{-1}	295.1	1.422
18	216.6	7565	1.2160×10^{-1}	295.1	1.422
20	216.6	5529	8.8910×10^{-2}	295.1	1.422
22	218.6	4047	6.4510×10^{-2}	296.4	1.432
24	220.6	2972	4.6940×10^{-2}	297.7	1.443
26	222.5	2188	3.4260×10^{-2}	299.1	1.454
28	224.5	1616	2.5080×10^{-2}	300.4	1.465
30	226.5	1197	1.8410×10^{-2}	301.7	1.475
32	228.5	889	1.3550×10^{-2}	303.0	1.486
34	233.7	663.4	9.8870×10^{-3}	306.5	1.514
36	239.3	498.5	7.2570×10^{-3}	310.1	1.543
38	244.8	377.1	5.3660×10^{-3}	313.7	1.572
40	250.4	287.1	3.9950×10^{-3}	317.2	1.601
42	255.9	220	2.9950×10^{-3}	320.7	1.629
44	261.4	169.5	2.2590×10^{-3}	324.1	1.657
46	266.9	131.3	1.7140×10^{-3}	327.5	1.685
48	270.6	102.3	1.3170×10^{-3}	329.8	1.704
50	270.6	80.0	1.0270×10^{-3}	329.8	1.704
52	269.0	62.2	8.0550×10^{-4}	328.8	1.696
54	263.5	48.3	6.3890×10^{-4}	325.4	1.668
56	258.0	37.4	5.0440×10^{-4}	322.0	1.64
58	252.5	28.7	3.9620×10^{-4}	318.6	1.612
60	247.0	22.0	3.0960×10^{-4}	315.1	1.584
62	241.5	16.7	2.4070×10^{-4}	311.5	1.555
64	236.0	12.6	1.8600×10^{-4}	308.0	1.526
66	230.5	9.5	1.4290×10^{-4}	304.4	1.497
68	225.1	7.1	1.0910×10^{-4}	300.7	1.467
70	219.6	5.2	8.2810×10^{-5}	297.1	1.438

(Continued)

H/km	T/K	$p/(\text{N/m}^2)$	$\rho/(\text{kg/m}^3)$	$a/(\text{m/s})$	$\mu \times 10^5/(\text{Pa}\cdot\text{s})$
72	214.3	3.8	6.2360×10^{-5}	293.4	1.408
74	210.3	2.8	4.6370×10^{-5}	290.7	1.387
76	206.4	2.0	3.4300×10^{-5}	2880.	1.365
78	202.5	1.5	2.5230×10^{-5}	285.3	1.343

2.3.2 Fundamental principles of fluid motion

Flight vehicles move in the air whose characteristics are similar to fluid. Therefore, the principles of fluid are discussed first.

2.3.2.1 Relative movement

As is known, an airplane is heavier than air, how can it overcome the gravitation and fly in the sky. It depends on the air force generated by the relative moment between air and plane. Without speed, there will be no aerodynamic forces. That is to say, the results depend on the relative movement. In order to learn the principles of flight, the law of relative movement must be known first. The concept of flight relative movement is shown as follows.

As shown in Figure 2.10, an airplane flies in static air with the speed V_∞ is just as the air moves against the airplane with the same speed. It depends on Newton's third law, the principles of acting and reacting. The movement of an airplane acts on the air and causes the air movement. On the contrary, the air reacts on the airplane.

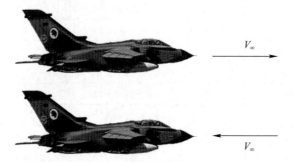

Figure 2.10 Aircraft relative movement

During the experimental study, usually, the plane holds still, the air flows over the surface of the plane in the reverse direction. The effects of air force generated on the plane are the same as those caused by the plane fly at the same velocity. That is the theory of relative movement.

2.3.2.2 Description of fluid motion

The space filled with moving fluid is usually called the flow field, and the physical quantities that characterize the fluid motion are called flow parameters, such as velocity, density, pressure, etc. So the flow field is also the field of the above physical quantities. Two methods are usually used in fluid mechanics to describe the motion of fluids, namely the Lagrangian method and the Eulerian method.

1) Lagrangian method

The Lagrangian method focuses on the masses of the fluid. That is the study of the flow field of each mass point of the law of motion parameters with time and the trajectory of the movement. By synthesizing the changes of the motion parameters of all fluid masses, the motion law of the whole flow field is obtained.

2) Eulerian method

Euler's method focuses on the spatial point in the flow field. That is, the study of the fluid mass point through the fixed point in space, the law of motion with time. By synthesizing the changes of motion parameters at all spatial points in the flow field, the law of motion of the entire flow field is obtained.

Lagrangian method and Eulerian method are just two different methods to describe the fluid motion, for the same flow field, both Lagrangian method and Eulerian method can be used. Euler's method is mostly used in aerodynamics.

2.3.2.3 Basic concepts of fluid motion

1) Path line

The set of all spatial points passed by the moving fluid mass point in the flow field over a period of time is called the path line of the fluid mass point.

2) Streamline

Assuming there is a spatial curve at a certain instant in the flow field, where the fluid mass velocity method at each point on the line coincides with the tangent line of the curve at that point, this spatial curve is called the stream line, which is shown in Figure 2.11.

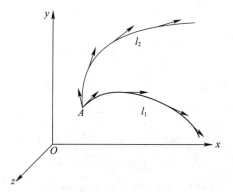

Figure 2.11   Stream lines at different moment in the flow field

3) Stream tube

Taking a closed curve C in the flow field that is not a streamline and making a stream line through each point on the curve C, the tubular surface formed by the collection of these stream lines is called a stream tube, which is shown in Figure 2.12.

2.3.2.4 Continuous equation

The continuity equation is a mathematical expression for the law of conservation of mass when applied to fluid flow.

Let us begin to consider the flow system of a continuum medium, which consists of fluid particles. A fluid particle that moves with a velocity V and has the density $\rho(x,t)$ at the position $x = x(x_0,t)$, is a representative object of the medium, having the mass of finite volume v. The mass of

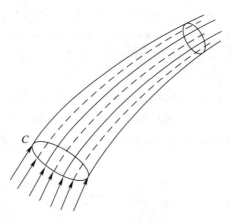

Figure 2.12 Stream tubes and streamlines in the flow field

the fluid particle can be obtained, using volume integral by

$$m = \int_v \rho \mathrm{d}v \qquad (2-30)$$

If we postulate that there are no sources or sinks in the medium, the mass of the fluid particle does not change in the position and time, i. e. the mass is conserved in space and time as follows:

$$\frac{\mathrm{D}m}{\mathrm{D}t} = \frac{\mathrm{D}}{\mathrm{D}t}\int \rho \mathrm{d}v = 0 \qquad (2-31)$$

where, v is the volume.

1) Integral form

$$\int \frac{\partial \rho}{\partial t}\mathrm{d}v + \int_A \rho V \mathrm{d}A = 0 \qquad (2-32)$$

For steady flow, $\partial \rho/\partial t = 0$, then

$$\int_A \rho V \mathrm{d}A = 0 \qquad (2-33)$$

For one-dimensional steady flow, there will be

$$\rho_1 V_1 A_1 = \rho_2 V_2 A_2 \qquad (2-34)$$

As seen in Figure 2.13 in every cross-section, the mass will be the same.

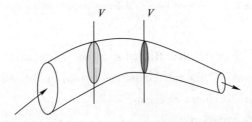

Figure 2.13 Continuous equation sketch

2) Differential form

The differential form can be written as,

$$\frac{\partial \rho}{\partial t} + \nabla \cdot (\rho V) = 0 \qquad (2-35)$$

2.3.2.5 Bernoulli equation

Bernoulli equation is about the conservation of energy. It means that energy can be transferred from one way to another way. Bernoulli equation is derived from the conservation of mechanical energy, so it is only applicable to the ideal fluid with negligible viscosity and not compressible.

Under the conditions of incompressible and ideal fluid, the law of conversation of energy can be expressed as follows.

$$p_1 + \frac{1}{2}\rho V_1^2 = p_2 + \frac{1}{2}\rho V_2^2 \tag{2-36}$$

where, p is the static pressure, $\frac{1}{2}\rho V^2$ is the Dynamic pressure.

According to this equation, with the increase of the fluid speed, the dynamic pressure will increase, and the static pressure will decrease.

For supersonic fluid, besides the kinetic and pressure energy, there is the internal energy produced by the molecular movement. Therefore, the energy equation for supersonic fluid is expressed by.

$$\frac{V^2}{2} + \frac{k}{k-1}\frac{p}{\rho} = \text{const.} \tag{2-37}$$

or

$$\frac{V^2}{2} + \frac{k}{k-1}RT = \text{const.} \tag{2-38}$$

where, k is the heat-insulating index, in air, $k = 1.4$.

2.3.3 Applications of the principles

2.3.3.1 Venturi tube

The Venturi tube (Figure 2.14) is named after Giovanni Battista Venturi (1746—1822), an Italian physicist. The principle of the Venturi effect is that the pressure will be reduced when a fluid flows through a constricted section of pipe. That is because the fluid speed in a constricted section of pipe is higher.

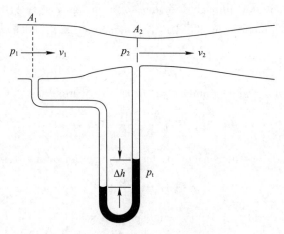

Figure 2.14 Venturi tube

A flow of air passes through a venturi tube, which is connected with a U-shape baresthesiometer. Then we can calculate the rate of flow.

$$Q = V_2 A_2 = A_2 \sqrt{2(p_1 - p_2)/[\rho(1 - A_2^2/A_1^2)]} \qquad (2-39)$$

2.3.3.2 Airspeed tube

An airspeed tube (Figure 2.15), also known as a pitot tube, is a pressure measurement instrument used to measure fluid flow velocity.

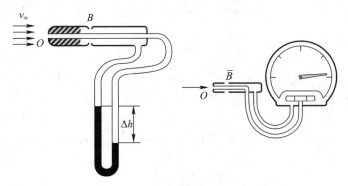

Figure 2.15 Airspeed tube

The principle of measuring the aircraft speed is that, when the aircraft flight forward, the air rush into the airspeed tube and the sensor in the tube can measure the dynamic pressure. If we compare the static air pressure and the dynamic pressure, we can know how fast the air rushed in, that is, how fast the aircraft flies.

$$V = \sqrt{\frac{2(P_0 - P)}{\rho}} \qquad (2-40)$$

Here, $(P_0 - P)$ is the differential pressure, and ρ is the density of air.

In addition, the air speed tube measured is not the aircraft speed relative to the ground, but only the speed relative to the atmosphere.

2.3.3.3 Laval nozzle

Considering the steady flow through a channel of a changing area A, the flow is isentropic and one-dimensional and has properties of the density ρ, temperature T, and local speed of sound a. The continuity equation is

$$\rho A V = \text{const.} \qquad (2-41)$$

The differential form of the continuity equation is also written as

$$\frac{d\rho}{\rho} + \frac{dA}{A} + \frac{dV}{V} = 0 \qquad (2-42)$$

While from the steady one-dimensional Euler equation with $\partial/\partial t = 0$, we have

$$V dV = -\frac{1}{\rho} d\rho \qquad (2-43)$$

The equation can be written in terms of the speed of sound as

$$V dV = -\frac{a^2}{\rho} d\rho \qquad (2-44)$$

Finally, combine Eq. (2-40) and Eq. (2-42), after introducing the Mach number,

$$\frac{dA}{A} = (Ma^2 - 1) \frac{dV}{V} \tag{2-45}$$

If $Ma = 1$, then $\frac{dA}{A} = 0$, the area of tube cross-section is the smallest at the throat.

As shown in Table 2.3, at subsonic, the change of speed is in inverse proportion with the variation of the area of the cross section. At supersonic, the change of speed is directly proportional to the variation of the area of the cross section.

Table 2.3 Relations among V, A and Mach number

Ma	<1	>1
$V\uparrow$ ($dV>0$)	$A\downarrow$ ($dA<0$)	$A\uparrow$ ($dA>0$)
$V\downarrow$ ($dV<0$)	$A\uparrow$ ($dA>0$)	$A\downarrow$ ($dA<0$)

From $Ma < 1$ to $Ma > 1$, the form of the nozzle must contract at first, and then expand. Such a tube is called Laval tube as shown in Figure 2.16.

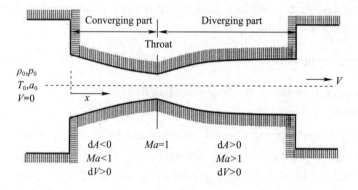

Figure 2.16 Laval nozzle and relations among V, A and Mach number

The Laval nozzle is the most commonly used component in rocket engines and aircraft engines, and it is an important part of the thrust chamber. The nozzle is first developed by Swedish inventor Gustaf de Laval in 1888 for applying in a steam turbine. It has two cone-shaped tubes, one is a shrink tube, and the other is an expansion tube. With the change of the cross-sectional area, the airspeed accelerates from subsonic to supersonic.

2.3.3.4 Airfoil and the generation of lift

The airfoil is a streamlined shape, capable of generating significantly more lift than drag. For any airfoil that generates lift, there must be a pressure imbalance, where the average air pressure on one side is lower than the other. As to an aircraft wing, the air flows at a faster speed and lower pressure at the top of the wing than at the bottom of the wing, which is shown in Figure 2.17.

According to the principle of mass continuity, for incompressible flow, the rate of volume flow must be constant within each stream tube. If a stream tube becomes narrower, the flow speed must increase in the narrower region to maintain the constant flow rate. From Bernoulli's principle, the pressure on the upper surface, where the flow rate is faster, is lower than the pressure on the lower surface, where the flow rate is slower. This pressure difference creates an aerodynamic force pointing

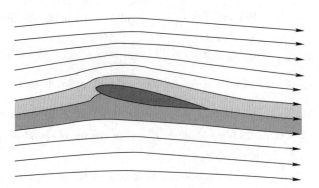

Figure 2.17   Streamlines around an airfoil

upward.

The above is only a brief explanation of the causes of lift in a two-dimensional plane. For an aircraft, the wing is a three-dimensional component consisting of many different airfoil interfaces, and the actual situation of lift in three-dimensional space needs to be considered. Also, under the influence of aircraft layout, the causes of lift are more complex. Interested students can continue to explore deeply in the future studies.

2.3.4 High-speed flow

2.3.4.1 Characteristics of the high-speed airflow

The airflow flowing at low speed is considered incompressible, the ρ of which is constant. Figure 2.18 shows the flowing behavior of the low-speed airflow at the variable-area channel. According to the Continuous equation and Bernoulli equation, when the pipe shrinks, i.e., $A_2 < A_1$, the velocity of the airflow will increase, $V_2 > V_1$, and the static pressure will decrease, $p_2 < p_1$, and vice versa.

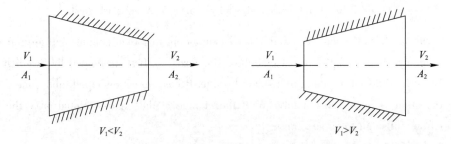

Figure 2.18   Flow characteristics of the low-speed airflow at variable-area channel

While the airflow flies at high speed, the variation of the density due to the change of velocity will bring a great change to the aerodynamics, which may even violate the principle of the flowing air. That is to say, the effects of high-speed flow cannot be neglected, which is the difference between the characteristics of high-speed airflow and low-speed airflow.

When the speed of airflow approaches the speed of sound, there will be a significant change in the pressure, the density, and the temperature of the airflow. Qualitative differences appear between the characteristics of high-speed airflow and low-speed airflow.

Figure 2.19 shows the flowing behaviors of the supersonic airflow at the variable-area channel. In contrary to the low-speed airflow, when the pipe shrinks, the velocity of the supersonic airflow will slow down, while the static pressure will increase. The reason for such difference is that the variation of the density caused by the change of cross-sectional area is larger than the variation of the velocity.

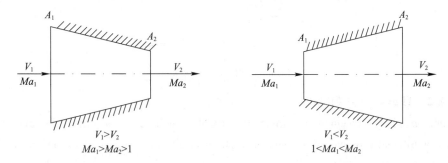

Figure 2.19 Flowing behaviors of the supersonic airflow at variable-area channel

As for the supersonic airflow, due to the density, which is not a constant anymore, the continuous equation of compressible fluid must be followed. Figure 2.19 shows that, at the variable-area channel, if $A_2 < A_1$, there will be $\rho_2 > \rho_1, v_1 > v_2, p_2 > p_1$. Contrarily, if $A_2 > A_1$, there will be $\rho_2 < \rho_1$, $V_2 > V_1, p_2 < p_1$.

Table 2.4 shows the variations of density when the velocity of airflow increases by 1% at different Ma numbers.

Table 2.4 Variations of density and velocity

Ma	0.2	0.4	0.6	0.8	1.0	1.2	1.4	1.6
$(\Delta v/v)$ %	1	1	1	1	1	1	1	1
$(\Delta \rho/\rho)$ %	-0.04	-0.16	-0.36	-0.64	-1	-1.44	-1.96	-2.56
$(\Delta A/A)$ %	-0.96	-0.84	-0.64	-0.36	0	0.44	0.96	1.56

It can be observed from the diagram that when $Ma < 1$, the velocity increases quickly, while the density decreases slowly. In this situation, to keep the flux of the air a certain value, the cross-section area of the pipe must decrease. For example, when $Ma = 0.8$, the velocity increases by 1%, and the density decreases by 0.64%, to keep the flux of air, the cross-section area of the pipe must decrease by 0.36%.

If $Ma > 1$, the velocity increases slowly, while the density decreases quickly. For instance, to keep the flux of the air a certain value, the cross-section area of the pipe must be increased. For example, when $Ma = 1.4$, the velocity will increase by 1%, and the density will decrease by 1.96%, to keep the flux of air, the cross-section area of the pipe must increase by 0.96%.

In a word, when the air flows at subsonic speed, in company with the increasing of the velocity, the cross-sectional area of the pipe must be decreased. For the supersonic airflow, the cross-section area of the pipe must be increased. In order to accelerate the airflow from subsonic to supersonic, besides the difference of pressure acting along the direction of flow, the shape of the Laval nozzle shown in Figure 2.20 is also needed. The airflow is accelerated while it flows towards the throat of the pipe, where it reaches a velocity of Mach 1. After the throat, the air continues to be accelerated, turning to be the supersonic airflow.

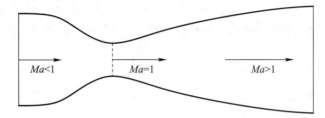

Figure 2.20　Laval nozzle

2.3.4.2　High-Speed Flight

Fighters, bombers, commercial transport aircraft, such as Concorde, fly at supersonic speed. Some additional physical phenomena in high-speed fly are not present in low-speed flight.

Particularly, the air compressibility becomes significant. As is presented in front that in the understanding of lift, the air is considered as an incompressible fluid because the forces involved are acting in low-speed flight.

It is to be noticed that the air is compressible at supersonic speed. All fluids, even water, are compressible to some extent. If air is compressible, its density will change. This is a phenomenon inherently different from that in low-speed flight, where the air density remains essentially unchanged as it passes over a wing, fuselage, or other parts of the airplane.

During the high-speed flight, another kind of drag, which is caused due to shock wave, will be considered. The drag is named wave drag as stated before. At one time, it became a huge obstacle to the planes' developments.

The level-flight speed of the piston-engine aircraft can reach the speed of 700km/h. When the velocity approaches the sonic speed, the plane will oscillate seriously and eventually run out of control. Sometimes, the structure of the aircraft will be destroyed, resulting in air accidents. The phenomenon is called the sound barrier as shown in Figure 2.21.

Figure 2.21　Sound barrier

Discovered from the research, the phenomenon of sound barrier is due to the shock wave and wave drag is created during the high-speed flight.

1) Shock waves

Compression of air can happen over such a small distance that it forms a shock front or shock wave. The dimensions of the air density change so small that the change is essentially instantaneous. In supersonic flight, a shock wave occurs when air must suddenly change speed and/or direction. Figure 2.22 gives the picture of shock waves on a space shuttle model in a supersonic wind tunnel.

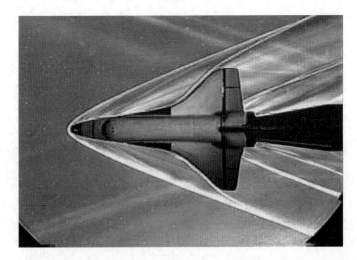

Figure 2.22 Shock wave on a model of the space shuttle

There are two types of shock waves involved with flight, normal (meaning perpendicular) shock waves and *oblique* (at an angle) shock waves. Normal shock waves are perpendicular to the direction of flight and are seen primarily on the surface of transonic wings or in pipes. They are caused by an abrupt change in density and pressure. Figure 2.23 shows what happens across a normal shock wave. Before the shock wave, the air is traveling at the Mach which is greater than 1. Behind a normal shock wave, the air is subsonic and the air density has increased.

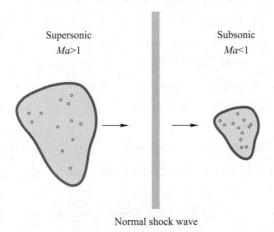

Figure 2.23 Density and Mach number change across a normal shock wave

Oblique shock waves are formed at an angle with respect to the oncoming air and occur when supersonic air must be turned. A supersonic airplane is traveling so fast, that the air has no chance to move out of the way as it does at subsonic speeds. Therefore, the moment when the air hits the leading edge of the wing must turn. The air turns almost instantaneously and forms the oblique shock wave. However, the shock waveforms at a given angle depending on the angle which must be turned. Figure 2.24 illustrates an oblique shock wave. As with a normal shock wave, the air density increases, and the air's velocity decreases across an oblique shock wave. The changes are not enough for the air to become subsonic, as in a normal shock. Therefore, the air behind an oblique shock wave remains supersonic relative to the aircraft. There are rare exceptions to this rule.

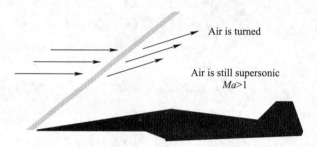

Figure 2.24 An oblique shock wave

All supersonic objects create shock waves. Normal shock waves cause a higher change in density than oblique shock waves. Therefore, supersonic aircraft are designed to avoid producing normal shock waves, since the greater the change of density across a shock wave is, the greater the energy loss and the drag are. This is accomplished by making the nose and wing leading edge sharp. Blunt noses lead to energy-consuming *bow shock*, which is a combination of a normal shock wave on the very nose joined to an oblique shock wave a little way back. Bow shocks are avoided by using sharp noses on supersonic airplanes.

As we known, the density, the pressure, and the velocity will be changed suddenly if there is the shock wave in the air. For a given shock angle, all of these properties can be found. Thus, for a given airplane geometry, the shock angles and pressures are easily determined. Supersonic flight is easier to analyze than subsonic flight. In low-speed aerodynamics, engineers must rely on complicated equations to solve the pressures over a vehicle. In supersonic aerodynamics, an engineer can use the published tables. However, being able to compute the pressures on the vehicle more easily does not translate into making the problem of supersonic flight easier. Now the penalty of supersonic flight will be discussed, that is, wave drag.

2) Wave drag

The airflow is banned by the shock wave so that the wave drag is formed. As the shock wave is a sort of strong compression wave, the wave drag is particularly large when the airflow passes the shock wave.

In any case, the wave drag created by the normal shock wave is bigger than that generated by the oblique shock wave. To the normal shock wave, the air is compressed at the highest degree, pressure, and density of air behind the shockwave reaching to the top, so the shock-wave intensity is the maximum. While supersonic airflow passes, the air is banned most intensively, its velocity drops down rapidly and the consumption of energy is very big, so the wave drag is relatively larger.

In addition, when flying at supersonic velocity, difform aircraft can generate different wave drags due to the differencesin shock waves. The more intensively the airflow banned by the flight vehicle, the much bigger the shock wave is, so as the wave drag.

To the blunt nose airfoil, a detached shock wave (intensive normal shock wave) is usually brought in front of the blunt nose. Detached shock wave which can hinder the airflow strongly will bring prodigious wave drags. As given in Figure 2.25 (a), to the acuminate airfoil, the oblique shock wave is ordinarily generated at front of the acumination. As given in Figure 2.25 (b), the inhibitory action of the shock wave is relatively weak. The more acuminate of the nose, the more incline of the shock wave, comparatively the wave drag is smaller. So noses of fuselage and wings are designed to be acuminate to decrease the intensity of shockwave and then to diminish the wave drag.

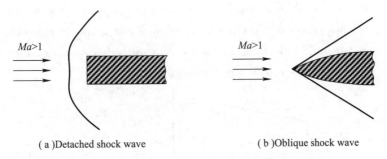

(a) Detached shock wave   (b) Oblique shock wave

Figure 2.25   Wave drag

3) Transonic flight

Commercial transports fly in the Mach 0.8 to 0.86 range, just below the speed of sound. This speed is not chosen arbitrarily. It is based on the presence of wave drag. While, if the airplane is flying at a speed less than the speed of sound, how can there be wave drag?

A wing diverts the air down. In bending the air down, it creates a lower pressure on the upper surface of the wing, which causes the air to accelerate. However, at speeds approaching the speed of sound, the air that is accelerated over the top of the wing becomes locally supersonic.

When air flows over the top of a subsonic wing, it accelerates to the point of greatest curvature of the air. At this point, the pressure is the lowest and the speed of the air is greatest. From that point to the trailing edge of the wing the airspeed decreases and the pressure increases in order to match the pressure of the air at the trailing edge. This is the trailing-edge condition.

The picture is quite different for the air flowing over the top of a transonic wing. The air accelerates as before, but by the time it reaches the point of maximum curvature, it is traveling at a greater speed than Mach 1. As the air continues to bend, because it is traveling faster than the communication speed of air, it is not able to effectively pull air down from above. Thus the density is substantially reduced, causing the pressure to continue to go down and the velocity to increase. This situation leaves the wing with the problem of how to meet the trailing-edge condition. The solution is the formation of a normal shock wave as shown in Figure 2.26.

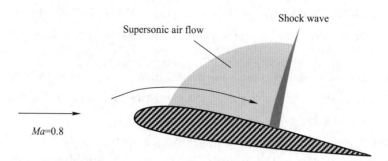

Figure 2.26 A transonic airfoil accelerates the air to supersonic speed and then forms a shock wave

For the shock wave, the pressure and the density increase abruptly and the velocity of the air goes below Mach 1. After the shock wave, the air can slow down further and the pressure continues to increase to meet the trailing-edge condition.

Figure 2.27 shows the airspeed, in units of Mach number, of the air over a subsonic and a transonic wing. The subsonic airfoil is traveling at a speed just below the *critical Mach number* such that the air never reaches supersonic. The transonic airfoil is just above the critical Mach number so the air becomes supersonic. The Mach number of the subsonic airfoil decreases after the peak while on the transonic airfoil it increases, until meeting the shock wave.

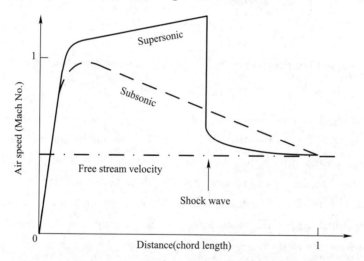

Figure 2.27 The airspeed over a transonic and a subsonic airfoil

How does the wing know where to put the normal shock wave? Let us first assume that the normal shock waveforms at the trailing edge in order to meet the trailing-edge condition. What would be found is that the force caused by the pressure differences across the shock wave is higher than the force of the wave drag. Thus, the shock wave will move forward on the wing. As it does, the pressure difference decreases until the wave moves to a place on the wing where the force from the pressure difference just equals the force due to wave drag. If the airplane were now to increase its speed, the force due to wave drag would increase and the shock wave would move toward the trailing edge. At some speed, the normal shock wave will reach the trailing edge.

Look again at the picture of the fighter flying at transonic speed in Figure 2.21. In the region where the air is supersonic with increasing speed, the pressure, density, and temperature are decreasing. At a point before the normal shock wave, the air has cooled enough to cause condensation, producing the cone of fog above and below the wing. The backside of the cone is a flat surface. This is the location of the normal shock wave where the pressure and temperature increase and the condensation disappears.

One may ask why there is a normal shock wave on the bottom of the wing. The fighter has almost symmetric wings and since the angle of attack is so small at transonic speeds, there is a reduction in pressure and acceleration of the air on both the top and the bottom of the wing. It is just that the acceleration and reduction in pressure are not as great on the bottom. Commercial jets that fly at transonic speeds are designed so that the normal shock wave forms only on the top of the wing.

In transonic (and supersonic) flight, the velocity of the air over the wing continues to increase until the normal shock wave is reached. Because of this, the center of the lift is farther back on the wing than in subsonic flight. For a typical wing in subsonic flight, the center of lift is about one-fourth chord length back from the leading edge of the wing. That means that the wing produces 50 percent of the lift by that point. The moment the wing becomes transonic the center of lift moves farther back. As the speed increases further, the center of the lift continues to move back. In very high-speed flight, the center of the lift can move as far as a half-chord length back from the leading edge.

4) Hypersonic flight

When the Mach number gets high (above about Mach 5), several things happen (Figure 2.28). First, the aerodynamics becomes Mach number independent. This means that for analysis purposes simple assumptions can be made and, in fact, the analysis of idealized, hypersonic flight is the easiest of all aerodynamic analyses.

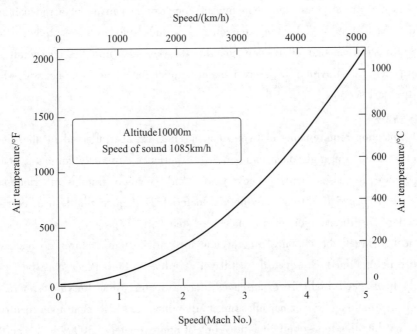

Figure 2.28 Air temperature as a function of airplane speed

Figure 2.29 shows an artist's conception of a hypersonic airplane. The only known existing hypersonic aircraft is the space shuttle during reentry. The X-15, shown in Figure 2.29, explored hypersonic flight in the 1960s, reaching an unofficial speed of Mach 6.7. After the record-making flight, the airplane was retired due to heat damage from the flight. Speculation abounds about the possibility of a super-secret hypersonic spy plane. However, there is only circumstantial evidence to support such a rumor.

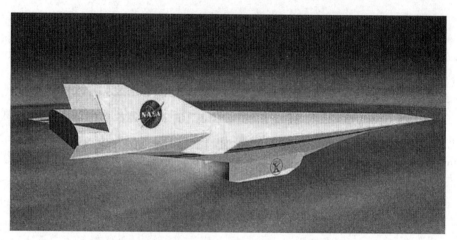

Figure 2.29   Depiction of Hyper-X hypersonic vehicle

The second change that occurs in hypersonic flight is that the energy transfer of the fast vehicle to the surrounding air becomes so great that the air chemistry begins to change. Oxygen and nitrogen molecules begin absorbing energy by breaking up or dissociating. This has many implications for the design of hypersonic aircraft. First, the changing air composition can affect aerodynamics. This was observed in the space shuttle, where the direction and attitude stability were predicted to be higher than it was. Fortunately, enough of a safety margin was designed into the spacecraft so that this shortcoming was not catastrophic. The second major implication is on skin heating, which will be discussed in the next section.

5) Skin heating

Thermal protection requirements of hypersonic aircraft are also affected by the dissociation of the air. Vehicles traveling at high Mach numbers will experience extremely hot gases. Some of this is due to temperature increases across shock waves and someare from skin friction. The high-temperature air will burn right through any normal material. The space shuttle uses ceramic tiles for thermal protection. The dissociation of air molecules helps in keeping the vehicle cooler. It takes energy from heat to break the chemical energy and the surface temperatures do not get as hot as would otherwise be predicted. However, they still get very hot, so the surface must be protected.

Extremely high-speed flight is experienced during reentry into the atmosphere. The Space Shuttle, as well as Apollo and Soyuz capsules, must all endure very high heat upon reentry. When the Space Shuttle first hits the atmosphere, it is traveling at approximately 23000km/h. The thin air that slams into the nose of the Space Shuttle convert kinetic energy to heat. In theory, the air that impacts the nose of the Space Shuttle will reach over 36000°F (20000°C), which is about three times the temperature of the sun.

When the air reaches a high temperature, it goes through complex changes. Some of the energy of impact goes into breaking chemical bonds rather than creating heat. Oxygen dissociates and ionizes. The impact is so great that the ionized gas that develops around a vehicle reentering the atmosphere could prevent radio communication with the outside. This is what is known as the *reentry blackout* experienced by all spacecraft since the first successful atmospheric reentry (the Russian Sputnik). Relative to having skin temperatures reaching 36000°F (20000°C) the temperature is closer to one-third that value, but still sun-like temperatures.

Figure 2.30 shows the temperature distribution of Apollo command modules. The temperature of the nose reaches about 2800°C.

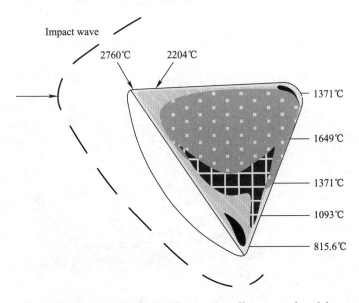

Figure 2.30 Temperature distribution of Apollo command modules

The Apollo command modules had a special carbon-based surface ontheir base. This surface burned off slowly as the craft reentered the atmosphere. The burning heat shield resulted in two effects. The first is that the burning consumes energy, and thus heat, from the air. The second is that the by-products were swept away, taking heat along with them. The astronauts were thus kept cool behind this heat shield. This form of skin cooling is called *ablation*. The problem with ablation is that it is not reusable.

Special tiles were designed for the Space Shuttle that is extremely poor heat conducts. The tiles absorb heat very slowly. When the surface of the tiles reaches the high reentry temperature, the tiles radiate heat out to maintain a constant surface temperature. The longer the Shuttle experiences the heat, the deeper the heat will penetrate the tiles. The tiles must be thick enough to prevent the heat from reaching the aluminum skin before the heat load is removed on landing.

The skin temperature distribution of a space shuttle is shown in Figure 2.31. According to the differences in skin temperatures, the skin can be divided into four regions. Different regions can adopt different kinds of tiles, as given in Figure 2.32.

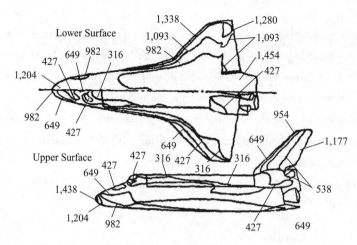

Figure 2.31　Skin temperature distribution of a space shuttle

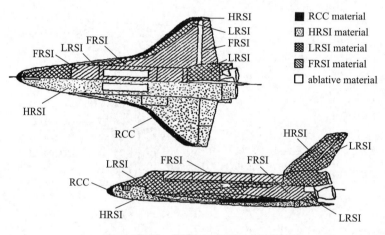

Figure 2.32　Different kinds of tiles

2.4　Reference system

As is known, the movement of a flight vehicle is caused by the forces and moments acted on it. In order to analyze the relations among the flight performances and the forces and moments, the forces and moments should be described in the reference systems, so does the flight performance. The reference systems are called coordinate systems, or reference frames.

Here, a way to define the reference frame is given.

2.4.1　Reference frames and transformations

The coordinate system is a system that uses one or more numbers or coordinates to uniquely determine the position of points or other geometric elements in space. The order of the coordinates is very important. Sometimes they are identified by their position in the ordered tuple, and sometimes they are identified by letters, such as "x coordinate", "y coordinate" and so on. In addition, there are usually many coordinate systems that can describe geometric figures, so it is very important to understand the conversion relationship between them.

2.4.1.1 General reference frames

The general equations of motion are developed for flight vehicles. Reference frames and coordinate transformations must be introduced first.

1) Right-hand rule and left-hand rule

Both the left-hand rule and the right-hand rule stem from the fact that three axes of cartesian coordinates in three-dimensional space have two possible directions. This can be seen by stretching out your hands together, palms up, fingers curled, and thumbs spread out. If the direction of the finger curl represents movement from the first or Y-axis to the second or Z-axis, then the third or X-axis can point along with either thumb. There will be left and right-handed rules when dealing with coordinate axes. Figure 2.33 illustrates the relations.

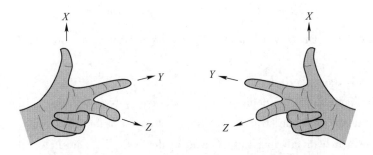

Figure 2.33 Light-hand rule and reft-hand rule

For the right-hand coordinate system, the right thumb points in the positive direction along the X-axis, and the curling direction of the fingers of the right hand represents the movement from the first or Y-axis to the second or Z-axis. When viewed from the top or X-axis, the system is counterclockwise.

For the left-hand coordinate system, the thumb of the left hand points in the positive direction along the X-axis, and the curling direction of the fingers of the left hand represents the movement from the first or Y-axis to the second or Z-axis. When viewed from the top or X-axis, the system is clockwise.

2) Typical reference frames

The often-used reference frame includes Earth-fixed reference frame $Axyz$, Body-fixed reference frame $Ox_1y_1z_1$, Flight path reference frame $Ox_2y_2z_2$, and Velocity reference frame $Ox_3y_3z_3$. Earth-fixed reference frame $Axyz$ is moved parallel, to be Quasi-Earth-fixed reference frame $Oxyz$ for us to research the relation between them.

(1) Earth-fixed reference frame

The definition of Earth-fixed reference frame is seen as Figure 2.34. Earth-fixed reference frame is connected with the earth. The origin A of the system is the point where the flight vehicle is launched. The direction of the axis Ax is usually to point to the target.

The direction of the axis Ay is upward and perpendicular to the axis Ax. The rest is arbitrary. Here, the axis Az is defined by the right-handed rule.

The Quasi-Earth-fixed reference frame $Oxyz$ is defined to move the Earth-fixed reference frame

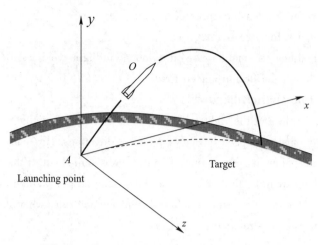

Figure 2.34 Earth-fixed reference frame $Axyz$

parallel from point A to point O which is the flight vehicle mass center so that the relation between the frames can be analyzed.

(2) Body-fixed reference frame

As shown in Figure 2.35, the definition of a Body-fixed reference frame is illustrated. The origin of the Body-fixed reference frame $Ox_1 y_1 z_1$ is at the mass center of the flight vehicle, in which the axis of Ox_1 is superposition with the longitudinal axis of the body, it is positive when it points to the head. The axis of Oy_1 which locates in the longitudinal symmetrical surface is perpendicular to the axis of Ox_1, it is positive when its direction is upward. The definition of the axis of Oz_1 whose direction is determined by the right-handed rule, being perpendicular to the plane of $Ox_1 y_1$. The system moves with the motion of the body.

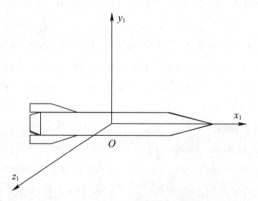

Figure 2.35 Body-fixed reference frame $Ox_1 y_1 z_1$

(3) Flight path reference frame

As shown in Figure 2.36, the Flight path reference form $Ox_2 y_2 z_2$ is given. The origin of the reference frame is the mass center of the flight vehicle, the axis of Ox_2 is superposition with the velocity vector of the mass center. The axis Oy_2 is in the vertical surface which contains the velocity vector V and is perpendicular to the axis of Ox_2, and it is positive when its direction is upward. Axis Oz_2 is defined by the right-handed rule.

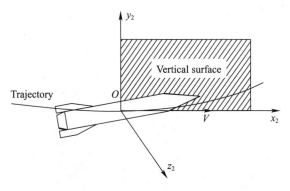

Figure 2.36 Flight path reference frame $Ox_2y_2z_2$

(4) Velocity reference frame

As seen in Figure 2.37, the Velocity reference frame $Ox_3y_3z_3$ is given. The frame $Ox_3y_3z_3$ is connected with the velocity vector V, the origin of the reference frame is the mass center of the flight vehicle, the axis of Ox_3 is superposition with the velocity vector of the mass center. The axis of Oy_3 which is in the longitudinal symmetrical surface containing the velocity vector V is perpendicular to the axis of Ox_3, and it is positive when its direction is upward. Axis Oz_3 is normal to the plane of Ox_3y_3, whose direction is determined by the right-handed rule. The frame $Ox_3y_3z_3$ is fixed with the velocity vector V.

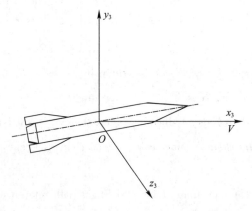

Figure 2.37 Velocity reference frame $Ox_3y_3z_3$

2.4.1.2 Coordinate transformation

1) Quasi-Earth-fixed reference frame $Oxyz$ and Body-fixed reference frame $Ox_1y_1z_1$

As stated before, the Quasi-Earth-fixed reference frame $Oxyz$ is used to describe the relation between different frames. The orientation of a flight vehicle is relative to Quasi-Earth-fixed reference frame $Oxyz$ as shown in Figure 2.38 by comparing it with Body-fixed reference frame $Ox_1y_1z_1$. Then three Euler angles ϑ, ψ, γ are defined to describe the attitude of the flight vehicle. These are pitching angle ϑ, yawing angle ψ, and slope angle γ.

Pitching angle ϑ is illustrated by the angle between the longitudinal axis of a flight vehicle Ox_1 and the horizontal surface Oxz.

Yawing angle ψ is described by the angle between the axis of Ox and the longitudinal axis'

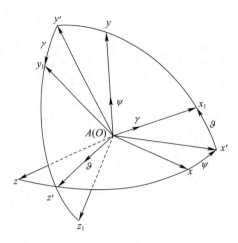

Figure 2.38 Quasi-Earth-fixed reference frame $Oxyz$ and Body-fixed reference frame $Ox_1y_1z_1$

projection on the horizontal surface Oxz.

Slope angle γ is given by the angle between the axis of Oy_1 and vertical surface including the longitudinal axis Ox_1 of a flight vehicle.

Coordinate transformations between Quasi-Earth-fixed reference frame $Oxyz$ and Body-fixed reference frame $Ox_1y_1z_1$ are as follows:

$$\begin{bmatrix} x_1 \\ y_1 \\ z_1 \end{bmatrix} = L(\gamma,\vartheta,\psi) \begin{bmatrix} x \\ y \\ z \end{bmatrix} \quad (2-46)$$

Here, transformation matrix $L(\gamma,\vartheta,\psi)$ is written by

$$L(\gamma,\vartheta,\psi) = \begin{bmatrix} \cos\vartheta\cos\psi & \sin\vartheta & -\cos\vartheta\sin\psi \\ -\sin\vartheta\cos\psi\cos\gamma + \sin\psi\sin\gamma & \cos\vartheta\cos\gamma & \sin\vartheta\sin\psi\cos\gamma + \cos\psi\sin\gamma \\ \sin\vartheta\cos\psi\sin\gamma + \sin\psi\cos\gamma & -\cos\vartheta\sin\gamma & -\sin\vartheta\sin\psi\sin\gamma + \cos\psi\cos\gamma \end{bmatrix}$$

$$(2-47)$$

2) Quasi-Earth-fixed reference frame $Oxyz$ and Flight path reference frame $Ox_2y_2z_2$

The relation between Quasi-Earth-fixed reference frame $Oxyz$ and Flight path reference frame $Ox_2y_2z_2$ is seen in Figure 2.39, and two angles are given to describe the relations of them. These are flight path angle θ and flight path deflection angle ψ_V.

Flight path angle θ is described by the angle between the velocity and horizontal surface Oxz.

Flight path deflection angle ψ_V is illustrated by the angle between the axis Ox and the velocity's projection on the horizontal surface Ox'.

According to Figure 2.39, coordinate transformations between Quasi-Earth-fixed reference frame $Oxyz$ and Flight path reference frame $Ox_2y_2z_2$ are written as follows:

$$\begin{bmatrix} x_2 \\ y_2 \\ z_2 \end{bmatrix} = L(\theta,\psi_V) \begin{bmatrix} x \\ y \\ z \end{bmatrix} \quad (2-48)$$

Chapter 2 Flight principles of flight vehicles

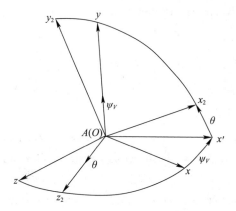

Figure 2.39 Quasi-Earth-fixed reference frame $Oxyz$ and Flight path reference frame $Ox_2y_2z_2$

Here, transformation matrix $L(\theta,\psi_V)$ is given by

$$L(\theta,\psi_V) = \begin{bmatrix} \cos\theta\cos\psi_V & \sin\theta & -\cos\theta\sin\psi_V \\ -\sin\theta\cos\psi_V & \cos\theta & \sin\theta\sin\psi_V \\ \sin\psi_V & 0 & \cos\psi_V \end{bmatrix} \qquad (2-49)$$

3) Velocity reference frame $Ox_3y_3z_3$ and Body-fixed reference frame $Ox_1y_1z_1$

The relation between Body-fixed reference frame $Ox_1y_1z_1$ and Velocity reference frame $Ox_3y_3z_3$ is shown in Figure 2.40. Two angles are given to describe their relationship. These are the angle of attack α, and angle of sideslip β.

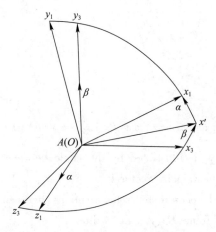

Figure 2.40 Speed reference frame $Ox_3y_3z_3$ and Body-fixed reference frame $Ox_1y_1z_1$

Angle of attack α is described by the angle between the axis of Ox_1 and the velocity's projection on the longitudinal symmetrical surface Ox_1y_1 of the body. If the axis Ox_1 is above the projection of velocity Ox', α is positive. Contrarily it is negative.

Angle of sideslip β is illustrated by the angle between the velocity V (Ox_3) and the longitudinal symmetrical surface Ox_1y_1. Observing from the flight direction, if the air flows to the flight vehicle from the right hand, β is positive. Contrarily it is negative.

From Figure 2.40, coordinate transformations between Velocity reference frame $Ox_3y_3z_3$ and

Body-fixed reference frame $Ox_1y_1z_1$ are written as follows:

$$\begin{bmatrix} x_1 \\ y_1 \\ z_1 \end{bmatrix} = L(\alpha,\beta) \begin{bmatrix} x_3 \\ y_3 \\ z_3 \end{bmatrix} \quad (2-50)$$

Here, transformation matrix $L(\alpha,\beta)$ is given by

$$L(\alpha,\beta) = \begin{bmatrix} \cos\alpha\cos\beta & \sin\alpha & -\cos\alpha\sin\beta \\ -\sin\alpha\cos\beta & \cos\alpha & \sin\alpha\sin\beta \\ \sin\beta & 0 & \cos\beta \end{bmatrix} \quad (2-51)$$

4) Flight path reference frame $Ox_2y_2z_2$ and Velocity reference frame $Ox_3y_3z_3$

The relation between Flight path reference frame $Ox_2y_2z_2$ and Velocity reference frame $Ox_3y_3z_3$ is seen in Figure 2.41. There is only one angle to describe their relationship. It is the slop angle of velocity γ_V.

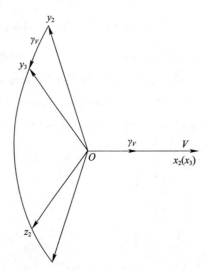

Figure 2.41 Flight path reference frame $Ox_2y_2z_2$ and Velocity reference frame $Ox_3y_3z_3$

Slope angle of velocity γ_V is illustrated by the angle between the axis Oy_3 and the vertical surface Ox_2y_2 which contains the velocity vector V.

According to Figure 2.41, coordinate transformations between Flight path reference frame $Ox_2y_2z_2$ and Velocity reference frame $Ox_3y_3z_3$ are written by

$$\begin{bmatrix} x_3 \\ y_3 \\ z_3 \end{bmatrix} = L(\gamma_V) \begin{bmatrix} x_2 \\ y_2 \\ z_2 \end{bmatrix} \quad (2-52)$$

Here, transformation matrix $L(\gamma_V)$ is given as follows:

$$L(\gamma_V) = \begin{bmatrix} 1 & 0 & 0 \\ 0 & \cos\gamma_V & \sin\gamma_V \\ 0 & -\sin\gamma_V & \sin\gamma_V \end{bmatrix} \quad (2-53)$$

2.4.2 European-American coordinate system and Soviet Union coordinate system

In the field of aircraft design related disciplines, there are two widely used coordinate system definition methods: One is used by the former Soviet Union and its related countries that introduced Soviet aviation technology to establish the aviation industry system (Abbreviated as Soviet coordinate system or Soviet system). Another is used in the United States and Europe (Abbreviated as Euro-American coordinate system or Euro-American system).

The ground coordinate system in the Soviet coordinate system is defined as: the origin O is fixed at a certain point on the ground, the Oy axis is vertically upward, and the Ox and Oz axes are in the horizontal plane and the Oy axis constitutes a right-handed rectangular coordinate system. The definition of the airframe coordinate system, velocity coordinate system, and track coordinate system is similar to that of the ground coordinate system, but with the center of mass of the aircraft as the origin.

The biggest difference between the definition of the European and American coordinate system and the Soviet coordinate system is the direction of the vertical axis. Let us take the Body-fixed reference frame as an example. As shown in the Figure 2.42, $Ox_b y_b z_b$ is the Body-fixed reference frame in the European and American coordinate system, and $Ox_1 y_1 z_1$ is the Body-fixed reference frame in the Soviet coordinate system. It is important to be aware that the positive direction of vertical axes of the two coordinate systems are opposite and the actual meaning of the y-axis and the z-axis are interchanged.

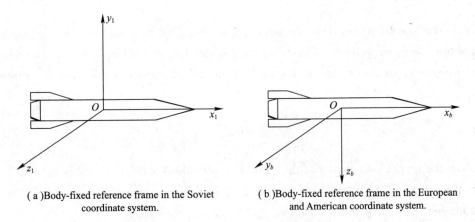

(a) Body-fixed reference frame in the Soviet coordinate system.

(b) Body-fixed reference frame in the European and American coordinate system.

Figure 2.42 The difference between the European and American coordinate system and the Soviet coordinate system

2.5 Flight mechanics description

In the environment of mechanics of flight vehicles, the external forces and moments exert on a flight vehicle while it flies in the air. The characteristics of those forces and the moments on them must be discussed at first. The forces and moments determine how fast the flight vehicle flies, how high flight vehicle can reach, how far flight vehicle can fly, and so forth. These are some of the elements of the study of the performance of flight vehicles.

2.5.1 Forces acted on flight vehicles

As shown in Figure 2.43, three kinds of forces, such as gravity, aerodynamics which includes lift, drag and lateral forces, and thrust act on a flight vehicle flying in the air.

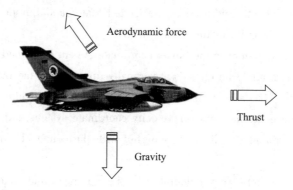

Figure 2.43 Forces acted on an airplane

2.5.1.1 Gravity

As known, there is the gravitational attraction between two bodies according to the law of universal gravitation discovered by Newton. That is,

$$F = G \frac{m_1 m_2}{r^2} \tag{2-54}$$

where, F is the gravitation between two bodies, G is the gravitation constant, m_1, m_2 express the masses of two bodies respectively, r stands for the distance between the centroids of two bodies.

If an object is on the surface of the earth, then the gravitation of the earth is expressed as follows:

$$F = G_e \frac{m}{R^2} \tag{2-55}$$

where, G_e is the gravitation constant of the earth, $G_e = 398600.5 \times 10^9 \text{m}^3/\text{s}^2$, R is the radius of the earth, in the equator of the earth, it is about 6378km. In the polar region, it is about 6357km.

At the sea level, for standard atmosphere, $F = mg_0$, $g_0 = 9.80665 \text{m/s}^2$.

In fact, the acceleration of gravity g changes with different altitudes, it can be written as follows:

$$g = g_0 \frac{R_0^2}{(R_0 + H)^2} \tag{2-56}$$

As shown in Figure 2.44, weight contains two parts.

$$G = G_0 + ma = mg + ma \tag{2-57}$$

where, mg is the gravity, ma is the centrifugal force, it is an inertial force produced by the rotation of the earth around its axis.

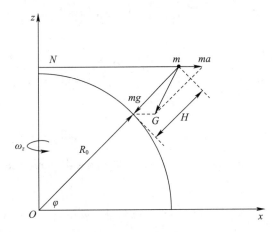

Figure 2.44 Gravity

$$a = (R_0 + H)\,\omega_z^2\cos\varphi \tag{2-58}$$

where, φ is the latitude angle of the earth, ω_z is the angular speed of rotation of the earth.

$\omega_z = 2\pi/8.616 \times 10^4 = 7.292 \times 10^{-5}$ rad/s. The value is too small so that this term ma is neglected, therefore, $G = mg$.

2.5.1.2 Thrust

As is known, engines produce thrust. Rocket engines and jet engines are used for some flight vehicles.

The thrust of a rocket engine is expressed by

$$P = \dot{m}u_e + A_e(p_e - p_a) \tag{2-59}$$

According to Eq. (2-59), there are two parts in it, one is momentum thrust produced by reacting force, another is pressure difference thrust.

A_e is the cross-section area of the tube, p_e is the gas pressure at the nozzle, p_a is the air pressure. There is no effect from the speed of a rocket, only by altitude.

But for a jet engine, P is affected not only by flight altitude but also by flight speed. So the power is stated as follows.

$$P = \dot{m}_a(u_e - V) + A_e(p_e - p_a) \tag{2-60}$$

where, V is the speed of a flight vehicle, $\dot{m}_a$ is the airflow quantity per second.

2.5.1.3 Aerodynamics

Aerodynamic force is exerted on a body by the air in which the body is immersed, and is due to the relative motion between the body and gas. We can get the formula of aerodynamic force by dimensional analysis. The aerodynamic force acted on a body can be divided into three parts, that is,

Aerodynamic force = drag + lift + side force (lateral force)

$$R = X + Y + Z \tag{2-61}$$

Usually, it is expressed by aerodynamic coefficients. That is,

$$R = C_R \frac{1}{2}\rho V^2 S \tag{2-62}$$

$$\begin{cases} X = C_x \dfrac{1}{2}\rho V^2 S \\ Y = C_y \dfrac{1}{2}\rho V^2 S \\ Z = C_z \dfrac{1}{2}\rho V^2 S \end{cases} \qquad (2-63)$$

where, C_R is the total aerodynamic coefficient, C_x is the drag coefficient, C_y is the lift coefficient, C_z is the side force coefficient, S is the reference area, $\dfrac{1}{2}\rho V^2$ is the dynamic pressure.

1) Lift

If an airplane moves in straight and level flight in the vertical plane, the 4 forces can be seen as Figure 2.45. That is to say, gravity, thrust, and aerodynamics (lift and drag) act on it. All forces made the airplane balance and movement in the air. Here, lift is one of the very important forces during the flight.

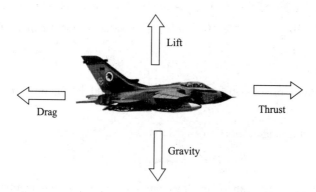

Figure 2.45 Forces acted on an airplane in straight and level flight

The sources for producing lift are from several parts of airplane configurations, such as wing, body, the combination of wing and body, nose tail, and so on. But the main part of them for lift is from the wing.

(1) Lift coefficient curve

Figure 2.46 shows the relations between the lift coefficient with angle of attack. Note that, at $0 - \alpha_{\text{linear}}$, the lift coefficient varies with the angle of attack α in a linear relation. That is $C_y = C_y^\alpha \alpha$, where $C_y^\alpha = \dfrac{\partial C_y}{\partial \alpha}$, means the partial derivative of lift coefficient to the angle of attack α. At the maximum lift coefficient $C_{y\max}$, the angle of attack is α_{cr} called the critical angle. If the angle of attack is bigger than α_{cr}, the lift coefficient is not increased, but decreased, at the time, there will be a stall for a flight vehicle.

(2) Lift coefficient from other parts

As the air flows around a flight vehicle body, a part of the lift is also produced. Besides the lift which is produced by wing and body respectively, there is lift caused by the interference between wing and body. It is the same as wing and tails.

Under the given conditions of configurations and size, the lift coefficient is expressed as follows.

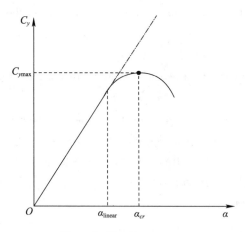

Figure 2.46 $C_y - \alpha$ curve

$$C_y = f(Ma, \alpha, \delta_z) \tag{2-64}$$

where, δ_z is the elevator deflection angle.

If α and δ_z are not too large, the lift coefficient can be expressed as follows.

$$C_y = C_{y0} + C_y^\alpha \alpha + C_y^{\delta_z} \delta_z \tag{2-65}$$

Here, C_{y0} means lift coefficient at $\alpha = \delta_z = 0$, at the same time, if the flight vehicle is symmetrical, then $C_{y0} = 0$.

2) Drag

Drag or resistance is composed of two parts, i.e., zero-lift drag and induced drag by lift.

$$X = X_0 + X_i \tag{2-66}$$

where X_0 is the drag at zero-lift. There are no relations with lift for X_0. And, X_i is the drag induced by lift.

It can be expressed by drag coefficient as follows.

$$C_x = C_{x0} + C_{xi} \tag{2-67}$$

where,

$$C_{xi} = C_{xi}^{\alpha^2} \alpha^2 + C_{xi}^{\beta^2} \beta^2 \tag{2-68}$$

β is the angle of the sideslip.

It is affected by,

$$C_x = f(Ma, Re, \alpha, \beta) \tag{2-69}$$

where Re stands for Reynolds number described the characteristics of a boundary layer.

(1) Zero-lift drag

Zero-lift drag is composed of friction drag, pressure drag, and wave drag at supersonic. It is mainly caused by the air viscous at low-speed flow. At supersonic flow, there will be wave drag.

I. Friction drag

It is due to the net effect of shear stress acted on.

II. Pressure drag

It is due to a net imbalance of surface pressure act on. When the air flows over the wing, for being held up at the front of the wing, its speed slows down but pressure increases. Owing to generating a vortex behind the wing edge for burbling, the air pressure decreases there. In this way, the imbalance of surface pressure is formed in the drag direction, the drag is given.

III. Wave drag

As stated in Section 2.3.4.2 1), shock waves are the dominant feature of the flow field around an airplane flying at supersonic. The presence of shock waves creates a pressure pattern around the supersonic airplane which leads to a strong pressure imbalance in the drag direction, and which integrated over the surface gives rise to wave drag.

Actually, wave drag is pressure drag.

(2) Induced drag

As stated before, besides the portion of the total flight vehicle drag, zero-lift drag, there is another part of the total drag, which is named induced-drag or drag due to lift, resulting from the lift. That is to say, the drag induced by lift is caused by the lift of a flight vehicle.

(3) C_x with Mach

The relation between drag coefficient C_x and Mach variations can be seen as Figure 2.47.

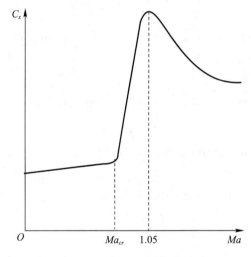

Figure 2.47 C_x-Ma curve

3) Lateral force

$$Z = C_z \frac{1}{2}\rho V^2 S \qquad (2-70)$$

Here, C_z is the lateral force coefficient.

Z is just like Y.

$$C_z = C_z^\beta \beta + C_z^{\delta_y} \delta_y \qquad (2-71)$$

where, β is the angle of the sideslip, δ_y is the rudder deflection angle.

4) Aerodynamic force decomposition

If we only consider the longitudinal symmetric plane, then the aerodynamics can be simplified as two models.

One is that, Aerodynamic force = axial force + normal force.

The direction of axial force is superposition with the body longitudinal axis while the direction of the normal force is perpendicular to it.

Another is that, Aerodynamic force = drag + lift.

The drag is in the opposite direction of the velocity of the flight vehicle, while the direction of lift is perpendicular to it.

Suppose that C_A is the axial force coefficient, C_N is the normal force coefficient, and C_D is the drag coefficient, C_L is the lift coefficient,

$$\begin{bmatrix} C_A \\ C_N \end{bmatrix} = L(\alpha) \begin{bmatrix} C_D \\ C_L \end{bmatrix} \qquad (2-72)$$

where, $L(\alpha) = \begin{bmatrix} \cos\alpha & \sin\alpha \\ -\sin\alpha & \cos\alpha \end{bmatrix}$, α is the angle of attack.

5) Two important aerodynamic concepts

(1) Polar curve

The relations between lift coefficient C_L & drag coefficient C_D can be described as Figure 2.48 at low-speed flow.

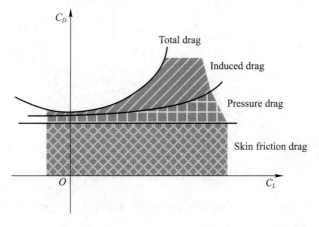

Figure 2.48   Polar curve

(2) Lift-drag ratio

One of the most important aerodynamic parameters is the lift-drag ratio, which is often referred to as lift Y over drag X and written as Y/X. Anyone interests in airplanes have likely heard these words at one time or another. The Y/X combines lift and drag into a single number that can be thought of as the airplane's efficiency of flight. Since lift and drag are both forces, Y/X has no dimensions, which means that it is just a number with no units. A higher value of Y/X means that the airplane is producing lift more efficiently. It can be calculated by C_y/C_x.

2.5.2 Moments acted on flight vehicles

2.5.2.1 Aerodynamic moments

As is known, moment equals force multiplied by the arm of force. For a flight vehicle, there are two important points, one is the center of gravity, the other is the pressure center which is the point acted on by aerodynamics. Generally, they are not in the same position. There is usually a distance between them. As a result, aerodynamic moments are produced.

For the convenience of analyzing the missile's rotary motions around its mass center, along each axis of the Body-fixed axis system, aerodynamic moments are divided into three components, which are named rolling moment M_x, yawing moment M_y, and pitching moment M_z. For example, as shown in Figure 2.49, the rolling moment M_x makes the missile roll around its longitudinal axis Ox_1. If the aileron deflection angle δ_x is positive (rear edge of right aileron turns down and rear edge of left aileron turns up), the rolling moment will be generated. Yawing moment M_y lets the missile roll around its vertical axis Oy_1. To the normal missile, when its rudder deflection angle is positive (rear edge of rudder turns right), then the yawing moment will be created. The pitching moment M_z makes the missile roll around its lateral axis Oz_1. To the normal missile, if the deflection angle of the elevator δ_z is positive (the rear edge of the elevator turns down), a pitching moment will be given to make its head down movement expressed.

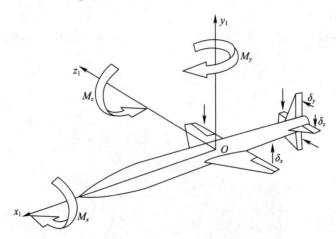

Figure 2.49 Moments acted on a flight vehicle

Like aerodynamic forces, aerodynamic moments could be expressed by their coefficients indicated as follows:

$$\begin{cases} M_x = m_x qSl \\ M_y = m_y qSl \\ M_z = m_z qSL \end{cases} \quad (2-73)$$

where, M_x is rolling moment around axis O_{x_1}, M_y is yawing moment around axis O_y, M_z is pitching moment around axis O_z, m_x, m_y, m_z are dimensionless moment coefficients, $q = \dfrac{1}{2}\rho V^2$ is dynamic pressure, S is reference area, l, L are reference length.

As stated before, lift acting on the axis-symmetrical missile could be approximately expressed as follows:

$$Y = Y^\alpha \alpha + Y^{\delta_z} \delta_z \quad (2-74)$$

Intersection of total aerodynamic force and the missile's longitudinal axis is called the pressure center.

The intersection of the longitudinal axis and the lift created by the attack angle is named the focus point of the missile. Lift caused by the deflection of the rudder acts on the pressure center of the rudder.

The position of pressure center X_P is realized as the range from the head to the pressure center. If the values and working positions of all lifts generated by each part of the missile are all known, the position of pressure center could be determined by the following equation.

$$X_P = \frac{\sum_{k=1}^{n} Y_k X_{Pk}}{Y} = \frac{\sum_{k=1}^{n} c_{yk} X_{Pk} \frac{S_k}{S}}{c_y} \quad (2-75)$$

To the winged missile, the lift generated by the wing is the main part of the total lift. Its position of pressure center depends on the relative position of wings to its bodies. Apparently, if the wing is installed farther away from the head, the value X_P is larger. Moreover, the position of the pressure center is affected by Mach number, attack angle α, the deflection angle of the rudder δ_z, and setting angles of wings and stabilizers. That is because when Ma, α, δ_z and setting angles change, pressure distribution on the missile will be altered. Figure 2.50 shows the relationship among the location of pressure center, Mach number, and attack angle. As seen in the picture, when Mach number approaches 1, the position of the pressure center changes acutely.

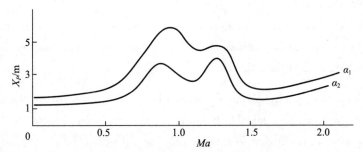

Figure 2.50 Change of pressure center and Mach number and α

Generally, the focus point will be superposition with pressure center only when the missile is axial symmetry ($c_{y0} = 0$) and $\delta_z = 0$.

The position of focus point x_F is considered as the range from the head to the focus point.

$$X_F = \frac{\sum_{k=1}^{n} Y_k^\alpha X_{Fk}}{Y^\alpha} = \frac{\sum_{k=1}^{n} c_{yk}^\alpha X_{Fk} \frac{S_k}{S}}{c_y^\alpha} \quad (2-76)$$

In this formula, Y_k^α is the derivative of lift generated by certain part to attack angle, X_{Fk} is the working position of lift generated by a certain part for attack angle.

1) Pitching moment

(1) Expression formula of pitching moment

To steer the head of a flight vehicle up and down, that is, to make the missile rotate around the axis O_z, there exists pitching angular speed ω_z. Furthermore, it will result in the variation of the angle of attack α.

$$M_z = Y \times \Delta = Y(X_G - X_P) \tag{2-77}$$

Definition to the direction of pitching moment, if the flight vehicle moves downward, then, M_z is negative. Otherwise, M_z is positive.

Generally speaking, pitching moment is affected by the following factors, Mach number, flight height, angle of attack, the deflection angle of elevator and their variation rate, pitching angular speed.

$$M_z = f(Ma, H, \alpha, \delta_z, \omega_z, \dot{\alpha}, \dot{\delta}_z) \tag{2-78}$$

If $\alpha, \delta_z, \omega_z, \dot{\alpha}, \dot{\delta}_z$ are small quantities, M_z can be expressed as follows. at certain flight height and velocity.

$$M_z = M_{z0} + M_z^\alpha \alpha + M_z^{\delta_z} \delta_z + M_z^{\omega_z} \omega_z + M_z^{\dot{\alpha}} \dot{\alpha} + M_z^{\dot{\delta}_z} \dot{\delta}_z \tag{2-79}$$

Here, M_{z0} is the pitching moment when $\alpha = \delta_z = \omega_z = \dot{\alpha} = \dot{\delta}_z = 0$. It is caused by the missile's dissymmetry to the plane Ox_1z_1.

$M_z^\alpha \alpha$ is the pitching moment produced by the angle of attack α, it is also called longitudinal static stability moment.

$M_z^{\delta_z} \delta_z$ is the pitching moment produced by the deflection angle of elevator δ_z, which is named as longitudinal control moment.

$M_z^{\omega_z} \omega_z$ is the pitching moment produced by pitching angle speed ω_z, which is named as longitudinal damping moment whose direction is opposite to the ω_z. The function of it is to prevent the missile from rolling around the axis Oz_1.

$M_z^{\dot{\alpha}} \dot{\alpha}, M_z^{\dot{\delta}_z} \dot{\delta}_z$ are the pitching moment produced by $\dot{\alpha}, \dot{\delta}_z$, which are also longitudinal damping moment. Sometimes, they can be neglected because of their small quantities.

If we use the moment coefficient to express, they can be written by

$$m_z = m_{z0} + m_z^\alpha \alpha + m_z^{\delta_z} \delta_z + m_z^{\omega_z} \omega_z + m_z^{\dot{\alpha}} \dot{\alpha} + m_z^{\dot{\delta}_z} \dot{\delta}_z \tag{2-80}$$

where, m_{z0} is the pitching moment coefficient when $\alpha = \delta_z = \omega_z = \dot{\alpha} = \dot{\delta}_z$.

m_z^α is the derivative of longitudinal static stability moment coefficient to α, usually, m_z^α is negative, for the coefficient m_z^α is associated with the longitudinal static stability. If $m_z^\alpha \big|_{\alpha = \alpha_\beta} < 0$, it is longitudinal static stable.

$m_z^{\delta_z}$ is the derivative of longitudinal control moment coefficient to deflection angle of elevator, which is called control efficiency.

$m_z^{\omega_z}$ is the derivative of the longitudinal damping moment coefficient to pitching angular speed.

$m_z^{\dot{\alpha}}$ and $m_z^{\dot{\delta}_z}$ is the coefficient of the pitching moment produced by $\dot{\alpha}, \dot{\delta}_z$.

Figure 2.51 shows the pitching moment. Figure 2.52 shows the pitching moment produced by α. Figure 2.53 shows the pitching moment produced by δ_z. Figure 2.54 shows the pitching moment produced by ω_z.

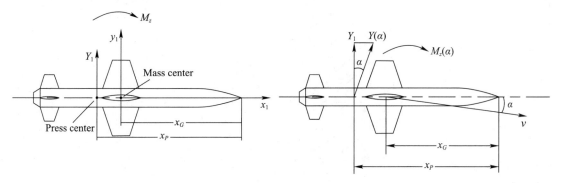

Figure 2.51 Pitching moment   Figure 2.52 Pitching moment produced by α

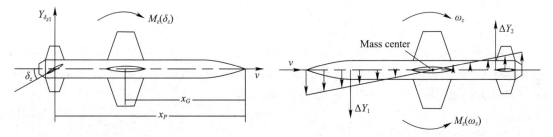

Figure 2.53 Pitching moment produced by δ_z Figure 2.54 Pitching moment produced by ω_z

(2) Concepts for pitching moment

Ⅰ. Steady of flight

If a flight vehicle moves in the air with constant parameters, we call it steady static flight, that is to say, the conditions of steady-state flight which mean $\omega_z = 0$, $\alpha(t)$ and $\delta_z(t)$ are regarded as constants. Under such assumption, the pitching moment and pitching moment coefficient can be written as follows:

$$M_z = M_{z0} + M_z^\alpha \alpha + M_z^{\delta_z} \delta_z \tag{2-81}$$

$$m_z = m_{z0} + m_z^\alpha \alpha + m_z^{\delta_z} \delta_z \tag{2-82}$$

If a flight vehicle is axis-symmetrical, the pitching moment coefficient can be stated by

$$m_z = m_z^\alpha \alpha + m_z^{\delta_z} \delta_z \tag{2-83}$$

There exist linear relations between α and δ_z at the steady-state flight.

Only when α and δ_z are small, the linear relationship above will come into existence. In company with the increase of α and δ_z, the linear relationship will be destroyed. If δ_z is supposed to be a certain value, the relationship between m_z and α could be described as a curve shown in Figure 2.55. As seen from the picture, when the attack angle exceeds certain scope, a linear relationship between m_z and M_x will not be kept anymore.

According to Figure 2.55, intersections of these curves and the abscissa axis satisfy that $m_z = 0$. Such intersections are named the equilibrium point. Here, the characteristics of missile motion are $\omega_z = \dot{\alpha} = \dot{\delta}_z = 0$. α and δ_z are kept to be a certain relationship to make the summation of pitching moments created by α and δ_z to be zero. This means the missile is at a longitudinal balance state.

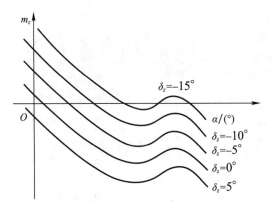

Figure 2.55 Curves of $m_z = f(\alpha)$

Ⅱ. Instantaneous balance assumption

As can be seen in Figure 2.56, at the state of $m_z = 0$, the quantities of all moment produced by lift relative to mass center equal to zero. That is to say, the flight vehicle locates at the balance state, also called the static balance state of longitudinal moment.

$$0 = m_z^\alpha \alpha + m_z^{\delta_z} \delta_z \tag{2-84}$$

It can be changed as follows:

$$\delta_{zB} = -\frac{m_z^\alpha}{m_z^{\delta_z}} \alpha_B \tag{2-85}$$

Thus, it is called balance relation formula of longitudinal moment, so the subscript B is used.

Furthermore, the "Instantaneous balance" assumption is given, which means if the rotation of a flight vehicle is neglected, there is no transient process for the variation of the motion parameters. That is, there is the relation by

$$0 = m_z^\alpha \alpha_B + m_z^{\delta_z} \delta_B \tag{2-86}$$

According to this formula, some parameters can be designed in the stage of the preliminary design of flight vehicle.

(3) Longitudinal static stability

According to the design of a flight vehicle, there are three types of static stability, which are statically stable, unstable and neutral.

The definition of static stability is introduced as follows. It is the tendency for a flight vehicle to return to its previous attitude after disturbed without controlling. Thus, restoring force and moment must exist. If the moment makes the flight vehicle depart from its initial attitude, even more, it is statically unstable. If the flight vehicle keeps on its present state once disturbed, the phenomenon is named as statically neutral stable state.

Let us discuss the longitudinal static stability first. Longitudinal static stability can be judged using a derivative $m_z^\alpha \big|_{\alpha = \alpha_B}$. If the flight vehicle flies in balance at a certain attack angle α_B, there is the increment of the angle of attack due to some reason (for instance, vertical gust), then the increment of lift acting on the focus point is created. When the deflection angle of the rudder δ_z keeps its original value, the increment of pitching moment could be written by

$$\Delta M_z(\alpha) = m_z^\alpha \big|_{\alpha = \alpha_B} \Delta \alpha q S L \tag{2-87}$$

If $m_z^\alpha \big|_{\alpha=\alpha_B} < 0$ shown in Figure 2.56 (a), and $\Delta\alpha > 0$, then $\Delta M_z(\alpha)$ is negative. For normal configurations of a flight vehicle, whose head will be lower to make attack angle return from ($\alpha_B + \Delta\alpha$) to α_B. This physical behavior is entitled to static stability. Aerodynamic moment generated after disturbance to make the flight vehicle return to its previous attitude is named as statically stable moment or restoring moment.

If $m_z^\alpha \big|_{\alpha=\alpha_B} > 0$, shown in Figure 2.56 (b), and $\Delta\alpha > 0$, then $\Delta M_z(\alpha)$ is positive. The increment of pitching moment will make the flight vehicle depart from its equilibrium position much more. Such an instance is named as statically unstable. Statically unstable aerodynamic moment is called turning torque.

If $m_z^\alpha \big|_{\alpha=\alpha_B} = 0$, as shown in Figure 2.56 (c), the situation of neural stable is established. When the flight vehicle departs from its equilibrium position, the increment of pitching moment is equal to zero. Added attack angle created by disturbance will neither be increased nor be decreased.

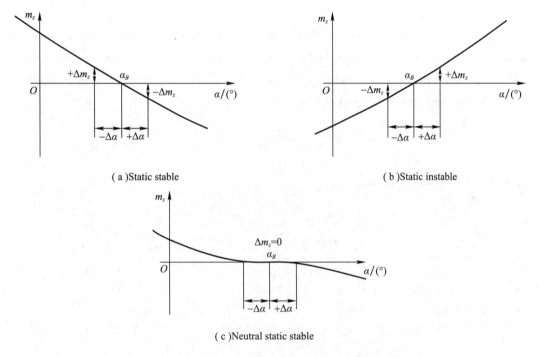

Figure 2.56 Three types of longitudinal static stability

Partial derivative m_z^α shows size and orientation of pitching moment coefficient created by attack angle per unit, it illustrates the quality of its longitudinal static stability.

Conditions of longitudinal static stability are summarized as follows:

If $m_z^\alpha \big|_{\alpha=\alpha_B} < 0$, it is longitudinal static stable.

If $m_z^\alpha \big|_{\alpha=\alpha_B} = 0$, it is longitudinal static neutral stable.

If $m_z^\alpha \big|_{\alpha=\alpha_B} > 0$, it is longitudinal static instable.

There is another expression of longitudinal static stability, which is shown as follows:

$$m_z^{C_y} = \frac{\partial m_z}{\partial C_y} = -\frac{X_F - X_G}{L} \tag{2-88}$$

Here, if $m_z^{C_y} < 0$, it is longitudinal static stable. That is, $X_F - X_G > 0$, it means that the mass center locates in front of the focus point. Furthermore, if a flight vehicle is statically unstable, it must be that the mass center locates after the focus point. If it is neutrally stable, the mass center and focus point must be at the same place.

Therefore, $m_z^{c_y}$ is also called longitudinal static stability. According to this definition, the value between focus point and mass center can be used to determine the margin of static stability.

Longitudinal static stability of a flight vehicle is associated with the flight quality. To achieve desirable longitudinal static stability, two methods are adopted during the design process. One is to change the configuration of the flight vehicle to alter the position of the focus point. Changing the configuration, area of wings and their relative positions to the body, transferring tail area, and adding opposite stabilizers are all belong to this method. The other is to change the positions of equipment in the flight vehicle to adjust the position of the mass center.

(4) Control moment

Towards the static stable flight vehicle, if it is a normal configuration, the deflection angle of the elevator should be negative (rear edge turns up) when it flies at a positive attack angle as shown in Figure 2.57 (a). If it is a canard configuration, the deflection angle of elevator should be negative to generate the required nose-up pitching moment as shown in Figure 2.57 (b).

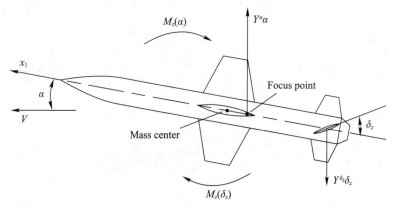

(a) normal configuration

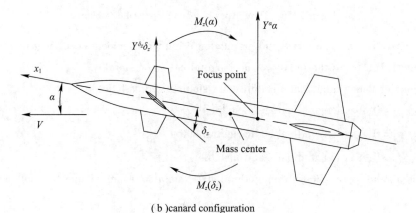

(b) canard configuration

Figure 2.57   Controlling moment

At the same time, the nose-down pitching moment formed by the lift $Y^\alpha \alpha$ against the mass center makes the flight vehicle at the state of moment balance. After deflexion of the elevator, the aerodynamic moment generated by the air force against the mass center is named as controlling moment expressed by

$$\begin{cases} M_z(\delta_z) = -c_y^{\delta_z}\delta_z qS (X_R - X_G) = m_z^{\delta_z}\delta_z qSL \\ m_z^{\delta_z} = -c_y^{\delta_z}(\bar{X}_R - \bar{X}_G) \end{cases} \quad (2-89)$$

In the formula, $\bar{X}_R = \dfrac{X_R}{L}$ is the relative value of the distance from the pressure center of the elevator to the head of missile, it is dimensionless.

$m_z^{\delta_z}$ is the control moment coefficient created by the elevator which has turned unit angle, it is called the efficiency of rudder. Towards normal configuration flight vehicle, the elevator is always behind the mass center so that $m_z^{\delta_z} < 0$. To tail-first flight vehicle, it is $m_z^{\delta_z} > 0$.

$c_y^{\delta_z}$ is the lift coefficient created by the elevator which has turned unit angle. The relationship between it and Mach number is shown in Figure 2.58.

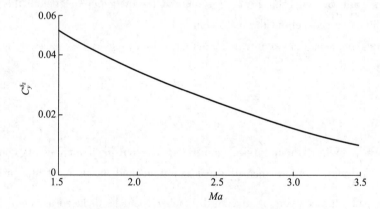

Figure 2.58 $c_y^{\delta_z}$ versus Mach number

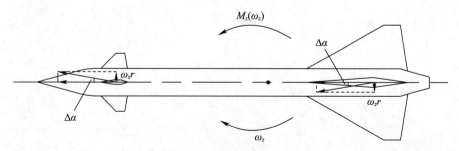

Figure 2.59 Pitching damping moment

(5) Pitching moment due to pitching angular speed

As seen in Figure 2.59, the damping moment is created for the flight vehicle to roll around the axis Oz_1, its magnitude is proportional to the spinning angular speed ω_z, and its direction is opposite to it. The function of the damping moment is to stop the missile from rolling around the axis Oz_1, thus it is named as pitching moment due to pitching angular speed (or longitudinal damping moment). Apparently, a damping moment won't exist without flight vehicle's rotating motion.

Supposed the mass center of the flight vehicle moves in velocity V, at the same time, it rolls around the axis Oz_1 in angular speed ω_z. Additional velocity whose size is $\omega_z r$ will be given to each point on the surface of the flight vehicle for rotation. Its direction is vertical to the radius vector r which connects the mass center and the point. When $\omega_z \neq 0$ the airflow direction relative to the flight vehicle body is changed. So, if $\omega_z > 0$, the attack angle of each point on the surface before mass center will decrease by an angle $\Delta\alpha(r)$ whose value is

$$\tan\Delta\alpha(r) = r\omega_z/V \qquad (2-90)$$

If the attack angle of each point is on the surface behind the mass center, then it will increase by an angle $\Delta\alpha(r)$. As the attack angle of each point is altered, additional lift $\Delta Y_i(\omega_z)$ has been generated on each point. Additional pitching moment $\Delta M_{zi}(\omega_z)$ caused by $\Delta Y_i(\omega_z)$ against mass center has also been produced. If $\omega_z > 0$, lift generated on each point before the mass center is downward. While added lift produced behind the mass center is upward, thus, added pitching moments created by such lifts have a direction opposite to ω_z. Total pitching moment due to pitching velocity $M_{zi}(\omega_z)$ could be got. Total $Y(\omega_z)$ is omitted for lift generated before the mass center is opposite to the lift produced behind the mass center.

Here, pitching moment due to pitching angular velocity could be replaced by its coefficient whose dimension is 1.

$$M_z^{\bar{\omega}_z} = m_z^{\bar{\omega}_z} qSL^2/V \qquad (2-91)$$

$$\bar{\omega}_z = \omega_z L/V \qquad (2-92)$$

where $m_z^{\bar{\omega}_z}$ is negative. Whose quantity size mainly depends on Mach number, its configuration, and the position of the mass center. Once the configuration and position of the mass center are determined, the relationship between $m_z^{\bar{\omega}_z}$ and Mach number could be deduced as shown in Figure 2.60. To write conveniently $m_z^{\bar{\omega}_z}$ is generally recorded as $m_z^{\omega_z}$.

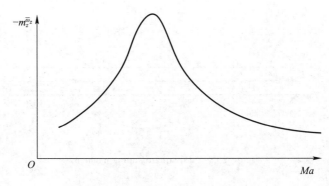

Figure 2.60 $m_z^{\bar{\omega}_z}$ versus Mach number

In general, comparing with pitching moment and controlling moment, pitching moment due to pitching angular velocity is relatively small. However, pitching moment due to pitching angular velocity which will urge the oscillation of the transition process to be lessened is an important factor to improve the quality of flight vehicle's transition process. Thus, it should be paid attention to.

(6) Additional pitching moment created by downwash

At the same time, parts of the pitching moment are produced using downwash. They can be considered as the damping moment as well. When $\dot{\alpha} \neq 0$ and $\dot{\delta}_z \neq 0$, two coefficients of additional pitching moment could be expressed as $\overline{m_z^{\dot{\alpha}}} \overline{\dot{\alpha}}$ and $\overline{m_z^{\dot{\delta}_z}} \overline{\dot{\delta}_z}$ respectively. To write conveniently, $\overline{m_z^{\dot{\alpha}}}$ and $\overline{m_z^{\dot{\delta}_z}}$ are replaced by $m_z^{\dot{\alpha}}$ and $m_z^{\dot{\delta}}$, whose dimension is 1.

2) Yawing moment

Yawing moment makes the flight vehicle turn around the axis Oy_1, which are mainly from a vertical tail. Just like pitching moment, yawing moment and the coefficient can be expressed by

$$M_y = M_y^\beta \beta + M_y^{\delta_y} \delta_y + M_y^{\omega_y} \omega_y + M_y^{\dot{\beta}} \dot{\beta} + M_y^{\dot{\delta}_y} \dot{\delta}_y + M_y^{\omega_x} \omega_x \qquad (2-93)$$

$$m_y = m_y^\beta \beta + m_y^{\delta_y} \delta_y + m_y^{\omega_y} \omega_y + m_y^{\dot{\beta}} \dot{\beta} + m_y^{\dot{\delta}_y} \dot{\delta}_y + m_y^{\omega_x} \omega_x \qquad (2-94)$$

where, $m_y^{\omega_x}$ is cross derivative, m_y^β is directional static stability.

If $m_y^\beta < 0$, then, it is named as directional static stable.

If $m_y^\beta = 0$, it is called directional static neutral stable.

If $m_y^\beta > 0$, it is called directional static instable.

Figure 2.61 shows the yawing moment. Figure 2.62 shows the yawing moment produced by vertical empennage.

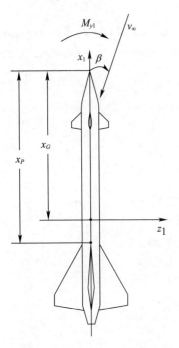

Figure 2.61 Yawing moment

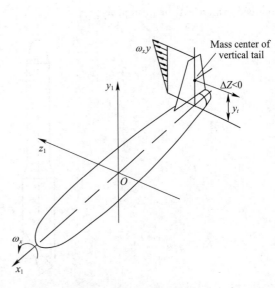

Figure 2.62 Yawing moment produced by vertical empennage

3) Rolling moment

The rolling moment makes the flight vehicle turn around axis Ox_1. It is produced by the unsymmetrical of a flight vehicle.

$$M_x = M_{x0} + M_x^\beta \beta + M_x^{\delta_x}\delta_x + M_x^{\delta_y}\delta_y + M_x^{\omega_x}\omega_x + M_x^{\omega_y}\omega_y \qquad (2-95)$$

$$m_x = m_{x0} + m_x^\beta \beta + m_x^{\delta_x}\delta_x + m_x^{\delta_y}\delta_y + m_x^{\omega_x}\omega_x + m_x^{\omega_y}\omega_y \qquad (2-96)$$

where m_x^β is called rolling static stability.

If $m_x^\beta < 0$, then it is rolling static stable.

As seen in Figure 2.63 and Figure 2.64, rolling damping moment is given by $M_x^{\omega_x}\omega_x$, and rolling control moment is expressed as $M_x^{\delta_x}\delta_x$.

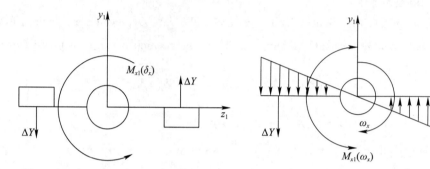

Figure 2.63   Rolling moment produced by δ_x Figure 2.64   Rolling moment produced by ω_x

2.5.2.2 Thrust moments

Where does the thrust act on?

Generally, the thrust acts on the geometry axis line, and the mass center locates at the line, then the arm of thrust is zero, so there is no thrust moment. $M_P = 0$.

However, as shown in Figure 2.65, the practical situation is that the thrust P does not pass through the mass center, and there is an angle between P and axis Ox_1, then the thrust moment M_P is produced, which can be written as,

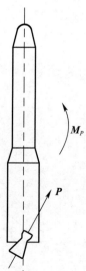

Figure 2.65   Thrust moment

$$M_P = R_P \times P \qquad (2-97)$$

where R_P is the radius vector.

2.6 Flight vehicle motion equation

2.6.1 Basic assumptions

To establish the equations of motion of a flight vehicle, basic assumptions for the flight vehicle equation of motion are given at first. An airplane is chosen as an example.

(1) The airplane is a rigid body.

(2) The Earth is regarded as flat and stationary in inertial space, thus the rotation velocity is neglected. And Newton's laws of motion are valid.

(3) The acceleration of gravity does not change with the flight altitude.

(4) The airplane is symmetrical plane with respect to oxy-plane. And the distribution of inner masses is symmetrical for the airplane, therefore, the product of inertia, $I_{xz} = I_{zx} = 0, I_{yz} = I_{zy} = 0$.

2.6.2 Fundamental principles

Equations of motion of a flight vehicle in inertial space are written according to the fundamental principles.

$$\begin{cases} \dfrac{\mathrm{d}(mV)}{\mathrm{d}t} = \Sigma F_i \\ \dfrac{\mathrm{d}H}{\mathrm{d}t} = \Sigma M_i \end{cases} \qquad (2-98)$$

where, m is mass of the airplane, ΣF_i is the total external forces acted on the airplane, V is speed of mass center, ΣM_i is the total external moments acted on the airplane, H is moment of momentum.

As we know, relations between the moving coordinate system (e.g. body-fixed reference frame) and the inertial coordinate system (i.e. Earth-fixed reference frame) for V and H are as follows:

$$\begin{cases} \dfrac{\mathrm{d}V}{\mathrm{d}t} = \dfrac{\delta V}{\mathrm{d}t} + \Omega \times V \\ \dfrac{\mathrm{d}H}{\mathrm{d}t} = \dfrac{\delta H}{\mathrm{d}t} + \Omega \times H \end{cases} \qquad (2-99)$$

where, V and Ω are the speed vector and angular speed vector of the airplane respectively, $\dfrac{\delta(\cdot)}{\mathrm{d}t}$ and $\dfrac{\mathrm{d}(\cdot)}{\mathrm{d}t}$ are the variation of V and H at the moving coordinate system and initial coordinate system respectively.

2.6.3 Flight vehicle motion equation establishment

1) Dynamics equation

(1) Dynamics equation of the movement of the missile's center of mass

The first equation of the system of Eq. (2-99) can be rewritten as

$$m\frac{dV}{dt} = m\left(\frac{\delta V}{\delta t} + \Omega \times V\right) = F + P \qquad (2-100)$$

$$\begin{cases} V = i_2 V_{x_2} + j_2 V_{y_2} + k_2 V_{z_2} \\ \Omega = i_2 \Omega_{x_2} + j_2 \Omega_{y_2} + k_2 \Omega_{z_2} \\ \dfrac{\delta V}{\delta t} = i_2 \dfrac{dV_{x_2}}{dt} + j_2 \dfrac{dV_{y_2}}{dt} + k_2 \dfrac{dV_{z_2}}{dt} \end{cases} \qquad (2-101)$$

where, i_2, j_2, and k_2 are the unit vectors of each axis of the flight path reference frame $Ox_2y_2z_2$; Ω_{x_2}, Ω_{y_2}, and Ω_{z_2} are the components of the rotational angular velocity Ω of the flight path reference frame relative to the Earth-fixed reference frame on each axis of $Ox_2y_2z_2$; V_{x_2}, V_{y_2}, V_{z_2} are the components of the missile's centroid velocity vector V on each axis of $Ox_2y_2z_2$.

According to the definition of the flight path reference frame

$$\begin{bmatrix} V_{x_2} \\ V_{y_2} \\ V_{z_2} \end{bmatrix} = \begin{bmatrix} V \\ 0 \\ 0 \end{bmatrix} \qquad (2-102)$$

Then

$$\frac{\delta V}{\delta t} = \frac{dV}{dt} i_2 \qquad (2-103)$$

$$\Omega \times V = \begin{vmatrix} i_2 & j_2 & k_2 \\ \Omega_{x_2} & \Omega_{y_2} & \Omega_{z_2} \\ V_{x_2} & V_{y_2} & V_{z_2} \end{vmatrix} = \begin{vmatrix} i_2 & j_2 & k_2 \\ \Omega_{x_2} & \Omega_{y_2} & \Omega_{z_2} \\ V & 0 & 0 \end{vmatrix} = V\Omega_{z_2} j_2 - V\Omega_{y_2} k_2 \qquad (2-104)$$

According to the conversion between the Flight path reference frame and the Earth-fixed reference frame, we can get

$$\Omega = \dot{\psi}_V + \dot{\theta} \qquad (2-105)$$

In the formula, $\dot{\psi}_V$ and $\dot{\theta}$ are respectively on the Ay axis of the Earth-fixed reference frame and the Oz_2 axis of the Flight path reference frame, so the following formulas can be obtained from Eq. (2-48) and Eq. (2-49)

$$\begin{bmatrix} \Omega_{x_2} \\ \Omega_{y_2} \\ \Omega_{z_2} \end{bmatrix} = L(\theta,\psi_V) \begin{bmatrix} 0 \\ \dot{\psi}_V \\ 0 \end{bmatrix} + \begin{bmatrix} 0 \\ 0 \\ \dot{\theta} \end{bmatrix} = \begin{bmatrix} \dot{\psi}_V \sin\theta \\ \dot{\psi}_V \cos\theta \\ \dot{\theta} \end{bmatrix} \qquad (2-106)$$

Substituting Eq. (2-106) into Eq. (2-104) can be obtained

$$\Omega \times V = V\dot{\theta} j_2 - V\dot{\psi}_V \cos\theta k_2 \qquad (2-107)$$

Substituting Eq. (2 – 103) and Eq. (2 – 107) into Eq. (2 – 100), expand it to get

$$\begin{cases} m\dfrac{dV}{dt} = F_{x_2} + P_{x_2} \\ mV\dfrac{d\theta}{dt} = F_{y_2} + P_{y_2} \\ -mV\cos\theta\dfrac{d\psi_V}{dt} = F_{z_2} + P_{z_2} \end{cases} \qquad (2-108)$$

In the formula, $F_{x_2}, F_{y_2}, F_{z_2}$ are the algebraic sum of all external forces (total aerodynamic force R, gravity G, etc.) on each axis of $Ox_2y_2z_2$ except thrust; $P_{x_2}, P_{y_2}, P_{z_2}$ are respectively the components of thrust P on each axis of $Ox_2y_2z_2$.

The expressions for the projection of total aerodynamic force R, gravity G and thrust P on the Flight path reference frame are listed below.

The total aerodynamic force R acting on the missile can be decomposed into drag force X, lift force Y and lateral force Z along the velocity coordinate system, which is

$$\begin{bmatrix} R_{x_3} \\ R_{y_3} \\ R_{z_3} \end{bmatrix} = \begin{bmatrix} -X \\ Y \\ Z \end{bmatrix} \qquad (2-109)$$

According to the conversion relationship between the Velocity coordinate system and the Flight path reference frame, Eq. (2 – 52) and Eq. (2 – 53) can be used to obtain

$$\begin{bmatrix} R_{x_2} \\ R_{y_2} \\ R_{z_2} \end{bmatrix} = \boldsymbol{L}^{\mathrm{T}}(\gamma_V)\begin{bmatrix} R_{x_3} \\ R_{y_3} \\ R_{z_3} \end{bmatrix} = \begin{bmatrix} -X \\ Y\cos\gamma_V - Z\sin\gamma_V \\ Y\sin\gamma_V + Z\cos\gamma_V \end{bmatrix} \qquad (2-110)$$

For short-range tactical missiles, gravity G can be considered as the negative direction along the Ay axis of the Earth-fixed reference frame, so it can be expressed as

$$\begin{bmatrix} G_x \\ G_y \\ G_z \end{bmatrix} = \begin{bmatrix} 0 \\ -mg \\ 0 \end{bmatrix} \qquad (2-111)$$

Project it onto the Flight path reference frame $Ox_2y_2z_2$, you can use Eq. (2 – 48) and Eq. (2 – 49) to get

$$\begin{bmatrix} G_{x_2} \\ G_{y_2} \\ G_{z_2} \end{bmatrix} = \boldsymbol{L}(\theta,\psi_V)\begin{bmatrix} G_x \\ G_y \\ G_z \end{bmatrix} = \begin{bmatrix} -mg\sin\theta \\ -mg\cos\theta \\ 0 \end{bmatrix} \qquad (2-112)$$

If the thrust P of the engine coincides with the longitudinal axis Ox_1 of the projectile, then

$$\begin{bmatrix} P_{x_1} \\ P_{y_1} \\ P_{z_1} \end{bmatrix} = \begin{bmatrix} P \\ 0 \\ 0 \end{bmatrix} \qquad (2-113)$$

Project it on the Flight path reference frame $Ox_2y_2z_2$, which can be obtained by Eq. (2-50) ~ Eq. (2-53)

$$\begin{bmatrix} P_{x_2} \\ P_{y_2} \\ P_{z_2} \end{bmatrix} = \boldsymbol{L}^{\mathrm{T}}(\gamma_V)\,\boldsymbol{L}^{\mathrm{T}}(\alpha,\beta) \begin{bmatrix} P_{x_1} \\ P_{y_1} \\ P_{z_1} \end{bmatrix} = \begin{bmatrix} P\cos\alpha\cos\beta \\ P(\sin\alpha\cos\gamma_V + \cos\alpha\sin\beta\sin\gamma_V) \\ P(\sin\alpha\sin\gamma_V - \cos\alpha\sin\beta\cos\gamma_V) \end{bmatrix} \qquad (2-114)$$

Substituting Eq. (2-110), Eq. (2-112) ~ Eq. (2-114) into Eq. (2-108), the scalar form of the dynamic equation of the movement of the center of mass of the missile is obtained as

$$\begin{cases} m\dfrac{\mathrm{d}V}{\mathrm{d}t} = P\cos\alpha\cos\beta - X - mg\sin\theta \\[6pt] mV\dfrac{\mathrm{d}\theta}{\mathrm{d}t} = P(\sin\alpha\cos\gamma_V + \cos\alpha\sin\beta\sin\gamma_V) + Y\cos\gamma_V - Z\sin\gamma_V - mg\cos\theta \\[6pt] -mV\cos\theta\dfrac{\mathrm{d}\psi_V}{\mathrm{d}t} = P(\sin\alpha\sin\gamma_V - \cos\alpha\sin\beta\cos\gamma_V) + Y\sin\gamma_V + Z\cos\gamma_V \end{cases} \qquad (2-115)$$

where, $\dfrac{\mathrm{d}V}{\mathrm{d}t}$ is the projection of the acceleration of the missile's center of mass along the tangential direction of the ballistic (Ox_2 axis), called the tangential acceleration; $V\dfrac{\mathrm{d}\theta}{\mathrm{d}t}$ is the acceleration of the missile's center of mass projected on the vertical plane (Ox_2y_2) along the ballistic normal (Oy_2 axis), which is called the normal acceleration; $-mV\cos\theta\dfrac{\mathrm{d}\psi_V}{\mathrm{d}t}$ is the acceleration of the missile's center of mass Projected in the horizontal component (that is, the Oz_2 axis), also known as the normal acceleration. The "-" sign at the left end of the formula indicates that the centripetal force is positive, and the corresponding $\dot{\psi}_V$ is negative; vice versa. It is determined by the definition of the sign of the angle ψ_V.

(2) Dynamics equation of the missile rotating around the center of mass

The scalar form of the dynamic vector equation of the missile rotating around the center of mass Eq. (2-98) is the simplest written in the Body-fixed reference frame.

The Body-fixed reference frame is a dynamic coordinate system, and the rotational angular velocity of the Body-fixed reference frame relative to the ground coordinate system is represented by $\boldsymbol{\omega}$.

In the same way, the dynamic equation of the missile rotating around the center of mass is established on the moving coordinate system (the Body-fixed reference frame). The second equation of Eq. (2-98) can be written as

$$\frac{\mathrm{d}\boldsymbol{H}}{\mathrm{d}t} = \frac{\delta \boldsymbol{H}}{\delta t} + \boldsymbol{\omega} \times \boldsymbol{H} = \boldsymbol{M} + \boldsymbol{M}_P \qquad (2-116)$$

Set i_1, j_1, and k_1 to be the unit vectors along each axis of the Body-fixed reference frame $Ox_1y_1z_1$; $\omega_{x_1}, \omega_{y_1}, \omega_{z_1}$ are the components of the rotational angular velocity of the Body-fixed reference frame relative to the Earth-fixed reference frame ω along each axis of the projectile coordinate system; The components of the moment of momentum H on each axis of the projectile coordinate system are $H_{x_1}, H_{y_1}, H_{z_1}$.

$$\frac{\delta H}{\delta t} = \frac{dH_{x_1}}{dt}i_1 + \frac{dH_{y_1}}{dt}j_1 + \frac{dH_{z_1}}{dt}k_1 \qquad (2-117)$$

Moment of momentum H can be expressed as

$$H = J \cdot \omega \qquad (2-118)$$

where, J is the inertia tensor.

The components of the moment of momentum H on each axis of the Body-fixed reference frame can be expressed as

$$\begin{bmatrix} H_{x_1} \\ H_{y_1} \\ H_{z_1} \end{bmatrix} = \begin{bmatrix} J_{x_1x_1} & -J_{x_1y_1} & -J_{x_1z_1} \\ -J_{y_1x_1} & J_{y_1y_1} & -J_{y_1z_1} \\ -J_{z_1x_1} & -J_{z_1y_1} & J_{z_1z_1} \end{bmatrix} \begin{bmatrix} \omega_{x_1} \\ \omega_{y_1} \\ \omega_{z_1} \end{bmatrix} \qquad (2-119)$$

where, $J_{x_1x_1}, J_{y_1y_1}, J_{z_1z_1}$ are the moment of inertia of the missile on each axis of the Body-fixed reference frame; $J_{x_1y_1}, J_{y_1z_1}, J_{z_1y_1}$ are the product of inertia of the missile on each axis of the Body-fixed reference frame.

For tactical missiles, they generally have an axisymmetric shape. At this time, it can be considered that the Body-fixed reference frame is its inertial principal axis system. Under this condition, the product of inertia of the missile on each axis of the Body-fixed reference frame is zero. For the convenience of writing, the above moments of inertia are represented by J_{x_1}, J_{y_1}, and J_{z_1} respectively, then Eq. (2-119) can be simplified as

$$\begin{bmatrix} H_{x_1} \\ H_{y_1} \\ H_{z_1} \end{bmatrix} = \begin{bmatrix} J_{x_1} & 0 & 0 \\ 0 & J_{y_1} & 0 \\ 0 & 0 & J_{z_1} \end{bmatrix} \begin{bmatrix} \omega_{x_1} \\ \omega_{y_1} \\ \omega_{z_1} \end{bmatrix} = \begin{bmatrix} J_{x_1}\omega_{x_1} \\ J_{y_1}\omega_{y_1} \\ J_{z_1}\omega_{z_1} \end{bmatrix} \qquad (2-120)$$

Substituting Eq. (2-120) into Eq. (2-117) can be obtained

$$\frac{\delta H}{\delta t} = J_{x_1}\frac{d\omega_{x_1}}{dt}i_1 + J_{y_1}\frac{d\omega_{y_1}}{dt}j_1 + J_{z_1}\frac{d\omega_{z_1}}{dt}k_1 \qquad (2-121)$$

$$\omega \times H = \begin{vmatrix} i_1 & j_1 & k_1 \\ \omega_{x_1} & \omega_{y_1} & \omega_{z_1} \\ H_{x_1} & H_{y_1} & H_{z_1} \end{vmatrix} = \begin{vmatrix} i_1 & j_1 & k_1 \\ \omega_{x_1} & \omega_{y_1} & \omega_{z_1} \\ J_{x_1}\omega_{x_1} & J_{y_1}\omega_{y_1} & J_{z_1}\omega_{z_1} \end{vmatrix}$$

$$= (J_{z_1} - J_{y_1})\omega_{z_1}\omega_{y_1}i_1 + (J_{x_1} - J_{z_1})\omega_{x_1}\omega_{z_1}j_1 + (J_{y_1} - J_{x_1})\omega_{y_1}\omega_{x_1}k_1 \qquad (2-122)$$

Substituting Eq. (2-121) and Eq. (2-122) into Eq. (2-116), the dynamic scalar equation of the missile rotating around the center of mass is

$$\begin{cases} J_{x_1}\dfrac{d\omega_{x_1}}{dt} + (J_{z_1} - J_{y_1})\,\omega_{z_1}\omega_{y_1} = M_{x_1} \\ J_{y_1}\dfrac{d\omega_{y_1}}{dt} + (J_{x_1} - J_{z_1})\,\omega_{x_1}\omega_{z_1} = M_{y_1} \\ J_{z_1}\dfrac{d\omega_{z_1}}{dt} + (J_{y_1} - J_{x_1})\,\omega_{y_1}\omega_{x_1} = M_{z_1} \end{cases} \quad (2-123)$$

where, $J_{x_1}, J_{y_1}, J_{z_1}$ are the moments of inertia of the missile with respect to each axis of the missile body coordinate system (i.e., the inertial principal axis system), which change continuously with the ejection of fuel combustion products; $\omega_{x_1}, \omega_{y_1}, \omega_{z_1}$ are the components of the rotational angular velocity ω of the Body-fixed reference frame relative to the ground coordinate system on each axis of the Body-fixed reference frame; $\dfrac{d\omega_{x_1}}{dt}, \dfrac{d\omega_{y_1}}{dt}, \dfrac{d\omega_{z_1}}{dt}$ are respectively are the components of the body rotation angular acceleration vector on each axis of the Body-fixed reference frame; $M_{x_1}, M_{y_1}, M_{z_1}$ are the components of the moment of the center of mass of all external forces (including thrust) acting on the missile on each axis of the Body-fixed reference frame.

For the convenience of writing later, the footnote "1" in Eq. (2-123) is omitted.

2) Kinematic equation

(1) Kinematics equation of missile's center of mass movement

To determine the trajectory (ballistic) of the missile's center of mass relative to the Earth-fixed reference frame, it is necessary to establish a kinematic equation for the movement of the missile's center of mass relative to the Earth-fixed reference frame. When calculating the aerodynamic force and thrust, it is necessary to know the height of the missile at any instant, and determine the position of the corresponding instant missile through ballistic calculation. Therefore, it is necessary to establish the position equation of the missile's center of mass relative to the Earth-fixed reference frame $Axyz$.

$$\begin{bmatrix} \dfrac{dx}{dt} \\ \dfrac{dy}{dt} \\ \dfrac{dz}{dt} \end{bmatrix} = \begin{bmatrix} V_x \\ V_y \\ V_z \end{bmatrix} \quad (2-124)$$

According to the definition of the Flight path reference frame, the Velocity vector of the missile's center of mass coincides with the Ox_2 axis of the Flight path reference frame, which is

$$\begin{bmatrix} V_{x_2} \\ V_{y_2} \\ V_{z_2} \end{bmatrix} = \begin{bmatrix} V \\ 0 \\ 0 \end{bmatrix} \quad (2-125)$$

Using the conversion relationship between the Earth-fixed reference frame and the Flight path reference frame, we can get

$$\begin{bmatrix} V_x \\ V_y \\ V_z \end{bmatrix} = \boldsymbol{L}^{\mathrm{T}}(\theta, \psi_V) \begin{bmatrix} V_{x2} \\ V_{y2} \\ V_{z2} \end{bmatrix} \quad (2-126)$$

Substituting Eq. (2-125) and the transposed matrix of Eq. (2-49) into Eq. (2-126) and substituting its result into Eq. (2-124), the kinematic equation of the movement of the center of mass of the missile is obtained as

$$\begin{cases} \dfrac{dx}{dt} = V\cos\theta\cos\psi_V \\ \dfrac{dy}{dt} = V\sin\theta \\ \dfrac{dz}{dt} = -V\cos\theta\sin\psi_V \end{cases} \quad (2-127)$$

(2) Kinematics equation of the missile rotating around the center of mass

To determine the attitude of the missile in space, it is necessary to establish a kinematic equation describing the attitude change of the missile body relative to the Earth-fixed reference frame, and to establish the relationship between the attitude angle ϑ, ψ, γ change rate and the rotation angular velocity components $\omega_{x_1}, \omega_{y_1}, \omega_{z_1}$ of the missile relative to the Earth-fixed reference frame.

We know that according to the conversion relationship between the Earth-fixed reference frame and the Body-fixed reference frame, we can get

$$\boldsymbol{\omega} = \dot{\boldsymbol{\psi}} + \dot{\boldsymbol{\vartheta}} + \dot{\boldsymbol{\gamma}} \quad (2-128)$$

Since $\dot{\psi}$ and $\dot{\gamma}$ respectively coincide with the Ay axis of the Earth-fixed reference frame and the Ox_1 axis of the Body-fixed reference frame, and $\dot{\vartheta}$ and Oz' axes coincide, there is

$$\begin{bmatrix} \omega_{x_1} \\ \omega_{y_1} \\ \omega_{z_1} \end{bmatrix} = \boldsymbol{L}(\gamma, \vartheta, \psi) \begin{bmatrix} 0 \\ \dot{\psi} \\ 0 \end{bmatrix} + \boldsymbol{L}(\gamma) \begin{bmatrix} 0 \\ 0 \\ \dot{\vartheta} \end{bmatrix} + \begin{bmatrix} \dot{\gamma} \\ 0 \\ 0 \end{bmatrix}$$

$$= \begin{bmatrix} \dot{\psi}\sin\vartheta + \dot{\gamma} \\ \dot{\psi}\cos\vartheta\cos\gamma + \dot{\vartheta}\sin\gamma \\ -\dot{\psi}\cos\vartheta\sin\gamma + \dot{\vartheta}\cos\gamma \end{bmatrix} = \begin{bmatrix} 0 & \sin\vartheta & 1 \\ \sin\gamma & \cos\vartheta\cos\gamma & 0 \\ \cos\gamma & -\cos\vartheta\sin\gamma & 0 \end{bmatrix} \begin{bmatrix} \dot{\vartheta} \\ \dot{\psi} \\ \dot{\gamma} \end{bmatrix}$$

$$(2-129)$$

After transformation, we can get

$$\begin{bmatrix} \dot{\vartheta} \\ \dot{\psi} \\ \dot{\gamma} \end{bmatrix} = \begin{bmatrix} 0 & \sin\gamma & \cos\gamma \\ 0 & \dfrac{\cos\gamma}{\cos\vartheta} & -\dfrac{\sin\gamma}{\cos\vartheta} \\ 1 & -\tan\vartheta\cos\gamma & \tan\vartheta\sin\gamma \end{bmatrix} \begin{bmatrix} \omega_{x_1} \\ \omega_{y_1} \\ \omega_{z_1} \end{bmatrix} \qquad (2-130)$$

After the above equation is expanded, the kinematic equation of the missile rotating around the center of mass is obtained:

$$\begin{cases} \dfrac{\mathrm{d}\vartheta}{\mathrm{d}t} = \omega_{y_1}\sin\gamma + \omega_{z_1}\cos\gamma \\ \dfrac{\mathrm{d}\psi}{\mathrm{d}t} = \dfrac{1}{\cos\vartheta}(\omega_{y_1}\cos\gamma - \omega_{z_1}\sin\gamma) \\ \dfrac{\mathrm{d}\gamma}{\mathrm{d}t} = \omega_{x_1} - \tan\vartheta(\omega_{y_1}\cos\gamma - \omega_{z_1}\sin\gamma) \end{cases} \qquad (2-131)$$

Similarly, for the convenience of writing later, the footnote "1" in Eq. (2-131) is omitted.

3) Mass change equation

During the flight of the missile, the mass of the missile continues to decrease due to the continuous consumption of fuel by the engine. Therefore, it is necessary to supplement the equation describing the mass change of the missile, which is

$$\frac{\mathrm{d}m}{\mathrm{d}t} = -m_c \qquad (2-132)$$

where, $\dfrac{\mathrm{d}m}{\mathrm{d}t}$ is the rate of change of missile mass which is a negative value because it is a decrease in mass; m_c is the missile mass consumption per unit time, also known as fuel mass flow per second. It is generally considered to be a known function of time, which may be a constant or a variable.

4) Geometric relation equations

From the definition of four common coordinate systems in Section 2.4.1, we can see that the relationship between them is connected by eight angles ($\vartheta, \psi, \gamma, \theta, \psi_V, \alpha, \beta, \gamma_V$), as shown in the Figure 2.66.

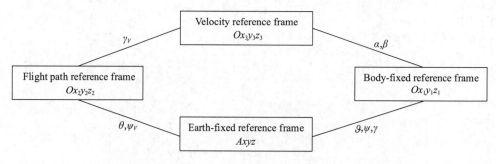

Figure 2.66 Eight angles between four coordinate systems

Because when a certain unit vector is projected onto the same axis of any coordinate system in different ways, the results should be equal. According to this principle, it can be known that these eight angles are not completely independent. After derivation, it can be seen that only five of the eight parameters are independent, and the remaining three angle parameters are represented by these five independent angle parameters. Therefore, there are three independent geometric relations between the eight angles. One form of expression is as follows:

$$\begin{cases} \sin\beta = \cos\theta[\ (\cos\gamma\sin(\psi-\psi_v) + \sin\vartheta\sin\gamma\cos(\psi-\psi_v)\ ] - \sin\theta\cos\vartheta\sin\gamma \\ \sin\alpha = \{\cos\theta[\sin\vartheta\cos\gamma\cos(\psi-\psi_v) - \sin\gamma\sin(\psi-\psi_v)\ ] - \sin\theta\cos\vartheta\cos\gamma\}/\cos\beta \\ \sin\gamma_V = (\cos\alpha\sin\beta\sin\vartheta - \sin\alpha\sin\beta\cos\gamma\cos\vartheta + \cos\beta\sin\gamma\cos\vartheta)/\cos\theta \end{cases}$$

$$(2-133)$$

5) Control relation equations

Manipulating the pitch, yaw, and tilt of the missile is to manipulate the missile's three degrees of freedom to change the magnitude and direction of the normal force, so as to achieve the purpose of changing the missile's flight direction. Regardless of the axisymmetric missile or the surface symmetric missile, the elevator is mainly used to control the pitch attitude of the missile; the rudder is mainly used to control the yaw attitude of the missile; the aileron is mainly used to control the tilt attitude of the missile. In addition, to change the speed, thrust control is usually used. It can be seen that, to realize the controlled flight of the missile, the missile should have four control mechanisms: the control mechanism of the elevator, the rudder, the aileron and the adjustment device of the engine thrust. Correspondingly, four constraints must be added to the missile, that is, there must be four control equations.

Set x_{*i} to study the instantaneous motion parameter value required by the guidance relation, x_i to the actual value of the same instantaneous motion parameter, and ε_i to the motion parameter error, then there is

$$\varepsilon_i = x_i - x_{*i} \quad (i=1,2,3,4) \quad (2-134)$$

Under normal circumstances, $\varepsilon_1, \varepsilon_2, \varepsilon_3$, and ε_4 can never be equal to zero. At this time, the control system will deflect the rudder surface and the engine adjustment device to eliminate the error. The magnitude and direction of the deflection angle of the rudder surface and the engine adjusting device depend on the value and the sign of the error ε_i. Generally speaking, it can be written as the following general control equation

$$\begin{cases} \phi_1(\cdots,\varepsilon_i,\cdots,\delta_i,\cdots) = 0 \\ \phi_2(\cdots,\varepsilon_i,\cdots,\delta_i,\cdots) = 0 \\ \phi_3(\cdots,\varepsilon_i,\cdots,\delta_i,\cdots) = 0 \\ \phi_4(\cdots,\varepsilon_i,\cdots,\delta_i,\cdots) = 0 \end{cases} \quad (2-135)$$

The equations can include the deflection angle of the rudder surface and the engine adjusting device, the motion parameter error, and other motion parameters.

Eq. (2-135) can be abbreviated as the following form

$$\phi_1 = 0, \quad \phi_2 = 0, \quad \phi_3 = 0, \quad \phi_4 = 0 \quad (2-136)$$

6) Missile motion equations

Synthesize Eq. (2 – 115), Eq. (2 – 123), Eq. (2 – 127), Eq. (2 – 131) ~ Eq. (2 – 133) and Eq. (2 – 136) obtained above to form a set of equations describing the space motion of the missile

$$\begin{cases} m\dfrac{\mathrm{d}V}{\mathrm{d}t} = P\cos\alpha\cos\beta - X - mg\sin\theta \\ mV\dfrac{\mathrm{d}\theta}{\mathrm{d}t} = P(\sin\alpha\cos\gamma_V + \cos\alpha\sin\beta\sin\gamma_V) + Y\cos\gamma_V - Z\sin\gamma_V - mg\cos\theta \\ -mV\cos\theta\dfrac{\mathrm{d}\psi_V}{\mathrm{d}t} = P(\sin\alpha\sin\gamma_V - \cos\alpha\sin\beta\cos\gamma_V) + Y\sin\gamma_V + Z\cos\gamma_V \\ J_x\dfrac{\mathrm{d}\omega_x}{\mathrm{d}t} + (J_z - J_y)\omega_z\omega_y = M_x \\ J_y\dfrac{\mathrm{d}\omega_y}{\mathrm{d}t} + (J_x - J_z)\omega_x\omega_z = M_y \\ J_z\dfrac{\mathrm{d}\omega_z}{\mathrm{d}t} + (J_y - J_x)\omega_y\omega_x = M_z \\ \dfrac{\mathrm{d}x}{\mathrm{d}t} = V\cos\theta\cos\psi_V \\ \dfrac{\mathrm{d}y}{\mathrm{d}t} = V\sin\theta \\ \dfrac{\mathrm{d}z}{\mathrm{d}t} = -V\cos\theta\sin\psi_V \\ \dfrac{\mathrm{d}\vartheta}{\mathrm{d}t} = \omega_y\sin\gamma + \omega_z\cos\gamma \\ \dfrac{\mathrm{d}\psi}{\mathrm{d}t} = \dfrac{1}{\cos\vartheta}(\omega_y\cos\gamma - \omega_z\sin\gamma) \\ \dfrac{\mathrm{d}\gamma}{\mathrm{d}t} = \omega_x - \tan\vartheta(\omega_y\cos\gamma - \omega_z\sin\gamma) \\ \dfrac{\mathrm{d}m}{\mathrm{d}t} = -m_c \\ \sin\beta = \cos\theta[(\cos\gamma\sin(\psi-\psi_V) + \sin\vartheta\sin\gamma\cos(\psi-\psi_V)] - \sin\theta\cos\vartheta\sin\gamma \\ \sin\alpha = \{\cos\theta[\sin\vartheta\cos\gamma\cos(\psi-\psi_V) - \sin\gamma\sin(\psi-\psi_V)] - \sin\theta\cos\vartheta\cos\gamma\}/\cos\beta \\ \sin\gamma_V = (\cos\alpha\sin\beta\sin\vartheta - \sin\alpha\sin\beta\cos\gamma\cos\vartheta + \cos\beta\sin\gamma\cos\vartheta)/\cos\theta \\ \phi_1 = 0 \\ \phi_2 = 0 \\ \phi_3 = 0 \\ \phi_4 = 0 \end{cases} \quad (2-137)$$

Eq. (2 – 137) is the missile space motion equations described in scalar form. It is a set of nonlinear ordinary differential equations. In these twenty equations, there are twenty unknowns: $V(t), \theta(t), \psi_V(t), \omega_x(t), \omega_y(t), \omega_z(t), x(t), y(t), z(t), \vartheta(t), \psi(t), \gamma(t), m(t), \alpha(t), \beta(t), \gamma_V(t), \delta_2(t), \delta_y(t), \delta_x(t)$ and $\delta_P(t)$. Therefore, Eq. (2 – 137) is closed. After the initial conditions are given, the numerical integration method can be used to solve the controlled trajectory and the corresponding changes of twenty parameters.

2.7 Flight performances

Usually, the flight performances of a flight vehicle are analyzed from three aspects, which are flight characteristics, utilization and cost. The main discussion is the flight characteristics.

2.7.1 Flight mission profile

The tasks of a flight vehicle can be described by the flight mission profile, which are usually expressed by the flight trajectory with flight parameters. Figure 2.67 presents a typical flight mission profile of an airplane.

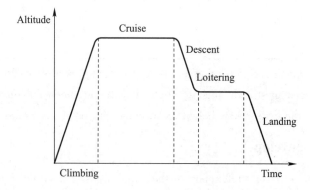

Figure 2.67   Flight mission profile

The airplane takes off at first, then climbs to a certain height, flies in cruise, and then descends to a certain lower height to loiter till it lands. The flight mission is finished.

Figure 2.68 illustrates a typical design mission for the same aircraft. Table 2.5 lists additional design constraints or requirements which may be specified for the aircraft. Note that the design mission and the constraint table specify performance which the customer requires from the aircraft. An aircraft design which fails to achieve these required performance levels will probably not be purchased by the customer.

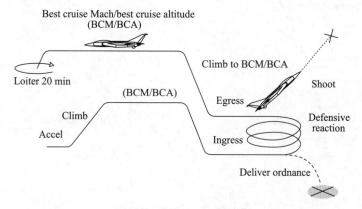

Figure 2.68   Typical design mission for a multi-role Jet Fighter

Table 2.5 Minimum performance requirements/constraints

Item	Requirement
Combat mission radius	400nm
Weapons payload	2 AIM-120 42000 lb MK-84 600 rounds 20mm ammunition
Takeoff distance	2000ft
Landing distance	2000ft
Max Mach number	$M = 1.8$ at optimum altitude
Instantaneous turn rate	18 (°) /s at $M = 0.9$, $H = 20000$ft
P_s	800ft/s, $M = 0.9$, $H = 5000$ft
Sustained g	$4 - g$ at $M = 1.2$, $H = 20000$ft $9 - g$ at $M = 0.9$, $H = 5000$ft

1ft = 0.3048m, 1 lb = 0.453kg

According to the typical design mission and requirements for an aircraft, the flight states, such as, ① flight at constant speed in straight line, ② flight at variable speed and in curve, ③ suspending flight, flight speed, and the ceiling how height an airplane can reach, as well as the range.

2.7.2 Flight envelope

The flight envelope of an aircraft refers to the capabilities of a design interms of airspeed and load factor or altitude. As seen in Figure 2.69, some limits for different types are given, e.g., stall limit, thrust constraints, dynamic pressure limit, pilot ejection, height restriction, use ceiling, maximum lift limit, and so on.

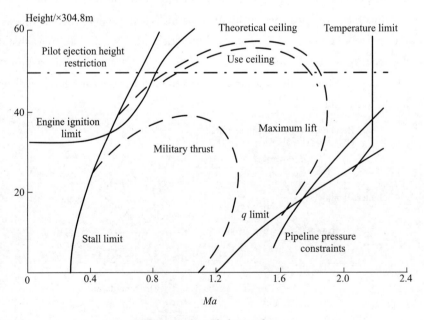

Figure 2.69 Flight envelope

2.7.3 Typical flight parameters

Some typical parameters of flight trajectory are used to describe flight performances of a flight vehicle. Some parameters are used for all flight vehicles, and the others are only for one type of flight vehicles.

1) Flight velocity

Flight velocity is one of the most important performance parameters of an aircraft. It includes minimum flying speed, maximum flying speed and cruising speed, and average speed.

(1) Minimum flying speed V_{min}

Minimum flying speed refers to the minimum velocity which the aircraft depends on to keep the straight-and-level flight at certain height. It is determined by the maximum lift coefficient C_{ymax}.

$$V_{min} = \sqrt{\frac{2G}{\rho C_{ymax} S}} \qquad (2-138)$$

where, G is the weight of the aircraft, ρ is the density of local air, S is the area of the wing.

It is important to the security of the aircraft which flies at low speed. With the increase of altitude, the minimum flying speed will increase as well as with the decrease of ρ.

(2) Max level speed V_{max}

Max level speed refers to the maximum flying speed an aircraft can reach in the straight-and-level flight. It indicates how fast an airplane can fly.

(3) Cruising speed V_{cr}

Cruising speed refers to the flight velocity, when the fuel expended by the engine is least per kilometer. Cruising speed is larger than minimum flying speed, but smaller than the max level speed. It produces the greatest economy when the plane flies at cruising speed.

2) Flight altitude

Ceiling indicates how high an airplane can fly. Steady-flight ceiling means the maximum altitude that an airplane can achieve when it does the straight-and-level flight. In company with the increase of the altitude, thrust of the engine will gradually fall off. When the airplane mounts up to a limit altitude, at which it can only do the straight-and-level flight rather than flying higher, then the altitude is named as the steady-flight ceiling.

3) Range

Range means flight distance. It indicates how far flight vehicle can fly, the minimum range and the maximum range. For a flight vehicle, the minimum and maximum heights will be reached.

4) Takeoff and landing

Regardless of an aircraft's design mission, it musttake off and land to start and finish its flight. Takeoff and landing present a particularly hazardous part of the flight of an aircraft. During the takeoff, the aircraft will be heavily loaded with fuel for the journey and the engines will be working at a high rate in order to take off in as short a distance as possible. The takeoff for a commercial airliner is further complicated by the need to adhere to appropriate noise abatement procedures. This may typically involve an initial climb at high angle in order to put the maximum distance between the aircraft and the ground at the boundary of the airport. This may then be followed by the need to reduce throttle setting as a populated area is reached.

Landing has also its difficulties. The pilot has to lead the aircraft to a precise touch-down point in three-dimensional space. Further more when touch-down is achieved, the aircraft must fly in the correct direction, aligned with the runway, and must be at a low air speed to facilitate bringing it to a halt in a reasonable distance while retaining a safe margin over the stalling speed.

(1) Takeoff distance

For an aircraft to takeoff, it uses excess thrust to accelerate to a safe flying speed. Normally, an air-speed 1.2 times the aircraft's stalling speed at its takeoff mass and configuration is considered safe to become airborne. This safe flying airspeed is called takeoff speed, V_{TO}. At reaching takeoff speed, the pilot raises the aircraft's nose to establish a pitch attitude and angle of attack called the takeoff attitude. Once the takeoff attitude is established and the aircraft has sufficient speed, it generates enough lift to start flying. Takeoff distance is the distance required for the aircraft to accelerate to takeoff speed and rotate. Some aircraft design requirements specify a rotation time, usually around three seconds, which must be allowed, and the distance covered by the aircraft added to the takeoff distance after it has reached takeoff speed. Design requirements may also specify required takeoff performance in terms of the distance required to accelerate, rotate, transition to a climb, and climb over an obstacle with a specified height. Figure 2.70 illustrates these steps or phases in a takeoff.

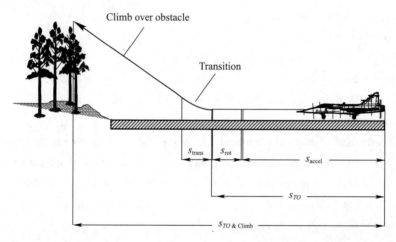

Figure 2.70 Sketch of takeoff

(2) Landing Distance

As Figure 2.71 illustrated, the landing maneuver is broken up into approximately the same steps as takeoff. As with takeoff, the details of the design requirements for landing distance vary. The landing speed, V_L is usually specified as $1.3V_{stall}$. The approach or descent to landing is also generally flown at V_L, or slightly faster. Some customers and/or regulatory agencies may specify landing distances over a fixed obstacle. Others may specify that the aircraft passes over the end or threshold of the landing runway at a specified height, or that it touches down a specified distance down the runway. The design specifications may require the landing analysis to include three or more seconds of free roll (deceleration only due to normal rolling friction and air drag) after touchdown before the brakes are applied. A landing analysis may also include the effects of reverse thrust or a drag parachute which is deployed at or slightly before touchdown. The simple case of no free roll allowance, so that $s_L = s_{\text{decel}}$, will be considered here.

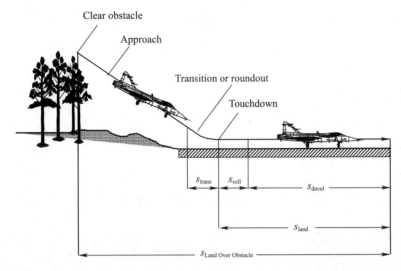

Figure 2.71 Sketch of landing

5) Turns

Turning performance is important to military fighter and other aircraft, which must maneuver in tight quarters, for instance, to takeoff and land at an airfield in a canyon or among skyscrapers in the center of a city.

The most important characteristics of turning performance, which are frequently specified as design requirements are turn rate and turn radius. This performance may be specified either as an instantaneous or a sustained capability.

As the names imply, a sustained turn rate or radius is the performance that the aircraft can maintain for a long period of time, minutes or even hours. An instantaneous turn rate or radius is the capability that the aircraft can achieve momentarily, but then the maximum performance may begin to decrease immediately.

The most commonly performed turning maneuver for an aircraft is the level turn. In this maneuver, the aircraft maintains a constant altitude (and in a sustained turn, a constant airspeed). Its velocity vector changes direction but stays in a horizontal plane. Figure 2.72 shows front and top views of an aircraft in a level turn.

6) V-n Diagrams

The turn analysis up to this point has said nothing of the limitations that the aircraft may have on its ability to generate the lift or sustain the structural loading needed to perform a specified turn. These limitations are often summarized on a chart known as a $V-n$ diagram.

Figure 2.73 is a $V-n$ diagram of a subsonic jet trainer. The maximum positive and negative load factors which the aircraft structure can sustain are shown as horizontal lines on the chart, since for this particular aircraft these structural limits are not functions of velocity. At low speeds, the maximum load factor is limited by the maximum lift the aircraft can generate, since,

117

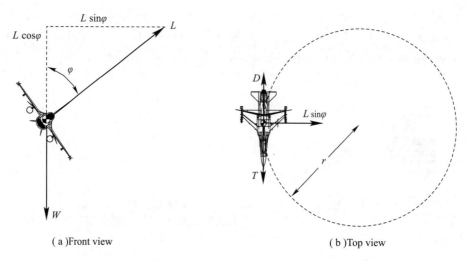

(a) Front view (b) Top view

Figure 2.72　Forces on an airplane in a level turn

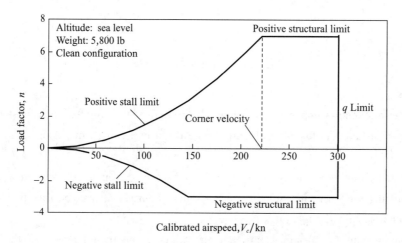

Figure 2.73　V-n Diagram for a subsonic jet trainer

$$L_{\max} = n_{\max} W = C_{L_{\max}} qS = C_{L_{\max}} / (2\rho V^2) \, S \qquad (2-139)$$

As a result, the maximum load factor is relative to the small limit.

7) Overload

The maneuverability of a flight vehicle describes the ability to alternate the direction and magnitude of velocity in a specific time. How to evaluate the maneuverability of flight vehicles? Obviously, we can utilize the normal acceleration and tangential acceleration to describe the ability. Here, the concept of overload or load factor is proposed.

The definition of the overload is given by

$$n = \frac{N}{G} \qquad (2-140)$$

where, $G = mg$, N stands for the total external forces except gravity acted on a flight vehicle.

Another definition will be introduced in the analysis of the ballistic kinematics of the flight vehicle. It is defined as the ratio of the combined force of all external forces acting on the missile (including gravity) to the weight of the vehicle. The definition is

$$n' = \frac{N+G}{G} \qquad (2-141)$$

Obviously, due to the different definitions of overload, the overload values in the same case are different.

Three concepts of overload need to keep in mind. They are called required overload n_R, available overload n_A, and limit overload n_L. The details are as follows:

n_R is the overload factor for requirements if the flight vehicle flies in some trajectory and in one way to guide and control.

n_A is the overload factor for available.

n_L is the overload factor to limit, that is to say, the flight vehicle can provide the limit loading factor.

The relations among them must satisfy the requirements of the following formula so that the flight vehicle can move in the air properly.

$$n_L > n_A > n_R + \Delta n \qquad (2-142)$$

where Δn is the increment.

The overloads are mainly affected by the flight trajectory curve, flight characteristics, forces acted on the flight vehicle, and so on.

The projections of the overload vector $\boldsymbol{n}$ on each axis of the Flight path reference frame $Ox_2y_2z_2$ are n_{x_2}, n_{y_2} and n_{z_2}. The projections of the overload vector $\boldsymbol{n}$ on each axis of the Velocity reference frame $Ox_3y_3z_3$ are n_{x_3}, n_{y_3} and n_{z_3}. The projection of the overload vector in the velocity direction n_{x_2} and n_{x_3} are called tangential overload, the projection in the direction perpendicular to the velocity n_{y_2}, n_{z_2} and n_{y_3}, n_{z_3} are called normal overload.

The projections of the overload vector $\boldsymbol{n}$ on each axis of the Body-fixed reference frame $Ox_1y_1z_1$ are n_{x_1}, n_{y_1} and n_{z_1}, n_{x_1} is called longitudinal overload, n_{y_1} and n_{z_1} are called lateral overload.

The maneuverability of the missile can be rated in terms of tangential and normal overloads. The greater the tangential overload, the greater the tangential acceleration the missile can produce, which means that the faster the value of the missile's velocity changes, the faster it can approach the target. The greater the normal overload, the greater the normal acceleration the missile can produce, and the greater the ability of the missile to change flight direction at the same speed, i.e., the missile can make a more curved ballistic flight.

2.8 Flight stability & control

As is known, the concepts of stability and control are very important for a flight vehicle design. They are a pair of contradictions. From the view of design, the two parts must be considered at the same time so that the balance of design can be reached.

2.8.1 Stability

2.8.1.1 Static stability

Static stability is the tendency for a flight vehicle to return to its previous attitude once disturbed. Thus, restoring force and moment must exist.

As shown in Figure 2.74 (a) the ball in the bowl illustrates a statically stable system. If the ball is displaced from the bottom, it will tend to return. An increase in the steepness of the sides of the bowl corresponds to increase instability.

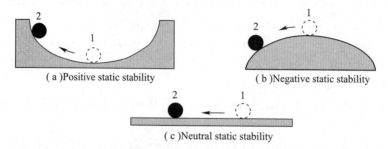

Figure 2.74　Simple systems with positive, negative, and neutral static stability

If one turns the bowl over, we have an illustration of static instability as seen in Figure 2.74 (b). If the ball is moved from the exact center of the bowl, it will continue to move away.

As seenin Figure 2.74 (c), the ball on the table illustrates the neutral stability. If the ball is moved, it will tend to stay in its new position.

As stated before, three types of statically stability, i.e., static stable, neutral, and unstable are discussed. A flight vehicle is said to be statically stable if it tends to return to its initial flight conditions, attitude, etc., after being disturbed by a gust or a small impulsive input from the controls. The more stable we make an aircraft, the less maneuverable it becomes. A very stable flight vehicle always tends to continue in its existing state, so excessive stability must be avoided.

Inthe space motion of a flight vehicle, we have defined three turning motions, i.e., pitching, yawing, and rolling motion. Pitching (nose-up/nose-down motion) is the longitudinal motion, while yawing and rolling belong to the lateral motion. So the static stability is divided into longitudinal static stability and lateral-directional static stability.

As stated before, longitudinal static stability can be judged using derivative $m_z^\alpha < 0$.

Lateral-directional (directional and rolling) static stability can be given by derivatives $m_y^\beta < 0$ (directional) and $m_x^\beta < 0$ (rolling).

2.8.1.2　Dynamic stability

Static stability indicates the trend of the stability, not the end of flight stability. We can have a statically stable airplane which is still not satisfactory in practice because it oscillates about the equilibrium position. If the amplitude of the oscillation grows with time then it is said to be negatively damped, and the aircraft is dynamically unstable.

Dynamic stability deals with how the motion caused by a disturbance changing with time. Figure 2.75 illustrates the concepts of statically and oscillatory dynamic stability for an aircraft flying with the wings level, without roll or bank. In this case, it is the longitudinal motion of the aircraft that is of interest to us. Let us consider a statically stable aircraft, which is slightly disturbed by increasing the angle of attack. The subsequent motion may take a wide variety of forms as shown in Figure 2.75 (a) ~ (c).

The first of these (Figure 2.75 (a)) shows a motion in which the aircraft simply returns to the

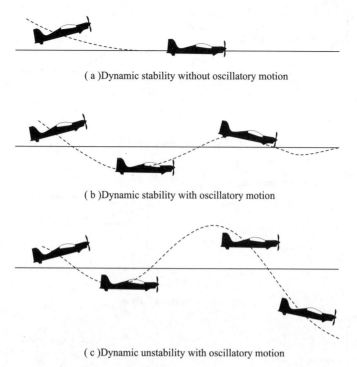

(a) Dynamic stability without oscillatory motion

(b) Dynamic stability with oscillatory motion

(c) Dynamic unstability with oscillatory motion

Figure 2.75 Dynamic stability and unstability of the motion

state it was in before the disturbance was applied. The motion is thus stable and the motion is not oscillatory.

Figure 2.75 (b) shows an oscillatory motion, but since the oscillations die out with time the motion is again dynamically as well as statically stable finally.

Figure 2.75 (c) shows the oscillations becoming greater with time rather than dying away. This motion is the dynamically unstable negatively damped motion. It is statically stable because whenever the aircraft is at a pitch angle that differs from its initial value, the moment acting on the aircraft is in the direction which tends to restore it to its original position.

From the view of design, statically stable is wanted. At the same time, the dynamic stability must be researched at the end. Usually, transport aircraft has static stability, but military airplanes are statically unstable, which use a control loop to increase the stability to change the stability characteristics.

Figure 2.76 gives the discussion of the three types of dynamic stability, i. e. , negative dynamic stability, neutral dynamic stability, and positive dynamic stability. Negative indicates that oscillations increase with time after disturbed. Neutral dynamic stability means that the amplitude of oscillations does not decrease with time. Positive dynamic stability indicates that the amplitude of oscillations decreases with time.

As known, the space motion of a flight vehicle can be decomposed to longitudinal motion and lateral-directional motion. The longitudinal motion is composed of short period motion and phugoid motion. The former means some parameters, e. g. $\Delta\alpha, \Delta\omega_z, \Delta\vartheta$, change and convergent very quickly in the first 10s. The latter indicates that motion parameters such as ΔV and $\Delta\theta$, which vary quite

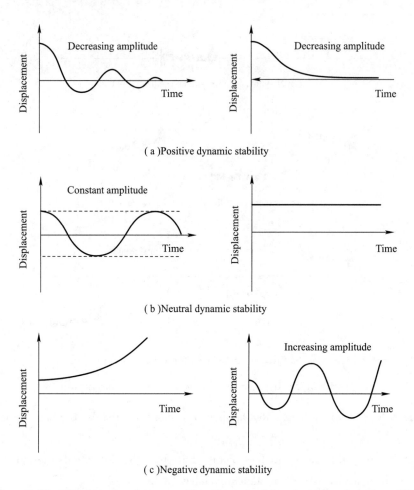

Figure 2.76 Time histories of systems with positive, neutral, and negative dynamic stability

slowly until they are controlled by pilots.

In lateral-directional motion, there are three modes, which are roll mode, Dutch roll mode, and spiral mode. The roll mode is static stability. Dutch roll mode motion couples rolling and yawing, whose name is from the Dutch speed skaters as they glide across ice-ground. Spiral is instability. Spiral divergence will lead to a downward spiral motion.

2.8.2 Control

2.8.2.1 Control forces

Control forces of winged missiles are mainly produced by the wings. Engine thrust can provide part of control forces, but only a little part. Generally, there are two methods to produce and change control forces, which are aerodynamics and thrust of the engine. Figure 2.77 ~ 2.79 show the control forces produced by aerodynamics in axisymmetric missiles and plane-symmetrical missiles. Figure 2.80 ~ 2.82 show the control forces produced by the thrust of engine. Furthermore, the methods in Figure 2.81 and Figure 2.82 only produce and change the control forces because their control units are fixed at the center of mass, while the methods in Figure 2.77 ~ 2.80 can produce and change both the control forces and control moments.

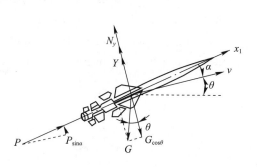

Figure 2.77 Control forces of axisymmetric missiles in longitudinal symmetrical surface

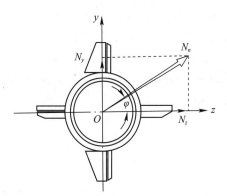

Figure 2.78 Control forces of axisymmetric missiles in any surfaces

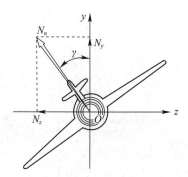

Figure 2.79 Control forces of plane-symmetrical

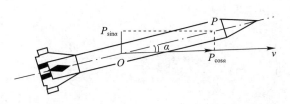

Figure 2.80 Control forces produced by thrust

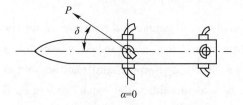

Figure 2.81 Control forces produced by rotational siphon-tube

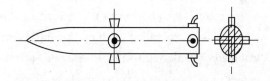

Figure 2.82 Control forces produced by lateral tube

2.8.2.2 Control components

For a flight vehicle, generally, three channels must be controlled. That is to say, pitching control, yawing control, and rolling control are made by the movement of elevators, rudders, and ailerons so that the flight vehicle can fly in terms of the design and technical requirements. The control for a flight vehicle should be fast and smooth so that the procedures and responses of flight parameters could be divergent and quickly.

From the view of the design of a flight vehicle, to meet the requirements of design and utilization, both control and stability should be considered. That is, on the one side, the stability should be satisfied. On the other side, the characteristics of control must also be considered.

Figure 2.83 shows three degrees of freedom in rotation, and the control surfaces which typically produce the moments to cause those rotations. Figure 2.83 (a) shows rotation about the aircraft's longitudinal Ox_1 axis, which is called rolling, and the maneuver is called a roll. Control surfaces on the aircraft's wings are called ailerons, which deflect differentially (one trailing edge up and one trailing edge down) to create more lift on one wing, less on the other, and therefore a net rolling moment is created.

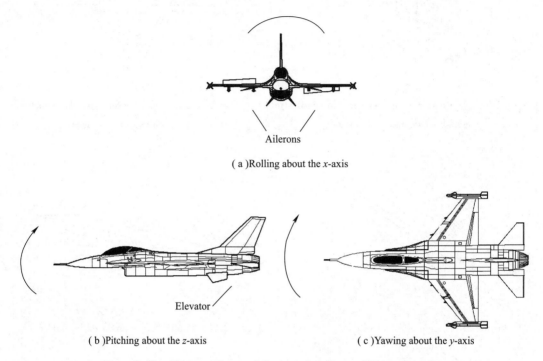

Figure 2.83 Three rotations and the control surfaces which produce them

Figure 2.83 (b) shows the aircraft in a pitch-up maneuver. Rotation of the aircraft about the lateral axis Oz_1 is called pitching. A control surface near the rear of the aircraft called an elevator or stabilator is deflected so that it generates a lift force, which, due to its moment arm from the aircraft center of gravity also creates a pitching moment. An elevator is a moveable surface attached to a fixed (immovable) horizontal stabilizer, a small horizontal surface near the tail of the aircraft which acts like the feathers of an arrow help keep the aircraft pointing in the right direction. A stabilator combines the functions of the horizontal stabilizer and the elevator. If the stabilator does not have a fixed portion, it is called all-moving.

Figure 2.83 (c) shows the aircraft yawing, rotating about the vertical axis Oy_1 so that the nose moves right or left. A moveable surface called a rudder which is attached to the aircraft's fixed vertical stabilizer deflects to generate a lift force in a sideways direction. The vertical stabilizer and rudder are toward the rear of the aircraft, having some distance from its center of gravity, so that the aerodynamic moment about the vertical axis causes the aircraft to yaw.

To the missiles, if a missile has four control surfaces as shown in Figure 2.84, one regards 1 and 3 as elevators and 2 and 4 as rudders even if the missile should roll subsequently.

If 1 and 3 are mechanically linked together such that a servo must impart the same rotation to both, then these surfaces are elevators pure and simple. The same argument applies to the

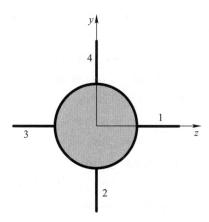

Figure 2.84 Control surfaces looking from the rear of the missile

rudders. When surface 1 and 3 each has their servo, they can act as ailerons. Looking in the direction z one surface is rotated $\delta°$ clockwise and the other surface $\delta°$ anti-clockwise, then a pure couple is imparted to the missile about the fore and aft axis, which tends to roll the missile. Such control surfaces are now called ailerons. We can double the power of the ailerons by doing the same thing to control surface 2 and 4. If now the aerodynamics are linear, i.e., the lifts are proportional to incidence, then the principle of superposition applies. Commands for elevator, rudder, and aileron movements can be added electrically resulting in unequal movements to opposite control surfaces. In this way, we have the means to control roll motion as well as the up/down (i.e. pitching) motion and left/right (i.e. yawing) motion. Alternative methods of control are shown in Figure 2.85.

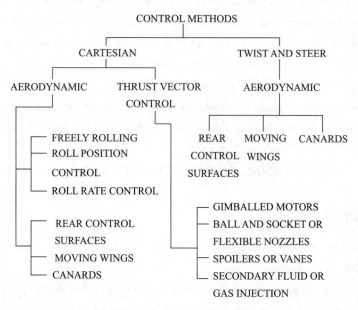

Figure 2.85 Control methods

2.8.2.3 STT & BTT Control

Skid-to-turn (STT) control is mostly used for missiles with axisymmetric aerodynamic layout, and is the main control mode for air-to-air missiles in service. For axisymmetric missiles, they are equipped with two pairs of wings evenly distributed along the circumference, relying on the manipulation of the rudder deflection angle of the elevator to adjust the angle of attack, which changes the direction and magnitude of lift to control the pitching motion of the missile. The change of the sideslip angle is achieved by changing the rudder deflection angle of the rudder, so that the direction and magnitude of the lateral force are changed to manipulate the yawing motion of the missile. During maneuvering, the STT missile is generally not controlled for the tilt state of the vehicle, but rather for the normal overload in the pitch and yaw directions, with the synthetic overload pointing in the direction required for the maneuver. The STT control method is now quite mature and has been widely used on in-service missiles. However, its main drawback is that the presence of a yaw angle in the case of a large angle of attack generates severe asymmetric vortices, which will limit the ultimate overload, so the STT missile cannot have good maneuverability.

Bank-to-turn (BTT) control technology, originally a common control method for aircraft, has also been increasingly used on missiles in recent years. When the BTT missile is required to maneuver, the roll control system rapidly adjusts the longitudinal plane of symmetry (the plane of maximum lift) to the desired direction of action, while the deflection of the pitch channel rudder will produce a normal overload directed at the target. Unlike STT control, which generates lateral overload by side slip angle, BTT control makes the missile get maneuvering overload in any direction by rolling to achieve maneuvering flight without sideslip angle. According to the different aerodynamic shapes of the missile and the power unit, BTT control is divided into BTT-180, BTT-90, and BTT-45. The difference between the three is that the maximum angle to control the missile roll during the maneuver is different, which is $\pm 180°$, $\pm 90°$ and $\pm 45°$ respectively. For axisymmetric missiles, both lifting surfaces can be used as the maximum lifting surface, and the BTT-45 is only suitable for controlling such axisymmetric missiles. The BTT-90 and BTT-180 are often used for symmetric missiles containing only one effective lift surface, and the BTT-180 missile must fly at a positive angle of attack when maneuvering. BTT-controlled missiles have many outstanding advantages compared with STT-controlled missiles, which can improve the aerodynamic efficiency of missiles, obtain better aerodynamic characteristics and achieve higher technical index requirements.

2.9 Space vehicles flight principles

2.9.1 Kepler laws

In astronomy, Kepler's laws gave a description of the motion of planets around the Sun, which is called Kepler's three laws, as follows.

(1) The orbit of each planet is an ellipse with the Sun at one of the two foci.

(2) A line joining a planet and the Sun sweeps out equal areas during equal intervals of time.

(3) The square of the orbital period of a planet is directly proportional to the cube of the semi-major axis of its orbit.

Kepler's laws are strictly only valid for a two-body system, which is not affected by the gravity of other planets. Nevertheless, Kepler's laws form a useful starting point to calculate the orbits of planets.

2.9.1.1 Kepler's first law

The orbit equation can be written as,

$$r + re\cos\theta = \frac{h^2}{\mu} \qquad (2-143)$$

or

$$r = \frac{h^2}{\mu} \frac{1}{1 + e\cos\theta} \qquad (2-144)$$

where, r is the distance between the orbiting body and the central body, θ is the direction of the orbiting body, called true anomaly, and the parameters, h is the specific relative angular momentum of the orbiting body, e is the eccentricity of the orbit, also called the orbital eccentricity, μ is the constant which divided by the distance squared, giving the magnitude of the acceleration, in the case of gravity, μ is the standard gravitational parameter. Figure 2.86 shows the geometry diagram of a typical elliptical orbit.

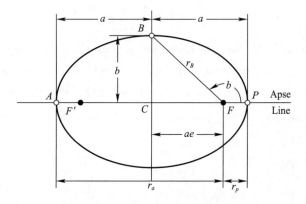

Figure 2.86 Elliptical orbit

The orbit equation defines the path of the body m_2 around m_1, relative to m_1. Remember that μ, h and e are constants. Observe as well that there is no significance to negative values of eccentricity, i.e., $e \geqslant 0$. Since the orbit equation describes conic sections, including ellipses, it is a mathematical statement of Kepler's first law, namely, that the planets follow elliptical paths around the sun. Two-body orbits are often referred to as Keplerian orbits.

2.9.1.2 Kepler's second law

The angular momentum of body m_2 relative to m_1 is the moment of m_2's relative linear momentum $m_2 \dot{r}$

$$\boldsymbol{H}_{2/1} = \boldsymbol{r} \times m_2 \dot{\boldsymbol{r}} \qquad (2-145)$$

where $\dot{\boldsymbol{r}} = v$ is the velocity of m_2 relative to m_1. Let us divide this equation through by m_2 and let $\boldsymbol{h} = \boldsymbol{H}_{2/1}/m_2$, so that

$$\boldsymbol{h} = \boldsymbol{r} \times \dot{\boldsymbol{r}} \qquad (2-146)$$

$\boldsymbol{h}$ is the relative angular momentum of m_2 per unit mass, that is the specific relative angular momentum. The unit of $\boldsymbol{h}$ is $\text{km}^2 \cdot \text{s}^{-1}$.

Taking the time derivative of $\mathbf{h}$ yields

$$\frac{d\mathbf{h}}{dt} = \dot{\mathbf{r}} \times \dot{\mathbf{r}} + \mathbf{r} \times \ddot{\mathbf{r}} \tag{2-147}$$

But $\dot{\mathbf{r}} \times \dot{\mathbf{r}} = 0$. Furthermore, $\ddot{\mathbf{r}} = -(\mu/r^3)\mathbf{r}$, so that

$$\mathbf{r} \times \ddot{\mathbf{r}} = \mathbf{r} \times \left(-\frac{\mu}{r^3}\mathbf{r}\right) = -\frac{\mu}{r^3}(\mathbf{r} \times \mathbf{r}) = 0 \tag{2-148}$$

Therefore,

$$\frac{d\mathbf{h}}{dt} = 0 \quad (\text{or } \mathbf{r} \times \dot{\mathbf{r}} = \text{const.}) \tag{2-149}$$

At any given time, the position vector $\mathbf{r}$ and the velocity vector $\dot{\mathbf{r}}$ lie in the same plane. Their cross product $\mathbf{r} \times \dot{\mathbf{r}}$ is perpendicular to that plane. Since $\mathbf{r} \times \dot{\mathbf{r}} = \mathbf{h}$, the unit vector normal to the plane is

$$\hat{\mathbf{h}} = \frac{\mathbf{h}}{h} \tag{2-150}$$

This unit vector is constant, thus, the path of m_2 around m_1 lies in a single plane.

As shown in Figure 2.87, since the orbit of m_2 around m_1 forms a plane, it is convenient to orient oneself above that plane and looks down upon the path. Let us resolve the relative velocity vector $\dot{\mathbf{r}}$ into components $\mathbf{v}_r = v_r \hat{\mathbf{u}}_r$ and $\mathbf{v}_\perp = v_\perp \hat{\mathbf{u}}_\perp$ along the outward radial from m_1 and perpendicular to it, respectively, where $\hat{\mathbf{u}}_r$ and $\hat{\mathbf{u}}_\perp$ are the radial and perpendicular (azimuthal) unit vectors (Figure 2.88). Then,

$$\mathbf{h} = r\hat{\mathbf{u}}_r \times (v_r \hat{\mathbf{u}}_r + v_\perp \hat{\mathbf{u}}_\perp) = rv_\perp \hat{\mathbf{h}} \tag{2-151}$$

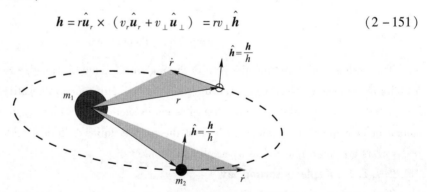

Figure 2.87 The path of m_2 around m_1 lies in a plane whose normal is defined by h

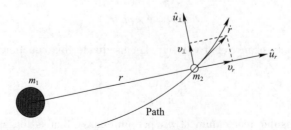

Figure 2.88 Components of the velocity of m_2, viewed above the plane of the orbit

That is,
$$h = rv_\perp \tag{2-152}$$

Clearly, the angular momentum depends only on the azimuth component of the relative velocity.

During the differential time interval dt the position vector $\boldsymbol{r}$ sweeps out an area dA. From the Figure 2.89, it is clear that the triangular area dA is given by

$$dA = \frac{1}{2} \times \text{base} \times \text{altitude} = \frac{1}{2} \times v dt \times r\sin\varphi = \frac{1}{2} r (v\sin\varphi) dt = \frac{1}{2} rv_\perp dt \tag{2-153}$$

Therefore,
$$\frac{dA}{dt} = \frac{h}{2} \tag{2-154}$$

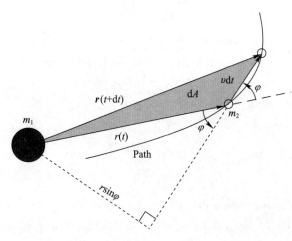

Figure 2.89 Differential area dA swept out by the relative position vector r during time interval dt

dA/dt is called the areal velocity, and it is constant. This result is known as Kepler's second law, i.e., equal areas are swept out in equal times.

2.9.1.3 Kepler's third law

As shown in Table 2.6, there are parameters of some planets, it is not hard to find out that $\dfrac{a^3}{T^2} \approx 3.3 \times 10^{18} \text{m}^3/\text{s}^2$ for all planets, in fact,

$$T = \frac{2\pi}{\mu} a^{\frac{3}{2}} \tag{2-155}$$

Table 2.6 Parameters of some planets

Planet	Semi-major axis/m	Period/s
Mercury	5.79×10^{10}	7.60×10^6
Venus	1.08×10^{11}	1.94×10^7
Earth	1.49×10^{11}	3.16×10^7
Mars	2.28×10^{11}	5.94×10^7
Jupiter	7.78×10^{11}	3.74×10^8
Saturn	1.43×10^{11}	9.30×10^8

This expression, which is identical to that of a circular orbit of radius a, reveals that like the energy, the period of an elliptical orbit is independent of the eccentricity. The equation embodies Kepler's third law: the period of a planet is proportional to the three-halves power of its semimajor axis.

2.9.2 Orbit equation & universal velocities

2.9.2.1 Orbit equation

The orbit equation defines the path of orbiting body m_2 around central body m_1.

According to Kepler's first law, the orbit equation in polar coordinates can be

$$r = \frac{h^2}{\mu} \frac{1}{1 + e\cos\theta} \qquad (2-156)$$

Note that h^2/μ is the semi-latus rectum of the conic section. This ratio, together with e, fully determines the geometry of the orbit. For a given orbit, the larger μ, the faster the orbiting body moves in it, twice as fast if the attraction is four times as strong.

2.9.2.2 Circular orbits

Setting $e = 0$ in the orbital equation yields

$$r = \frac{h^2}{\mu} \qquad (2-157)$$

That is, r = constant, which means the orbit m_2 around m_1 is a circle. Since $\dot{r} = 0$, it follows that $v = v_\perp$ so that the angular momentum formula $h = rv_\perp$ becomes simply solving for v yielding the velocity of a circular orbit.

$$v_{\text{circular}} = \sqrt{\frac{\mu}{r}} \qquad (2-158)$$

2.9.2.3 Escape velocity

If the eccentricity equals 1, then the orbit equation becomes

$$r = \frac{h^2}{\mu} \frac{1}{1 + \cos\theta} \qquad (2-159)$$

As the true anomaly θ approaches $180°$, the denominator approaches zero, so that r tends towards infinity. The energy of a trajectory for which $e = 1$ is zero, so that for a parabolic trajectory the conservation of energy,

$$\frac{v^2}{2} - \frac{\mu}{r} = 0 \qquad (2-160)$$

In other words, the speed anywhere on a parabolic path is

$$v = \sqrt{\frac{2\mu}{r}} \qquad (2-161)$$

If the body m_2 is launched on a parabolic trajectory, it will coast infinity, arriving there with zero velocity relative to m_1. It will not return. Parabolic paths are therefore called escape trajectories. At a given distance r from m_1, the escape velocity is given.

$$v_{esc} = \sqrt{\frac{2\mu}{r}} \quad (2-162)$$

Let v_0 be the speed of a satellite in a circular orbit of radius r. Then we have

$$v_{esc} = \sqrt{2} v_0 \quad (2-163)$$

That is, to escape from a circular orbit requires a velocity boost of 41.4 percent. However, remember our assumption that m_1 and m_2 are the objects in the universe. A spacecraft launched from the earth with velocity v_{esc} (relative to the Earth) will not coast to infinity (i.e. leave the solar system) because it will eventually succumb to the gravitational influence of the sun and end up with the same orbit as the earth.

As we know, the first universal velocity is 7.91km/s. If the flight vehicle rotates around the Earth, the first universal speed should be reached. The second universal velocity is 11.18km/s. If the flight vehicle moves from the sun, the second universal velocity must be achieved. The third universal velocity is 16.63km/s, which must be attained by the flight vehicle to fly away from the solar system.

2.9.2.4 Orbit description

To determine an orbit requires specifying six independent quantities. These can be the six classical orbital elements or the total of six components of the state vector, r and v, at a given instant.

The traditionally used set of orbital elements is called the set of Keplerian elements, after Johannes Kepler and his laws. The Keplerian elements are the six as follows:

a is the semimajor axis,

e is the eccentricity,

i is the inclination,

Ω is the longitude of the ascending node,

ω is the argument of periapsis,

τ is the mean anomaly at epoch.

In principle, once the orbital elements are known for a body, its position can be calculated forwards and backward indefinitely in time. However, gravity is from an assumed point source, and the orbital elements change over time.

Keplerian elements describe an orbit's physical attributes. Satellites orbit in elliptical path's with the center of the earth at one of the ellipse's focus points. The position in the orbit closest to the earth is called perigee, the position in the orbit, which is furthest away from the earth is called apogee.

1) Orbital size and shape

The semi-major axis and the eccentricity are used to describe the orbit's size and shape. The semi-major axis describes the size of the ellipse. The semi-major axis is half of the distance between apogee and perigee. Based on Kepler's third law of planetary motion, the orbital period from the semi-major axis can be calculated.

Eccentricity affects the shape of the ellipse, which is computed as the linear eccentricity (the distance from the center of the ellipse to the center of the earth) divided by the semi-major axis. A zero eccentricity describes a circular orbit. An eccentricity approaching one describes a highly elliptical orbit.

2) Orbital orientation

The remaining three orbital elements orient the orbit with respect to the Earth.

The plane of the orbital ellipse is oriented by two angles, the inclination and the right ascension of the ascending node.

Inclination describes the orbital plane's tilt angle in respect to the equator, and it also specifies the highest latitudes (North and South) over which the satellite directly overflies. A zero inclination describes an equatorial orbit, and a 90° inclination describes a polar orbit. Inclination greater than 90° describes orbits that move against earth rotation (called retrograde).

An orbit's ascending node is its South-to-North equatorial crossing. The right ascension of the ascending node is the angle measured eastward from the Vernal Equinox to the ascending node. The Vernal Equinox is the Sun's apparent ascending node making the beginning of the Northern hemisphere's spring.

Argument of perigee is the angle measured in the direction of satellite motion from the ascending node to perigee.

True anomaly is the angle measured in the direction of motion from perigee to the satellite's position at some defined epoch time. Mean anomaly describes what the satellite's true anomaly would be if it were in a circular orbit. Mean anomaly from the orbit's true anomaly and eccentricity can be computed. The commonly available Keplerian elements use mean anomaly.

2.9.3 Satellite orbits

Gravitational force is only one main force acting on a satellite or flight vehicle when it is in orbit. This force is continuously pulling the satellite towards the Earth, and it does not fall directly down to the earth because of its high velocity. During a period of satellite orbit, there is a perfect balance between the gravitational force and the centripetal force to maintain the orbit of the satellite.

Satellites can fly in several types of orbit for different purposes. The most common orbits for environmental satellites are geostationary and polar, but some instruments also fly in inclined orbits. Other types of orbits are possible, such as the Molniya orbits commonly used for Soviet spacecraft.

2.9.3.1 Geosynchronous orbit

A synchronous orbit is an orbit in which an orbiting body (usually a satellite) has a period equal to the average rotational period of the body being orbited (usually a planet), and in the same direction of rotation as that body.

A geosynchronous orbit is an orbit around the Earth with an orbital period that matches the Earth's rotation period.

$$T = 23\text{h}56\text{min}4\text{s} = 86164\text{s} \tag{2-164}$$

$$\omega_E = \frac{2\pi}{86164\text{s}} = 7.292115 \times 10^{-5}\,\text{rad/s} \tag{2-165}$$

$$a = \left(\frac{\mu T^2}{4\pi^2}\right)^{1/3} = 42164 \text{km} \qquad (2-166)$$

As a result, all earth geosynchronous orbits have a semimajor axis of 42164km.

Geostationary orbit is a special case of a geosynchronous orbit, it is a geosynchronous orbit that is circular and at zero inclination, that is, directly above the equator.

In the special case of a geostationary orbit, $h = r - R = 35786$km.

A satellite in such an orbit is at an altitude of approximately 35786km above mean sea level. And the velocity of satellite is as follows

$$v = \sqrt{\frac{\mu}{r}} = 3074.6 \text{m/s} \qquad (2-167)$$

Geostationary satellites provide a great field of view, enabling coverage of more earth surface to forecast the weather events or early warn the forest fire danger. Because a geostationary orbit must be in the same equatorial plane as the earth's rotation, it cannot provide clear images of the Polar regions with poor spatial resolution.

2.9.3.2 Sun-synchronous orbit

A sun-synchronous orbit is a geocentric orbit that combines altitude and inclination in such a way that an object on that orbit ascends or descends over any given point of Earth's surface at the same local mean solar time.

Sun-synchronous orbits (Figure 2.90) are those whose orbital plane makes a constant angle α with radial from the Sun. To get such an orbit, the orbital plane must rotate in inertial space with the angular velocity of the Earth in its orbit around the Sun, which is 360° per 365.26 days, or 0.9859° per day.

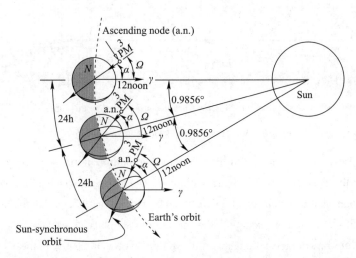

Figure 2.90　Sun-synchronous orbit

2.9.3.3 Molniya orbit

Molniya orbit (Figure 2.91) is a kind of highly elliptical orbit with an inclination of 63.4°, an argument of perigee of $-90°$, and an orbital period of precisely one half of a sidereal day. As shown in Figure 2.92, Molniya orbits are named after a series of Soviet Molniya communications satellites, which have been using this type of orbit since the mid-1960s.

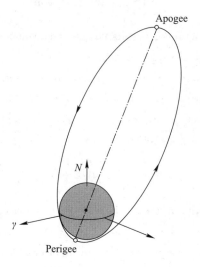

Figure 2.91　Typical Molniya orbit

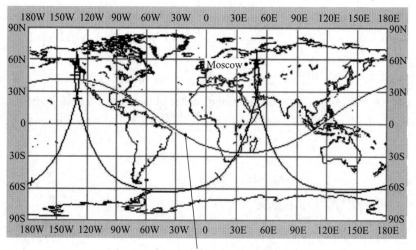

Molniya is visible from Moscow when the track is north of this curve.

Figure 2.92　Ground track of Molniya satellite

If a satellite is launched into an orbit with an inclination of 63.4° (prograde) or 116.6° (retrograde), then the apse line will remain stationary. The Russian space program made this a key element in the design of the system of Molniya communications satellites. All the Russian launching sites are above 45° latitude. For example, the northernmost place Plesetskis located at 62.8°N. Launching a satellite into a geostationary orbit would involve a costly plane change maneuver. Furthermore, a geostationary satellite cannot view effectively the far northern latitudes into which Russian territory extends.

The Molniya telecommunications satellites are launched from Plesetskis into 63° inclination orbits with a period of 12 hours. The apse line of these orbits has a length of 53000km. Perigee (typically 500km altitude) lies in the southern hemisphere, while apogee is at an altitude of

40000km above the northern latitudes, farther out than the geostationary satellites. A Molniya "constellation" consists of eight satellites in planes every 45°. Each satellite is above 30° north latitude for over eight hours, coasting towards and away from apogee.

2.9.4 Lunar exploration orbit

Lunar exploration activities have been begun since the end of the 1950s. For over half a century, even after more than ten years of rest periods, the exploration and research on the Moon has never stopped.

The Moon moves around the Earth in an elliptical orbit. The average inclination angle of the Moon's orbit to the Earth's orbit is 5°09′. The Moon rotates while orbiting around the Earth, with a period of 27.32166 days. The final orbit of the lunar probe is "the orbit around the Moon", not the "orbit of the Moon." When the probe becomes a lunar satellite orbiting around the Moon, it will also accompany the Moon in an orbit around the Earth.

A spacecraft flying around the Earth in a large elliptical orbit, when its apogee is exactly in the direction of the Moon, can also probe the Moon nearby. However, ordinary lunar probes either fly around the Moon for observation, or land on the Moon for detection. Because the Moon itself is within the gravitational range of the Earth, a probe flying to the Moon does not have to reach the second cosmic speed. As long as the initial speed is greater than 10.848m/s, it can fly to the Moon. The probe that flies to the Moon is mainly affected by the gravity of the Earth before 66000 kilometers away from the Moon, and its flight orbit is an elliptical orbit relative to the Earth. After 66000 kilometers from the Moon, it is mainly affected by the Moon's gravity, and the flight orbit is hyperbolic.

The lunar exploration orbit is roughly composed of five consecutive segments: launch orbit, Earth parking orbit, transfer orbit, lunar orbit, and Moon landing orbit. In specific lunar exploration activities, depending on the purpose of the exploration, the probe may only fly part of its orbit. Among these five orbits, the Earth-Moon transfer orbit is the one with the longest flight time and the most difficult to design and optimize.

The basic Earth-Moon transfer orbit adopts the Homan transfer idea. The transfer process is as follows. The first velocity pulse Δv_1 generate at any point P of circle C_1 with radius r_1, Transfer to ellipse E, its perigee is P, and a second velocity pulse Δv_2 is generated at apogee A of E, making the orbit transfer to a circular orbit C_2 with a radius of r_2. This double-pulse transfer process is called Hohmann orbital transfer shown in Figure 2.93.

Symbolic representation of this process:

$$(C_1) \xrightarrow{\Delta v_1} P(E) \cdots A(E) \xrightarrow{\Delta v_2} P(C_2)$$

In order to find the necessary Δv_1 and Δv_2, the relational equations are

$$\begin{cases} v_{r1} + \Delta v_1 = v_{ep} \\ v_{ra} + \Delta v_2 = v_{r2} \end{cases} \qquad (2-168)$$

as well as:

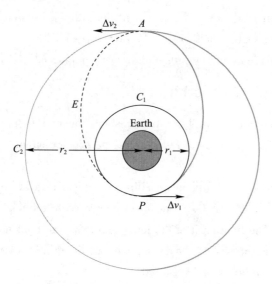

Figure 2.93 Hohmann orbital transfer

$$\begin{cases} v_{r1} = \sqrt{\mu/r_1}, v_{r2} = \sqrt{\mu/r_2} \\ v_{ep} = \sqrt{2\mu \dfrac{r_2}{r_1(r_1+r_2)}} = v_{r1}\sqrt{\dfrac{2r_2}{r_1+r_2}} \\ v_{ea} = \sqrt{2\mu \dfrac{r_1}{r_2(r_1+r_2)}} = v_{r2}\sqrt{\dfrac{2r_1}{r_1+r_2}} \end{cases} \quad (2-169)$$

So the required characteristic velocity is

$$\begin{cases} \Delta v_1 = v_{r1}\left(\sqrt{\dfrac{2r_2}{r_1+r_2}} - 1\right) \\ \Delta v_2 = v_{r2}\left(1 - \sqrt{\dfrac{2r_1}{r_1+r_2}}\right) \end{cases} \quad (2-170)$$

And the total characteristic speed is

$$\Delta v_\Sigma = \Delta v_1 + \Delta v_2 \quad (2-171)$$

The time t_{tr} of the Hohmann transition is equal to half of the period of the transition elliptical orbit:

$$t_{tr} = \dfrac{\pi}{\sqrt{\mu}}\left(\dfrac{r_1+r_2}{2}\right)^{3/2} \quad (2-172)$$

In order to save energy, probes that fly to the Moon generally enter a parking orbit around the Earth at first, and then change to a transition orbit to fly to the Moon. When approaching the Moon, they either fly around the Moon for detection, or descend from the orbit around the Moon, and landing on the lunar surface. Some probes that land on the Moon do not enter the orbit orbiting the Moon, but directly descend from the transition orbit and land on the lunar surface. No atmosphere on the Moon can be used for deceleration, so the landing probe will either directly hit the lunar surface (hard landing) or use a rocket to decelerate to achieve a soft landing.

2.9.5 Mars exploration orbit

Mars exploration has always been one of the hot spots of space exploration. Both Mars and the Earth are revolving around the Sun, with different orbital period (the earth is 365 days, and Mars is 687 days), so the distance between the two planets is changing with time, and the closest distance is 55 million km every 780 days (equivalent to about 26 months). When the distance is the furthest, Mars and the Earth are on both sides of the Sun, 400 million kilometers apart. What's more complicated is that because the orbits of Mars and the Earth are elliptical orbits and have different eccentricities, the distance between Mars and the Earth is different each time. The shortest rendezvous distance occurs once every 15 to 17 years, which is called "Mars Great Opportunity."

It is impossible for countries to wait for the opportunity of the "Mars Great Opportunity" that occurs once in more than ten years. Instead, countries use the time window that occurs every 26 months or so to launch Mars probes. In addition, since it takes at least a few months for the probe to enter the orbit of Mars, how far in advance should the probe "set off" from the Earth. It also needs to be calculated very accurately so that the probe can "enter into" the orbit of Mars when Mars is relatively close to the Earth under the conditions of limited fuel and limited number of orbit maneuvers. If the launch is too late, it is possible that when the probe runs out of fuel, Mars has already moved past the position closest to the Earth, causing the probe to miss the Mars. If the launch is too early, Mars may "have not arrived" when the probe arrives at the rendezvous position, causing the probe to fly over the target ahead of time and enter the solar system.

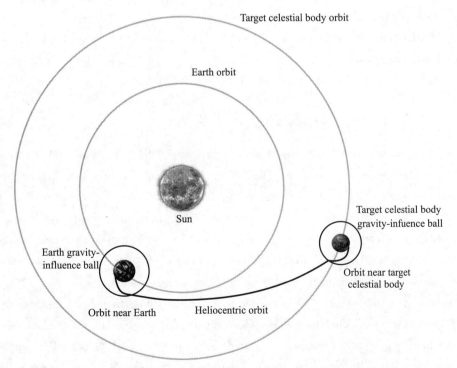

Figure 2.94 Conic splicing diagram of interplanetary transfer orbit

There are three stages from launch to Mars orbit, as shown in Figure 2.94. In order to reduce the complexity of orbit calculation and the difficulty of technical implementation, most of the Mars exploration missions in history have used the "Hohmann transfer orbit" to plan the flight path, and "Tianwen-1" is no exception.

This kind of orbit requires the least number of orbit changes, has the advantages of short time-consuming, simple operation, etc. The basic process includes three stages.

In the first stage, the Mars probe starts from the Earth and accelerates in low-Earth orbit until it reaches the second cosmic speed that can escape the gravity of the Earth.

The second cosmic velocity is

$$v_p = \sqrt{V_{\infty E}^2 + \frac{2\mu_E}{r_p}} \qquad (2-173)$$

The speed increment is

$$\Delta V = v_p - v_c = \sqrt{V_{\infty E}^2 + \frac{2\mu_E}{r_p}} - \sqrt{\frac{\mu_E}{r_p}} \qquad (2-174)$$

At this stage, it can be simplified that the probe is mainly affected by the gravity of the Earth, and the trajectory is hyperbolic with respect to the Earth, and the "Earth-centered inertial coordinate system" can be used to calculate and simulate the movement of the probe.

In the second stage, after the probe escapes the gravity of the Earth and before entering the gravitational influence of Mars, it can be simplified that the probe is only affected by the gravity of the Sun, and its trajectory is elliptical with respect to the Sun. The movement process can be simulated in the "heliocentric inertial coordinate system".

In the third stage, the probe applies a deceleration maneuver to make it enter the Martian gravitational influence range.

The deceleration maneuvering speed increment should satisfy:

$$\Delta V > v_p - v_h = \sqrt{\frac{2\mu_A}{r_p} + V_{\infty A}^2} - \sqrt{\frac{2\mu_A}{r_p}} \qquad (2-175)$$

It can be simplified that only subject to the gravitational force of Mars, and its trajectory is hyperbolic or elliptical relative to Mars. The "Mars inertial coordinate system" can be used to simulate and calculate its motion process. Finally, the Patched-conic method is used to join the trajectory of these three stages, the approximate orbit from the Earth to Mars can be obtained. Between different stages, the probe realizes the orbit transfer through acceleration/deceleration, which is the so-called orbit maneuver process. Of course, the speed direction and size of the probe can also be adjusted in the middle of the same stage.

For example, on July 23, the CZ-5 carrier rocket lifted China first Mars exploration mission "Tianwen-1" probe and ignited at the Wenchang Space Launch Site in China. And on October 9, 2020, the main engine of "Tianwen-1" ignited for about 480s and successfully completed the "deep space maneuver." It leaved the Earth's orbit and flew towards Mars. On February 10, 2021, Tianwen-1 probe implemented Mars capture braking. Tianwen-1 arrived in Mars orbit as a single spacecraft. It successfully landed on the Mars on May 15, 2021.

Questions

1. Please deduce the equation of the motion of a rocket.
2. Suppose that a rocket takes off vertically with an initial mass of 2500kg, including fuel of 2200kg, and the gas flow speed $u_e = 2000$m/s, when the fuel variation ratio is 50kg/s. Please get the maximum speed that the rocket can reach.
3. Suppose that a rocket takes off vertically at $t = 0$s with an initial mass of 1000kg, including fuel of 900kg, when the fuel variation ratio is 10kg/s, gas flow speed $u_e = 2100$m/s. Please get the acceleration speed and flight speed at $t = 0$s and $t = 45$s of the rocket.
4. Please compute the ratio of density and pressure at the height of 10km to the sea level.
5. Why can a flight vehicle fly in the air?
6. Describe the mechanical environment of a flight vehicle.
7. Please briefly describe how the aerodynamics is generated.
8. Which methods may be used to generate larger lift?
9. What is the composition of the drag?
10. Describe the moments acted on a flight vehicle.
11. Explain the following terms:
    (1) longitudinal static stability,
    (2) longitudinal dynamic stability.
12. Try to analyze the longitudinal static stability characteristics with diagrams.
13. Try to discuss the relation of focus point and mass center with the static stability.
14. Try to find the balance relations of the forces of a missile flying in vertical plane with constant altitude under the following conditions:
    (1) $\alpha = 0$,
    (2) $\alpha \neq 0$.
15. How to establish the equations of motion of a flight vehicle?
16. What are the differences between Soviets y stem and Euro-American system?
17. What are the definitions of angle of attack and angle of sideslip?
18. What is the definition of flight path reference frame?
19. Please derive the transformation formula between quasi-Earth-fixed reference frame and body-fixed reference frame.
20. How to simplify the equation of motion into longitudinal movement and lateral movement?
21. To analyze the forces and moments acting on a missile and draw a sketch in vertical plane, then (1) establish the equations of motion of the missile; (2) please get the loading factors at Body-fixed reference frame and Velocity reference frame.
22. Suppose that a missile is launched horizontally, which has a thrust of 4,905N, and an initial mass of 25kg, including 5kg of electronic devices. Try to get the loading factor and the inner force of the electronic devices acted on at the moment of takeoff. ($g = 9.81$m/s^2)
23. How to control a flight vehicle?

24. How to realize the TURN of a flying vehicle?
25. What is the three-channel control?
26. How to describe Kepler's orbit (equation)?
27. What're the Keplerian elements?
28. Describe the characteristics of the geosynchronous orbit.

Words and phrases

ablation　消融,切除
abound　多,大量存在,富于,充满
acronym　首字母缩略词
acumination　尖头
apogee　远地点
Bernoulli equation　伯努利方程
blunt　钝的
burbling　气流分离,起泡,附面层分离
camber　拱形
cartesian coordinates　笛卡儿坐标系
ceramic　陶器的
chord　翼弦
Concorde　协和式飞机
condensation　浓缩
cone　锥形物,圆锥体
continuous medium　连续介质
cosmic ray　宇宙线
COSPAR,Committee on Space Research　空间研究委员会
critical Mach Number　临界马赫数
cruising speed　巡航速度
curvature　弯曲,曲率
depiction　描写,叙述
dimensionless　无量纲的
dissociate　分离,游离,分裂
drag-induced by lift　诱导阻力
dynamic pressure　动压
electromagnetic radiation　电磁辐射
ellipse　椭圆
energetic particle radiation　高能粒子辐射
energy conservation　能量守恒
exosphere　外大气层,外逸层

exponential　指数的,幂数的
friction　摩擦,摩擦力
galactic cluster　银河系
gas state equation　气体状态方程
geospace　地球空间
heat-insulating index　绝热指数
hinder　阻碍,打扰
incendiary agent　燃烧剂
inhibitory　禁止的,抑制的
internal energy　内能
ionize　使离子化,电离
ionosphere　电离层
kinetic energy　动能
Laval tube　拉瓦尔喷管
lift coefficient　升力系数
magnetic　磁的,有磁性的,有吸引力的
mass conservation　质量守恒
mesosphere　中间层
micrometeoroid　微流星体
Molniya orbit　闪电轨道
NACA　(美国)国家航空咨询委员会
NASA　(美国)国家航空航天局
Nitrogen　氮
oscillate　振荡
parabola　抛物线
parasite drag　附加阻力
penalty　处罚
perigee　近地点
piston-engine　活塞发动机
plasma　等离子体
pressure energy　压力能
propagation　(声波、电磁辐射等)传播

ramification 分支,分叉,衍生物,支流	superposition 重叠,重合,叠合
reentry 再入,再进	synchronous 同步的,同时的
Reynolds number 雷诺数	tangent 切线的,相切的
rumor 流言,谣言,传闻	thermosphere 高温层
servo 伺服系统	tile 瓦片,瓷砖
shock wave 激波	trailing-edge 机翼后缘
slam 砰,撞击,猛击,冲击	troposphere 对流层
speculation 思索	true anomaly 真近点角
Sputnik (苏联)人造地球卫星	USSR 苏联
stable plasma jet 等离子体流(太阳风)	vacuum 真空
Standard Atmosphere 标准大气	Vernal Equinox 春分点
stickiness 黏性	vertical ascending gust 垂直向上的阵风
sticking coefficient 黏着系数	vortex 漩涡,旋风,涡流
stratosphere 平流层	zero-lift drag 零升阻力

References

[1] Tang Sheng-jing. Analysis methods for novel cargo aircraft concepts[M]. Munich:Verlag Dr. Hut,2002.
[2] Torenbeek E. Synthesis of subsonic airplane design[M]. Holland:Delt University Press,1982.
[3] 谢础,贾玉红. 航空航天技术概论[M]. 北京:北京航空航天大学出版社,2005.
[4] 赵育善,吴斌. 导弹引论[M]. 西安:西北工业大学出版社,2002.
[5] 钱杏芳,林瑞雄,赵亚男. 导弹飞行力学[M]. 北京:北京理工大学出版社,2000.
[6] 王春利. 航空航天推进系统[M]. 北京:北京理工大学出版社,2004.
[7] 罗格. 世界航天器与运载火箭集锦[M]. 北京:宇航出版社,2000.
[8] 樊启发. 世界制导兵器手册[M]. 北京:兵器工业出版社,1996.
[9] 邢继发. 世界导弹与航天发动机大全[M]. 北京:军事科学出版社,1999.
[10] 王保国,刘淑艳,刘艳明,于勇. 空气动力学基础[M]. 北京:国防工业出版社,2009.
[11] 钱翼稷. 空气动力学[M]北京:北京航空航天大学出版社,2004.
[12] 李福昌. 运载火箭工程[M]. 北京:宇航出版社,2002.
[13] 苗瑞生,居贤铭,吴甲生. 导弹空气动力学[M]北京:国防工业出版社,2006.
[14] 陈克俊,刘鲁华,孟云鹤. 远程火箭飞行动力学与制导[M]北京:国防工业出版社,2014.
[15] 张晓今,张为华,江振宇. 导弹系统性能分析[M]北京:国防工业出版社,2014.

Chapter 3 Flight vehicle system composition

A system is defined as one synthesized body that is composed of several parts interacted and mutual depended. In this chapter, in order to have a more comprehensive and in-depth understanding of the aircraft, it is a meaningful attempt to try to introduce the aircraft and its basic components and functions from a system perspective.

3.1 Flight vehicle system

A flight vehicle system is an overall body for specific tasks. For the design and research of a flight vehicle, the method of system engineering should be applied. To analyze the overall system, the compositions of the system should be considered.

As discussed in Chapter 1, rocket, missile, aircraft and spacecraft are the most common kinds of flight vehicles. In the following parts, it is going to analyze their system compositions in detail respectively.

3.2 Missile system

3.2.1 Missile system composition

A missile weapon system refers to one that completes the scheduled task, and it is at the higher level above the missile system. Its composition usually includes a missile system, fire and control equipment, and auxiliary equipment. Figure 3.1 gives a typical composition of the missile weapon system.

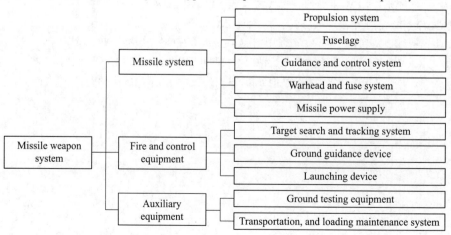

Figure 3.1 A typical missile weapon system

The fire and control equipment track the target and controls the launch of the missile. It mainly includes a target search and tracking system, ground guidance device, and launching device, etc.

The auxiliary equipment provides missile storage, transportation, maintenance, and other services. It usually includes ground testing equipment, transportation, loading maintenance system and training devices, etc.

The main body of the missile system is the missile which is mainly composed of five parts, i. e., propulsion system, fuselage, guidance and control system, warhead and fuse system, and missile power supply.

3.2.1.1 Propulsion system

Propulsion system is the device that takes the engine as the core to provide sufficient power for a missile. Guaranteeing required range and velocity, the system is also named propulsion subsystem.

Rocket engines such as solid and liquid rocket engines, jet engines, turbojet engines and ramjet engines, and combination engines are used on the missiles.

Some missiles, for example, the ground to air missile and ship to air missile, two engines, or a dual-thrust motor should be adopted. One engine is used to boost the missile to make it depart from the launching device and accelerate rapidly, so this engine is called boosters; the other one is used as the main engine which is able to maintain the missile flying at a certain velocity so that it can track the plane or tank, so the engine is named sustainer motor. Long-range miss.les and intercontinental missiles whose velocities are required to reach thousands of meters per second often adopt step rocket. Each step uses one or several rocket engines.

3.2.1.2 Guidance and control system

Guidance and control system is the general terms of instruments, devices, and equipment that could lead and control the missile to fly towards the target. In order to direct the missile to hit the target accurately, on one hand, the relative moment between the missile and the target should be detected and measured. Furthermore, the leading command is formed for a missile to attack a target according to the moving deviations of them. On the other hand, according to the leading command, the missile should be controlled to alter its flight movement to fly in the required direction and trajectory to hit the target. A guidance system is composed of a lead system and a control system, which could achieve the above requirements. There are many kinds of guidance systems, correspondingly, their principles of operation are diverse.

A guidance system could be fixed on the missile, for example, an automatic homing guidance system is disposed of in this way. But for many other missiles, only the control system is installed on the missile, and the leading system is sited on the command station which could be located on the land, ship, or plane.

The control system is a small loop that is in the whole guidance system. Generally, it is on the missile body which is also called an autopilot to control the attitudes and displacements of the missile.

3.2.1.3 Warhead and fuse

This part of the missile is used to destroy and damage the target directly. For it is always at the head of the missile, we name it warhead customarily.

Considering the different targets, a variety of warheads such as blast warhead, fragmenting warhead, shaped-charge warhead, chemical warhead, biological agent warhead, and nuclear warhead are produced to satisfy different situations.

The warhead is detonated by a fuse. That is to say, the function of fuses is to control when the warhead detonates safely and reliably.

3.2.1.4 Fuselage

The fuselage is the main body of the missile. Usually, it is a revolving body composed of cabins, segments, and aerodynamic surfaces such as wings, control planes, and stabilizers (tails). Warhead and fuse, guidance and control system, propulsion system and missile power supply are all installed in it.

3.2.1.5 Missile power supply

The missile power supply is the device that provides electrical energy to the subsystems on the missile. In addition to the battery, devices of power distribution and power transformation are usually included. Silver-zinc battery is generally used, which has relatively more energy per unit weight and could be stored for a long time. For some missiles, a small turbo-generator is utilized to supply energy for parts of them. Some cruise missiles adopt small generators which are driven by the turbofan engine.

3.2.2 Typical examples of the missile system

1) Javelin missile system

The Javelin, as shown in Figure 3.2 and Figure 3.3, is a man-portable, fire-and-forget, antitank missile employed by dismounted infantry to defeat current and future threats from armored combat vehicles. Javelin is intended to replace the Dragon system in the Army and the Marine Corps because it has significant improvements over Dragon.

The range of Javelin is 2500m, which is more than twice of its predecessor, the Dragon. The Javelin has secondary capabilities against helicopters and ground-fighting positions. It is equipped with an imaging infrared (I2R) system and a fire-and-forget guided missile. The Javelin's normal engagement mode is top-attack to penetrate the tank's most vulnerable armor. It also has a direct-attack capability to engage targets with an overhead cover or in bunkers. Figure 3.2 indicates a typical scene of the Javelin missile in use. Its "soft launch" allows employment within buildings and

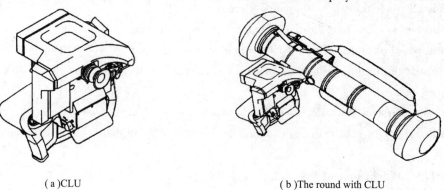

(a) CLU　　　　　　　　　　　(b) The round with CLU

Figure 3.2　Javelin missile system

Figure 3.3 Javelin missile in use

enclosed fighting positions. The soft launch signature reduces the gunner's exposure to the enemy, thus increases survivability. Javelin is also much more lethal than Dragon. It has a top attack dual warhead capability which enables it to defeat all known enemy armor systems.

(1) System composition

The Javelin system consists of the Command Launch Unit (CLU) and the round. The CLU incorporates a passive target acquisition and fire control unit with an integrated day sight and a thermal imaging sight. The gunner's controls for the missile system are on the CLU. The day sight is equipped with ×4 magnification and the night sight with ×4 and ×9 magnification optics.

The round consists of the Javelin missile and the Alliant Techsystems (ATK) Launch Tube Assembly. Javelin is a fire-and-forget missile with lock-on before launch and automatic self-guidance. The missile is equipped with an imaging infrared seeker which is based on a cadmium mercury telluride (CdHgTe) 64 × 64 staring focal plane array in the eight-to-twelve-micron waveband. The tandem warhead is fitted with two shaped charges: the precursor warhead to initiate explosive reactive armor and the main warhead to penetrate base armor. The propulsion system is a two-stage solid propellant design that provides a minimum smoke soft launch.

There are two states for the Javelin missile system. One is for battle, that is, the former cover of the launching tube is open, CLU and round are combined together, and everything is ready for launch. The other is for moving state; all the compositions are separated for packing so that two soldiers could take them to move to other places.

As shown in Figure 3.1, launch and guidance equipment consists of the Command Launch Unit, i.e. CLU, and the round where a missile is in.

It consists of the observation and detection equipment, battery, cooler for the imaging system, launching tube, and so on.

The launching tube is composed of round, components for mechanical-electronic insurance, seal components, and soft launcher, in case of avoiding and decreasing the sound, flame, and smoke, and the backward force from a firearm.

(2) Missile construction

As seen in Figure 3.4, the total missile is composed of three main sections, in turn from the head to the end are, seeker section, warhead section, and propulsion section.

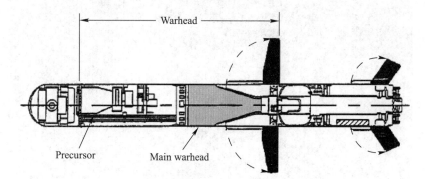

Figure 3.4 Javelin missile configuration

It includes body, wings, and tails, seeker, pre-warhead, main warhead, guidance electronic cabin, launching engine and flight engine, executing mechanism, and so on.

(3) System working principles

The system could be deployed and ready to fire in less than 30s and the reload time is less than 20s. The missile is mounted on the Command Launch Unit and the gunner engages the target using the sight on the CLU, by placing a curser box over the image of the target. The gunner locks on the automatic target tracker in the missile by sending a lock-on-before-launch command to the missile. When the system is locked-on, the missile is ready to fire and the gunner cannot carry out post launch tracking or missile guidance. Unlike conventional wire guided, fibre-optic cable guided, or laser beam riding missiles, Javelin is autonomously guided to the target after launch, leaving the gunner free to reposition or reload immediately.

Ejecting the missile from the launch tube with a soft launch gives a low-recoil shoulder. The soft launch enables firing from inside buildings or covered positions. Once the missile is clear, the larger propellant in the second stage is ignited and the missile is propelled towards the target. The weapon has two attack modes, direct or top attack. The gunner selects direct attack mode to engage covered targets, bunkers, buildings, and helicopters. The top attack mode is selected against tanks, in which case the Javelin climbs above and strikes down on the target to penetrate the roof of the tank where there is the least armor protection. The missile is launched at an 18° elevation angle to reach a peak altitude of 150m in top attack mode and 50m in direct fire mode.

2) Patriot missile system

The Patriot is aground-to-air missile used by the United States Army and several allied nations. It is manufactured by the Raytheon Company of the United States. The Patriot System replaced the Nike Hercules system as the U.S. Army's primary High to Medium Air Defense system and replaced the Hawk system as the U.S. Army's medium tactical air defense system. In addition to these roles, Patriot has been given the function of the U.S. Army's anti-ballistic missile (ABM) system, which is now the

Chapter 3　Flight vehicle system composition

Patriot's primary mission. The most recent upgrade of Patriot is called PAC-3.

(1) System composition

The Patriot system has four major operational functions: communications, command and control, radar surveillance, and missile guidance. The four functions combine to provide a coordinated, secure, integrated, mobile air defense system. The heart of the Patriot battery is the fire control section, consisting of the AN/MPQ-53 or-65 Radar Set, the AN/MSQ-104 Engagement Control Station (ECS), the OE-349 Antenna Mast Group (AMG), and the EPP-III Electric Power Plant. The system's missiles are transported on and launched from the M901 Launching Station, which can carry up to four PAC-2 missiles or up to sixteen PAC-3 missiles.

Figure 3.5 shows the composition of the Patriot missile system and the main functions of each part. The real radar set, antenna mast group, and launch station are given in Figure 3.6.

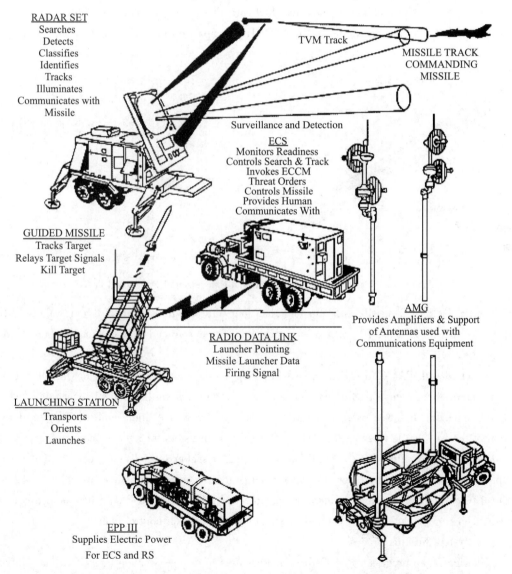

Figure 3.5　The composition of Patriot ground-to-air missile system

147

Figure 3.6　The radar set, antenna mast group, and launch station of Patriot

(2) Missile construction

There are two kinds of Patriot missile which are PAC-2 and PAC-3. The latter one which is shown in Figure 3.7 (a) was deployed in 2002. The comparison of PAC-2 and PAC-3 is shown in Figure 3.7 (b).

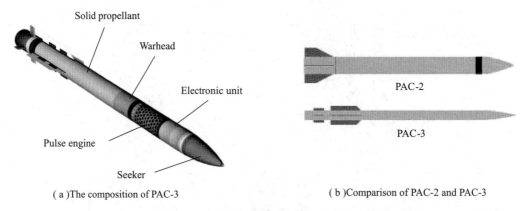

(a)The composition of PAC-3　　　　　　(b)Comparison of PAC-2 and PAC-3

Figure 3.7　The composition of PAC-3 and comparison of PAC-2 and PAC-3

The Patriot PAC-2 missile is 5.2m long, has four tailfins with a span of 87cm, weighs 914kg, and has a range of at least 70km. It uses a proximity-fuzed warhead that detonates when it comes near the target.

Compared with PAC-2, the PAC-3 is 5.2m long, has a diameter of 25cm, has a tailfin system with a span of 48cm, weighs 318kg, and has a range of 20km. The tailfin system, consisting of a set of cruciform fixed fins followed by maneuvering fins, is used for coarse flight control, while the forward section of the fuselage is ringed with 180 small solid-fuel thrusters for agile maneuvering.

The PAC-3 is equipped with a millimeter-wave radar seeker linked to a processor that not only computes the interceptor's trajectory but also matches the incoming warhead with a library of known warhead types to determine the optimum attack procedure. Although the PAC-3 has its own guidance system, it has an RF data link to maintain contact with the launch control center.

(3) System working process

Following is the process a PAC-2 firing battery uses to engage a single target (an aircraft) with a single missile:

A hostile aircraft is detected by the AN/MPQ-53 Radar. The radar examines the track's size, speed, altitude, and heading, then decides whether it is a legitimate track or "clutter" created by RF interference. Based on many factors, including the track, IFF response, or its presence area, the ECS operator reports it to the superiors and waits for feedback.

Once the engagement command is received, a firing battery will be selected to take the shot. The system's launchers are from "standby" into "operate". When the fire signal is sent to the selected launcher, the patriot missile will be fired automatically by the system.

The AN/MPQ-53 Radar, which has been continuously tracking the hostile aircraft, "acquires" the just-fired missile and begins feeding it interception data. The Radar also "illuminates" the target for the missile's semi-active radar seeker.

The monopulse receiver in the missile's nose receives the reflection of illumination energy from the target. The track-via-missile uplink sends this data through an antenna in the missile's tail back to the AN/MPQ-53 set. In the ECS, computers calculate the maneuvers that the missile should perform in order to maintain a trajectory to the target and the TVM uplink sends these to the missile.

Once in the vicinity of the target, the missile detonates its proximity fused warhead.

3.3 Rocket system

3.3.1 Multiple rocket launcher system

3.3.1.1 Multiple rocket launcher system composition

A multiple rocket launcher (MRL) is an unguided rocket artillery system. Like other rocket artillery, multiple rocket launchers are less accurate and have a much lower (sustained) rate of fire than batteries of traditional artillery guns. However, they have the capability of simultaneously dropping many hundreds of kilograms of explosives, with devastating effects. Figure 3.8 indicates the main components of a typical MRL system.

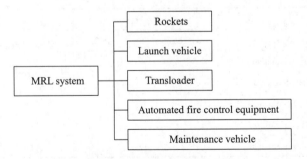

Figure 3.8　A typical MRL system

3.3.1.2 Typical examples of the multiple rocket launcher system

The BM-30 or 9A52 which is usually called Tornado is a Soviet heavy multiple rocket launcher. The system is designed to defeat personnel, armored, and soft-skinned targets in concentration areas, artillery batteries, command posts, and ammunition depots. It was created in the early 1980s and entered service in the Soviet Army in 1989.

The main components of the system are Rockets 9M55 or 9M528 (in containers), BM 9A52 − 2

Launch vehicle shown in Figure 3.9, TZM 9T234-2 Transloader with an 850kg crane and 12 spare rockets, automated fire control equipment in the command post 1K123 "Vivary", maintenance vehicle PM-2 – 70 MTO-V.

Figure 3.9 BM9A52-2 launch vehicle

Thereare several kinds of rockets for the Tornado MRL system, which are listed in Table 3.1. Rocket 9M55K is one kind of cluster munition and is the mainly used rocket for Tornado. It is 7.6m long, weighs 800kg, has a minimum range of 20km and maximum range of 70km. 72 sub-munitions are mounted in warheads whose total weight is 243kg. Figure 3.10 shows the construction of 9M55K.

Table 3.1 Various rockets for Tornado

Variant		Rocket		Warhead	Range	
Name	Type	Weight/kg	Length/m	Weight/kg	Min./km	Max./km
9M55K	Cluster munition	800	7.6	243	20	70
9M55K5	HEAT/HE-Fragmentation			243		
9M55F	Separable HE-Fragmentation			258		
9M55S	Thermobaric			243		
9M528	HE-Fragmentation	815		243	25	100

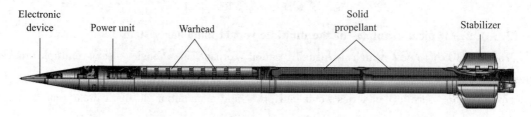

Figure 3.10 The composition of 9M55K

3.3.2 Carrier rocket system

3.3.2.1 Carrier rocket system composition

Carrier rocket, also referred to as launch vehicle, is designed to send the payloads such as satellite, manned spaceship, space station, space probe, or space shuttle into the preliminary orbit. It is developed on the basis of the intercontinental ballistic missile. But requirements for the carrier rocket are different from those of ballistic missiles, reliability, carrying capacity, versatility, and economy are stressed.

Carrier rocket is a complex system, which is composed of several interrelated, interacting, and interdependent subsystems based on a combination with a specific function. A carrier rocket system includes the carrier rocket or launch vehicle, the launch pad, the telemetry tracking and command (TT&C) system, and other infrastructure. Figure 3.11 shows the basic components of a carrier rocket system.

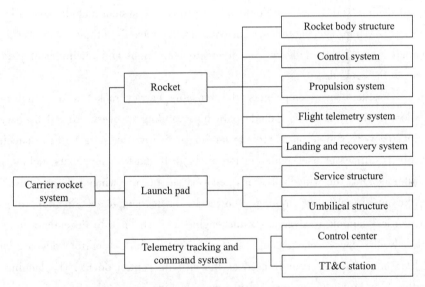

Figure 3.11 The basic components of a carrier rocket system

A launch pad is the area and facilities where rockets or spacecraft lift off. A typical launch pad consists of the service and umbilical structures. The service structure provides an access platform to inspect the launch vehicle prior to launch. Most service structures can be moved or rotated to a safe distance. The umbilical structure has propellant loading, gas, power, and communication links to the launch vehicle.

A telemetry tracking and command system includes a control center and several TT&C stations locating along the flight path of rockets. The TT&C system using continuous wave, single pulse measurement radar system, live television system, telemetry receivers, remote control devices, and optical telescopes, completes the continuous monitoring and control of rockets cooperating with the device installed in rockets. Figure 3.12 shows the launch pad and space control center.

Carrier rocket generally consists of the following components: the rocket body structure, control system, propulsion system, flight telemetry system, and additional systems. The rocket body structure's

Figure 3.12　A launch pad and a space control center

task is bearing forces and connecting all the parts together. The structure of a liquid-propellant rocket includes payload fairing, instrument cabinet, propellant tanks, inter-stage section, inter-tank section, the engines, tail section, and so on. The liquid rocket propulsion system includes rocket engines and a propellant delivery system. The control system controls the rocket's attitude stability in accordance with its intended orbit. Flight telemetry system records the flight and environment parameters and transmits them to the ground.

In addition to the above components, the reusable carrier rocket also includes a recovery system. Nowadays, recovery methods include parachute landing recovery, winged fly-back recovery, vertical landing recovery, and so on. Parachute landing recovery uses parachutes to decelerate in the air and then opens airbags or uses buffer motors to land. Its basic components include parachutes, airbags, or buffer motors. The winged fly-back recovery adopts the winged flying body, which lands horizontally like an aircraft after braking in orbit. Its basic components include wings and their controls. The vertical landing recovery uses the engine to push back to decelerate and land on the ground vertically. Its basic components include the attitude control device and landing mechanism. RCS or grid fins is usually used as the attitude control device. The landing mechanism includes landing legs, buffer, footpad, and deployment locking mechanism.

3.3.2.2　Typical examples of the carrier rocket system

1) Saturn V

Saturn launch vehicle is the U.S. heavy-lift rocket designed for the Apollo program, which is the U.S. manned lunar landing mission. It includes Saturn Ⅰ, Saturn ⅠB, and Saturn Ⅴ. Among these, the Saturn Ⅴ is the most capable, which could send the Apollo spacecraft weighing up to 50 tons into the Moon orbit.

Saturn V has three stages. The first stage uses five F1 LOX/kerosene engines. The second stage uses five J-2 engines. And the third stage uses one J-2 engine. Figure 3.13 shows the composition of the Saturn V.

2) CZ-2F

CZ-2F is the China carrier rocket designed for China manned space project. It includes two types, for spaceship launching and for space laboratory launching. The LEO launch capability of each type is 8100kg and 8600kg respectively.

Chapter 3 Flight vehicle system composition

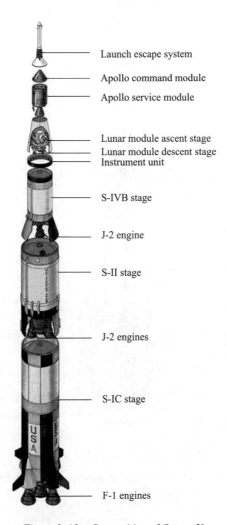

Figure 3.13 Composition of Saturn V

CZ-2F is a two-stage liquid rocket. It also has four boosters and an escape tower used for emergencies while manned flight. In the event of a major failure, the spacecraft carrying astronauts will safely escape from the danger zone.

The CZ-2F carrier rocket consists of the structure system, power plant system, control system, propellant system, fault detection and treatment system, escape system, telemetry system, external safety system, additional system and ground equipment system. Figure 3.14 shows the composition of CZ-2F.

3.4 Air vehicle system

3.4.1 Airplane system

3.4.1.1 Airplane system composition

An airplane is a complex system. In the design stage and the flight and maintenance manuals

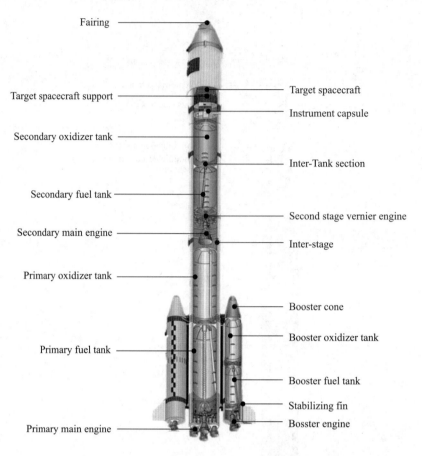

Figure 3.14 Composition of CZ-2F

(used by pilots and maintenance technicians), it is broken down into simpler systems that carry out homogeneous functions. Figure 3.15 indicates the typical composition of the airplane system.

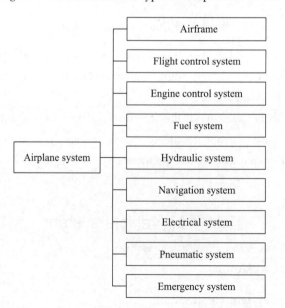

Figure 3.15 Typical composition of the airplane system

In the following section, only some basic systems are discussed.

1) Flight control system

Though flight controls have advanced considerably throughout the years, all aircraft are governed by the same basic principles of flight control, no matter the vehicle is the most sophisticated high-performance fighter or the simplest model aircraft.

Generally, the flight control systems consist of primary flight control and second flight control. The primary flight control provides control surfaces that allow the aircraft to maneuver in pitch, roll, and yaw. These control surfaces include the elevator, aileron, and rudder. In the second flight control, additional control surfaces are required for the specific purposes of controlling the high lift devices required during the approach and landing phases of flight. High lift control is provided by a combination of flaperons and leading-edge slats.

The system has also to be designed so that it provides stable control for all parts of the aircraft flight envelope; this requires a thorough understanding of the aerodynamics and dynamic motion of the aircraft. Most modern fighter aircraft of any sophistication now possess a fly-by-wire system due to the weight savings and considerable improvements in handling characteristics that may be achieved. Many aircraft embody automatic flight control systems to ease the burden of flying the aircraft and reduce pilot workload.

2) Engine control systems and fuel system

The basic control action of the engine system is to control the flow of fuel and air to the engine to operate at its optimum efficiency over a wide range of forward speeds, altitudes, and temperatures whilst allowing the pilot to handle the engine without fear of malfunction. So, control can be exercised over the following aspects of engine control: fuel flow, air flow, and exhaust gas thrust. With the development of digital technology and serial data transmission systems, as well as higher performance electronic devices, the FADEC (Full Authority Digital Electronic Control) was introduced into the engine control system.

In early jet aircraft, the control of fuel to the combustors was performed by pneumatic and hydromechanical flow control devices. This soon proved to be totally unsatisfactory, since the wide range of ambient conditions encountered in flight meant that continual throttle adjustments were needed. The task of handling engines was eased by the introduction of electronic control in the form of magnetic amplifiers in early civil and military aircraft. As the technique developed, the control system became more sophisticated with additional engine condition sensors and multiple servo-loops. With the modern FADEC system, there are no mechanical control rods or mechanical reversions, and the pilot can perform carefree handling of the engine throughout the flight envelope.

The fuel system is an essential element in the overall suite of systems required to assure safe flight, whose components include fuel transfer pumps, fuel booster pumps, fuel transfer valves, and non-return valves. The purpose of an aircraft fuel system is primarily to provide a reliable supply of fuel to the engines. Modern aircraft fuels are hydrocarbon fuels like those used in the automobile. Piston-engined aircraft uses a higher-octane fuel called AVGAS in aviation parlance. Jet engines use a cruder fuel with a wider distillation cut and with a lower flashpoint.

3) Hydraulic system

Hydraulic systems now play an important role in all modern aircraft, both military and civil. In choosing an aircraft system, the principal requirements are low weight, low volume, low initial cost, high reliability, and low maintenance. Hydraulic systems meet all these requirements reasonably well and have additional attractions. With hydraulic power, the pilot can move the control surfaces with ever-increasing speeds and demands for maneuverability. And the system can also be used for undercarriage, braking, and anti-skid systems. Though more electric systems are being considered to replace hydraulically powered systems in some areas, the hydraulic system remains the most effective source of power for both primary and secondary flying controls. The operation principle of the hydraulic system is as follows.

The primary source of power on an aircraft is the engine, and the hydraulic pump is connected to the engine gearbox. The pump causes a flow of fluid at a certain pressure, through stainless steel pipes to various actuating devices. A reservoir ensures that sufficient fluid is available under all conditions of demand.

4) Navigation system

In the early time, there were two types of navigation systems. One was mechanical navigation. Navigation information was displayed on a group of instruments called the basic or primary six, which included the attitude indicator, a vertical speed indicator showing the rate of climb and descent, airspeed indicator, turn-and-bank coordinator, a heading indicator showing the magnetic compass course, and the altimeter. These instruments are still used.

The other was Radio navigation. The earliest radio navigation aid was the four-course radio range. Four towers set in a square transmitted the letters A and N in Morse code. A pilot flying along one of the four beams toward the square would hear only an A or N in the dashes and dots of the code. The dashes and dots grew louder or fainter as he flew, depending on if he was flying toward or away from one of the corners. Turning right or left, he would soon hear a different letter being transmitted, telling him which quadrant he had entered. The beams flared out so that at certain points they overlapped. Where the A or N signals meshed, the Morse code dashes and dots sounded a steady hum, painting an audio roadway for the pilot. Unfortunately, mountains, mineral deposits, railroad tracks, and even the atmospheric disturbance of the setting sun could distort the signals.

Today's aircraft are tracked as computer-generated icons wandering across radar display screens, with their positions, altitude, and airspeed updated every few seconds. Pilots and controllers communicate using both voice and data transmitting radios, with controllers relying on radar tracking to keep aircraft on course. Today, cockpit navigation information is increasingly displayed on a monitor, but the position of information and its formatis nearly identical to the basic six instruments of early and simpler aircraft. New technologies also provide new kinds of navigation methods, like the Global Positioning System (GPS), the Inertial Navigation System (INS), and the Inertial Guidance System (IGS) which is one technology that allows pilots to accurately determine their position anywhere on the Earth within seconds.

5) Electrical system

Electrical systems have made significant advances over the years as aircraft have become more dependent upon electrically powered services.

The generic parts of a typical alternating current (AC) aircraft electrical system comprise the following: power generation, primary power distribution and protection, power conversion and energy storage, and secondary power distribution and protection.

In most aircraft, the system generates a constant frequency of 115 VAC (Voltage of Alternating Current) at 400Hz. In U. S. military circles, great emphasis is being placed by the U. S. Air Force and the U. S. Navy on the development of 270 VDC systems.

6) Pneumatic system

The aircraft pneumatic system is sometimes called vacuum pressure systems, which is similar to hydraulic systems but with air instead of fluid. The pneumatic system provides compressed air for various requirements for an aircraft. It is one of the three power sources, the other two are electrical power and hydraulic power. The easy availability of high-pressure air from the modern engine is key to the use of pneumatic power as means of transferring energy or providing the motive power on the aircraft.

The modern turbofan engine is an effective gas generator and this has led to the use of engine bleed air for a number of aircraft systems, either for reasons of heating, provision of motive power, or as a source of air for cabin conditioning and pressurization systems. Bleed air is extracted from the engine compressor and after cooling and pressure reduction/regulation it is used for a variety of functions. Other areas of the aircraft use pneumatic principles for sensing the atmosphere surrounding the aircraft for instrumentation purposes. The sensing of air data is crucial to ensuring the safe passage of the aircraft in flight.

7) Emergency system

Usually, the emergency systems are designed to operate once and only when there is an emergency, which may be the final means of survival for the aircraft, crew, and passengers. So, the integrity of these systems must be high. Hence there is a need to separate them from the aircraft primary systems so that failures are not propagated from the primary systems into the emergency systems. Emphasis is placed on separate sources of power, alternative methods of operation, and clear emergency warning indications.

The emergency systems need to be handled in a very specific and rapid manner. The main principle of an emergency system involves two stages: informing the operator and performing a mitigating action. First of all, for these stages, various types of warning systems have been designed to facilitate immediate notification of the emergency, including location and severity. In some cases, the mitigating actions are carried out automatically by built-in subsystems. In others, pilots must perform actions based on a checklist or previous training routines.

3.4.1.2 Typical examples of the airplane system

1) J-10

J-10 is China's third-generation fighter independently developed by Avic Chengdu aircraft industrial (group) Co. LTD. It is a multi-purpose fighter that could carry out air defense, air control, attack, and other combat tasks.

J-10 uses a three-axis four-redundancy digital fly-by-wire flight control system. It has a high thrust turbofan engine and two hydraulic systems. The core of the electrical system is a high-power alternator. Besides, it also has an emergency power unit. Figure 3.16 shows the appearance of J-10.

Figure 3.16   J-10

2) A380

A380 aircraft is a large civil jet aircraft developed by Airbus Corp. It has a layout of 555 and 850 seats.

A380 aircraft uses four Rolls-Royce Trent 400 engines, similar to engines used on Boing 747 and 767. It uses the Integrated Modular Avionics (IMA) System developed by Thales Cop. For safety, it has four independent flight control systems, each of these can control the aircraft. It also has two conventional hydraulic systems. Figure 3.17 shows an A380 aircraft flying in the sky.

Figure 3.17   A380 in flight

3.4.2   UAV system

3.4.2.1   UAV system composition

Unmanned Aerial Vehicle (UAV) system usually has the same elements as manned aircraft system, except the elements which support the aircrew aboard. Basically, a UAV system comprises a flying platform, autopilot, navigation, communication, power plant, and payload. It usually has a control station and launch-and-recovery equipment on the ground.

1) Flying platform

The flying platform, also called UAV, is the unmanned aerial vehicle that is operated by radio remote control equipment and onboard preset control device. Based on how the lift is generated, UAV can be divided into fixed-wing UAV, rotor UAV (including unmanned helicopter and multi-rotor UAV), bionic UAV, unmanned airship, and so on. Some typical UAVs are shown in Figure 3.18.

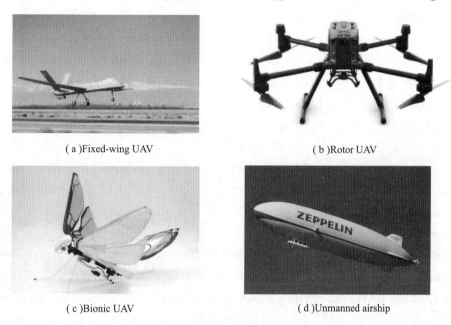

(a) Fixed-wing UAV (b) Rotor UAV

(c) Bionic UAV (d) Unmanned airship

Figure 3.18 Some typical UAVs

The main function of the flying platform is to carry the payload to where it works and carry various sub-systems that maintain the operation of the UAV system, including control system, navigation system, communication system, power plant, etc. The type and performance of the flying platform are determined by the mission it undertakes. When designing a flying platform, Flight distance, speed and time are the most concerned index.

2) Control station

The control station is the command center of the UAV system, which is usually based on the ground (GCS), aboard ship (SCS), or possibly airborne in a "parent" aircraft (ACS).

An operator would monitor the flight attitude and change the flight parameters of the UAV through the control station. What the control station provides are changing the attitude and track of the UAV, operating the various types of payload that the UAV carries, controlling the communication process, launching and retrieving the UAV.

3) Autopilot

The autopilot, also known as the automatic flight control system (AFCS), is the central part of the UAV system. Its main function is to direct the flight platform to complete the corresponding task under the command of the control station or to autonomously control the flight platform to complete its task.

The hardware of the autopilot includes the central processing unit (CPU), various communication interfaces, and micro-electromechanical system (MEMS) sensors. The CPU is the core component of the autopilot, running the core algorithm of the autopilot. AVR, ARM, FPGA, and DSP are the most commonly used processor of the UAV by now. The function of the communication interface is to exchange data between the processor and the sensor. The function of the MEMS sensor is to measure the flight attitude of the UAV. Compared with the mechanical sensor, the MEMS sensor is small in size and low in price, which is very suitable for a small UAV system.

4) Navigation

The navigation system allows the UAV and the operator to know where the aircraft is at any moment in time. Nowadays, the most commonly used navigation system of the UAV is the global navigation satellite system (GNSS), which accesses positional information from a system of earth-satellites. The GNSS now available can give a continuous positional update of the UAV and are extremely light in weight, compact, and also quite cheap. The accuracy can be further improved by using differential global positioning system (DGPS).

Besides, for the nonautonomous UAV system which needs continuous communication between the UAV and the control station, or where the GPS is blocked, radar tracking, radio tracking or direct reckoning, etc. can be used alternatively.

5) Communication

The communication system exchanges data between the UAV and the control station through the data link. The UAV data link system includes an uplink, also called a command link, for the control station to control the UAV and equipment onboard and a downlink for the control station to receive signals or data from the UAV and equipment onboard. The data link can also be used to measure the distance and azimuth of the ground antenna relative to the aircraft. This information can be used for the navigation of the UAV to improve the accuracy of the measurement of the target position by onboard sensors.

The communication system consists of the airborne part and the ground part. The airborne part includes the airborne data terminal (ADT) and the antenna, while the ground part includes the ground data terminal (GDT) and the antenna. Both the airborne and the ground data terminal include radio frequency receivers, transmitters, and modems.

6) Power Plant

Different uses give different requirements for the power plant of the UAV, but small size, low cost, and reliable operation are the common goals. The widely used UAV power plant at present includes the electric motor, piston engine, turboshaft engine, turbofan engine, and turbojet engine.

The electric motor is usually used on miniature UAV (e.g. multi-rotor UAV), whose take-off mass is less than 10kg and the endurance time is less than one hour. The piston engine is usually used on low-speed-low-altitude small UAV and the turboshaft engine is usually used on low altitude unmanned helicopter. Generally, for single-use target drones, suicide unmanned aerial vehicles or missiles, turbojet engines are used because of the high thrust-to-weight ratio requirement but a low requirement for the service life (1~2h).

7) Payload

Equipment carried by the UAV to complete a specific task while without affecting its flight after removal, are referred to as the payload of the UAV, or the mission payload.

Divided by functions, the payload includes the photoelectric payload (e. g. visible light imaging camera, infrared thermal imager, etc.), the delivery payload (e. g. weapons, etc.), the acquisition payload (e. g. air sampling, etc.), and other types of payload (e. g. communication relay, experiment, etc.).

8) Launch-and-Recovery equipment

(1) Launch equipment

Launch equipment is required for those UAVs which do not have a vertical flight capability, nor have access to a runway of suitable surface and length. This usually takes the form of a ramp along which the aircraft is accelerated on a trolley, propelled by a system of rubber bungees, by compressed air or by rocket, until the aircraft has reached an airspeed at which it can sustain airborne flight.

(2) Recovery equipment

Recovery equipment is also required for UAV without a vertical flight capability when there is no suitable surface for landing. It usually uses a parachute installed within the UAV. An alternative form of recovery equipment is a large net into which the aircraft is flown and caught.

Figure 3.19 shows a typical launch and recovery equipment.

(a) UAV launch equipment

(b) UAV recovery equipment

Figure 3.19 Typical UAV launch and recovery equipment

3.4.2.2 Typical examples of the UAV

1) RQ-4

RQ-4, also called Global Hawk, is a single-engine turbofan high-altitude long-endurance reconnaissance UAV developed by Northrop Grumman for the U. S. air force. It is used to replace the U-2 reconnaissance aircraft to perform long-range, high-altitude, wide-area, and long-lasting intelligence, surveillance and reconnaissance (ISR) missions. It is the largest and heaviest UAV in the world now.

The Global Hawk UAV system consists of two RQ-4A UAVs and two ground control stations. Each ground control station is composed of a task control unit operated by four people and a launch and recovery unit operated by two people. The body of the UAV is a single wing configuration with a high aspect ratio, and the wing is made of carbon fiber reinforced plastic. RQ-4A is powered by the AE3007 turbofan engine developed by Rolls Royce. The payload is Raytheon's integrated sensor system (ISS) including photoelectric/infrared sensors and I/J-band synthetic aperture radar. RQ-4A adopts a wheel take-off and landing mode. Figure 3.20 shows an RQ-4 in flight.

Figure 3.20　RQ-4 in flight

2) Matrice 300 RTK

Matrice 300 RTK is a quadrotor UAV developed by DJI Cop. The UAV is oriented to industrial applications and could complete different tasks such as emergency fire fighting, police law enforcement, energy inspection, surveying, and mapping, etc.

The flying platform of the Matrice 300 RTK is a foldable quadrotor structure. The power system adopts TB60 rechargeable intelligent battery. Its payload includes the laser rangefinder, zoom camera, wide-angle camera, etc. The payload can be installed on the PTZ of the body, whose maximum load is 2.7kg. Figure 3.21 shows the main structure of Matrice 300 RTK.

Figure 3.21　Main structure of Matrice 300 RTK

3.5　Spacecraft system

3.5.1　Space engineering system

Since the first launch of the man-made satellite, hundreds of spacecrafts have been developed and used in scientific, military, and commercial missions. The spacecraft becomes far more complex than any previously implemented auto machine as the craft also now had to be capable of including more functions and enduring the hostile environment of space.

Space engineering system is the systems at the higher level above the spacecraft system. To perform its function, the spacecraft should be launched by a launch vehicle on the launch pad, measured and controlled by the telemetry, tracking, and command (TT&C) system, and cooperated by the ground application system.

Space engineering systems can be divided into manned space flight systems and unmanned space flight systems. Figure 3.22 indicates the typical composition of an unmanned space flight system.

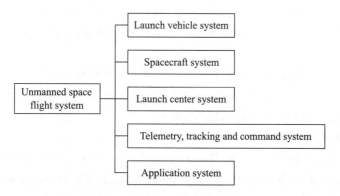

Figure 3.22 Typical composition of an unmanned space flight system

Besides the engineering system which the unmanned space flight system consists of, the manned space flight system includes the astronaut system, the landing system, etc. The following parts are going to introduce the composition of the manned space flight system by the example of "Shenzhou VII" spaceship.

1) Launch vehicle system

Launch vehicle system is to develop a powerful launch vehicle capable of sending a manned spaceship into orbit. The CZ-2F launch vehicle is improved from its previous series by adding fault detection and an escape system. The successful launch record makes it one of the most reliable launch vehicles in the world.

2) Spaceship system

The main task of the spaceship system is to develop Shenzhou spaceships. The Shenzhou VII spaceship consists of three modules. After manned mission, the orbiting capsule can be remained in orbit for half a year, working as a low earth orbit satellite, and can be modified as a docking module when needed.

3) Launch center system

Launch center system is to complete spaceship's and launch vehicle's general assembling, testing and preparation before launch, and carry out launch operation. A new launch pad for the manned space mission is constructed at Jiuquan Launch Center.

4) Telemetry, tracking and command system

Telemetry, tracking and control system's aim is to complete tracking and control of manned spaceflight during the mission and realize telecommunication requirements such as receiving the onboard TV signal and astronaut voice data on the ground. Based on the existing Chinese network, an S-band tracking and communication system is set up, capable of connecting with an international network. This new land and sea-based network for manned spaceflight missions include 3 ground tracking and control centers, dozens of earth stations, and 4 Yuanwang tracking ships.

5) Astronaut system

The main task of the astronaut system is to select and train astronauts, and give medical supervision and guarantee to astronauts during their training courses and flight tests. A Beijing Astronauts Selection and Training Center is established, large astronaut ground simulation training facilities are developed, and so is the personal equipment such as rescue device, spaceflight suits, and space food. Members of China's astronaut team are selected and get trained.

6) Application system

The main task of the Application System is to develop spaceship payload and provide equipment and instruments for earth observation and space science experiments, such as space material, space life, space physics, and remote sensing application tests.

7) Landing system

The main task of the landing system is to recover the reentry capsule and rescue astronauts. The main landing site in Inner Mongolia and an auxiliary landing site nearby Launch Center are set up. Several land and sea rescue areas under flight orbit trace are established in China in order to assure the safe return of astronauts if an accident happens.

3.5.2 Typical examples of the spacecraft system

3.5.2.1 Satellite

Any satellite comprises various subsystems, dependent upon mission profile. In general, a satellite system may be divided into two principal elements, the payload and the platform. The payload refers to all kinds of scientific instruments, experiment equipment, which is the motivation for the mission. For example, for a communication satellite, the payload is the communication subsystem, which has one or more antennas to carry out the communications mission (receiving and transmitting information).

The platform provides the support and certain resources to ensure the normal operation of the onboard payload. It comprises the service subsystem such as power, communications, attitude and orbit control, guidance, navigation and control (GNC), command and data handling (CDH or C&DH), thermal control (TCS), propulsion, and structures.

Figure 3.23 indicates the main composition of the satellite system.

Payload referring to the instruments, equipment, and subsystems directly conduct the mission of the spacecraft. The payload is dependent upon the mission of the spacecraft, and is typically regarded as the part of the spacecraft "that pays the bills". Typical payloads could include scientific instruments (cameras, telescopes, or particle detectors, for example).

Guidance, navigation, and control (GNC) system is a key subsystem that determines the performance of the space mission. Navigation refers to the calculation of the motion parameters, such as position, attitude, and velocity. Guidance means to steer the spacecraft to the desired target based on the difference between the current motion parameters and the scheduled parameters. Control means adjusting the spacecraft's orbit and attitude to keep it around the designed value by an

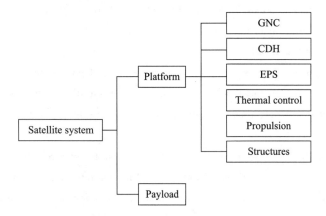

Figure 3.23 Main composition of the satellite system

actuator.

The CDH subsystem is the "brain" of a spacecraft, which plays an important role in information exchange between payload and subsystem, communication of user on the ground and the spacecraft. It performs validation and decoding of the commands after receiving from the ground station and distributes the commands to the appropriate subsystems and components. The CDH also collects onboard state data and science data from the relative subsystems and components and packages the data for storage on a data recorder or transmits to the ground via the communications subsystem.

Electric Power System (EPS) is an important service subsystem of spacecraft, which is used to generate, store, convert, modulate and distribute electrical power for various spacecraft subsystems and payloads. Most of the spacecrafts gain power from the sun by converting light energy to electric energy according to the theory of the photoelectric effect. But for some spacecraft located in a special environment, new electric power generators or energy-storing devices are needed. For example, to supply sufficient power on the surface of the moon during a long-time night, the lunar rover uses Radioisotope Thermoelectric Generator (RTG) to generate electrical power.

The thermal control system controls the process of the heat exchanger of the spacecraft inside and outside and keeps the temperature of thermal stability predefined range. Spacecraft must be withstanding very hard thermal environment, during the flight procedure of the mission with temperatures potentially ranging across hundreds of degrees Celsius. Without the thermal control system, some components will be damaged and the space mission will lead to failure.

Spacecraft may or may not have a propulsion subsystem, depending upon whether the mission profile calls for propulsion or not. The Swift spacecraft is an example of a spacecraft that does not have a propulsion subsystem. Typically though, LEO spacecraft (for example Terra (EOS AM-1) includes a propulsion subsystem for altitude adjustments (called drag make-up maneuvers) and inclination adjustment maneuvers. A propulsion system is also needed for spacecraft that performs momentum management maneuvers. Components of a conventional propulsion subsystem include fuel, tankage, valves, pipes, and thrusters.

Spacecraft must be engineered to withstand launch loads imparted by the launch vehicle and must have a point of attachment for all the other subsystems. Depending upon mission profile, the structural subsystem might need to withstand loads imparted by entry into the atmosphere of another planetary body, and landing on the surface of another planetary body.

1) FY-3 meteorological satellite

FY-3 meteorological satellite is China's second generation of polar orbit meteorological satellite. It is the development and improvement of FY-1 meteorological satellite technology. It could obtain all-weather three-dimensional multispectral global quantitative atmospheric, surface characteristic parameters.

The FY-3 system includes four satellites, among which the FY-3D satellite was launched by China's CZ-4C carrier rocket in 2017 at Taiyuan Satellite Launch Center. The satellite is equipped with ten payloads, including the microwave thermometer, the microwave hygrometer, the microwave imager, the space environment monitoring instrument package, the global navigation satellite occultation detector, the infrared hyperspectral atmosphere detector, the near infrared hyperspectral greenhouse gas monitor, the wide-angle aurora imager, the ionospheric photometer, and the resolution spectral imager. Figure 3.24 shows the main structure of FY-3D.

Figure 3.24　Main structure of FY-3D

2) ChinaSat-16 communication satellite

ChinaSat-16 communication satellite, also known as SJ-13, is China's first high-throughput communication satellite with high orbit, developed by China Academy of Space Technology.

ChinaSat-16 uses Dongfanghong-3B satellite platform, and the payload can provide 26 user beams, covering China and offshore areas. It could be used for remote education, medical treatment, Internet access, airborne and ship communication, emergency communication, and other fields. Figure 3.25 shows the main structure of ChinaSat-16.

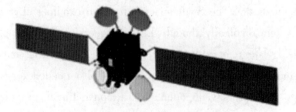

Figure 3.25　Main structure of ChinaSat-16

3.5.2.2 Manned spacecraft system

In general, a manned spacecraft system may also be divided into two principal elements, the payload and the platform. The payload and the platform are similar to the satellite system depended on its mission. Besides, a manned spacecraft also includes the onboard service module, the docking module, the returning module cabin to provide the place where the astronauts live and execute scheduled astronautic missions and help the astronaut to return to the ground. Figure 3.26 indicates the main composition of the manned spacecraft.

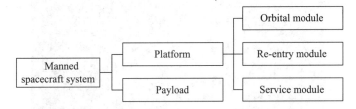

Figure 3.26 Main composition of the manned spacecraft

1) Apollo 9

The world-famous Apollo program began in May 1961 and ended in December 1972, lasting 11 years and 7 months. Its purpose is to achieve a manned landing on the moon and man's field investigation on the moon. Apollo 9 is the third manned flight of the Apollo program and the first manned flight including the lunar module (LM). Its goal is to simulate the maneuvers that will be performed in actual lunar missions. Other goals include a comprehensive inspection of launch vehicle and spacecraft systems, personnel, and procedures.

Apollo 9 was composed of a command module (CM), a command service module (CSM), a lunar module, and an instrument unit (IU), and was launched by a Saturn V rocket. The command module is a cone-shaped craft about 390cm in diameter at the large end, served as a command, control, and communications center. The command service module was a cylinder 390cm in diameter, providing the main propulsion and maneuvering capability. It was jettisoned just before CM reentry. The lunar module was a two-stage vehicle that accommodated two men and could transport them to the lunar surface. It had its own propulsion, communication, and life support systems. Figure 3.27 shows the Apollo 9 in orbit.

2) Shenzhou 6

Shenzhou spacecraft is an important part of China's manned space flight. Shenzhou-6 spacecraft is the second manned spacecraft of the Shenzhou series. It was successfully launched at Jiuquan Satellite Launch Center in October 2005. It circled the earth 77 times and orbited the earth for more than four days.

Shenzhou 6 spacecraft is composed of the orbital module, the re-entry module, and the propulsion module, of which there-entry module is located in the middle.

During the autonomous flight of spacecraft, the orbital module is used as the life cabin for astronauts to live, rest and sleep in orbit. After the spacecraft completes its mission, the re-entry module returns to the ground, and the orbital module continues to stay in orbit for various tests of the payload subsystem.

Figure 3.27 Apollo 9 in orbit

  The re-entry module is the cockpit of the astronauts. During the launch and return stage, the astronauts must ride in the re-entry module, wear the spacesuit inside the module, and be bound on the seat, to bear greater overload, impact and rotation. In case of fatal failure, the re-entry module should also be used as the astronaut's emergency rescue module.

  The propulsion module provides thrust for the spacecraft in the process of autonomous flight and braking stage, to change the flight trajectory and control the attitude of the spacecraft. The propulsion module is also equipped with the power supply, the environment control system, the guidance navigation and control system, the measurement and control system to support the work of the spacecraft. Figure 3.28 shows the composition of Shenzhou 6.

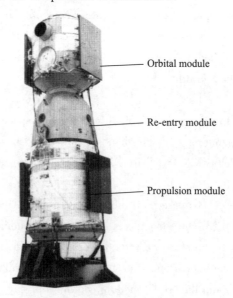

Figure 3.28 Composition of Shenzhou 6

3.5.2.3 Space station system

The main components of a space station system are similar to that of a manned spacecraft system. But the space station is much more complicated than the manned spacecraft.

The space station consists of one or more modules such as the pressurized astronaut living module, the working experiment module, the scientific instrument module, and the non-pressurized service module, etc. Outside the cabins are the solar panels and the docking ports to ensure the power supply and dock with other spacecraft. The living module is the main living place where astronauts eating relaxing and sleeping. The working experiment module and scientific instrument module are mainly used for scientific experiments and material processing. In the service module are the power system and various gas sources, etc.

Besides these components, the space station may also include trusses connecting each cabin section, repair shop, and fuel chamber, etc. A repair shop is a place for repairing spaceships. The fuel chamber is where the fuel is stored.

Figure 3.29 indicates the main components of the space station system.

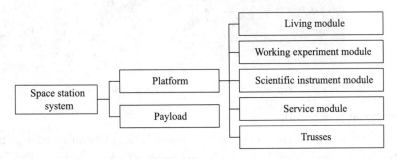

Figure 3.29 Main components of the space station system

1) Mir space station

Mir (means peace) space station is the third generation manned space station of the Soviet Union and the ninth space station in human history. Its design work began in 1976 and was launched on February 20, 1986. At 14:00:12 Beijing time on March 232001, it successfully crashed into the South Pacific Ocean.

The Mir space station consists of five function modules.

Kvant-1 (means quantum) module, also known as the astrophysics module of Mir space station, is mainly used for manned/automatic astronomical observation. It consists of a laboratory module, a transport module, and a non-pressurized payload module.

Kvant-2 module is the first radial module of the Mir space station. Its main function is to expand the Mir space station and serve as a passageway and entrance for extravehicular activities. Kvant-2 consists of the airlock cabin, the scientific equipment cabin, and the service storage cabin.

Kristall (means crystal) is the second radial module of Mir. It serves as the dock for the shuttle.

Spektr (means optics) is mainly used for atmospheric physics and biomedicine research.

Priroda (means nature), also known as the environmental monitoring module, is equipped with multispectral, microwave and infrared scanners, lidar, ozone sensors, and other environmental monitoring sensors. It is mainly used for land, ocean, and atmosphere remote sensing and material science, life science, and biotechnology research.

Figure 3.30 shows the Mir space station in space.

Figure 3.30　Mir space station in space

2) International space station

The international space station (ISS) is the largest space platform in orbit at present. It is a space laboratory with modern scientific research equipment, which can carry out large-scale, multi-disciplinary basic and applied scientific research. It provides a large number of experimental loads and resources for scientific experimental research in the microgravity environment and supports people to stay in earth orbit for a long time. The international space station project is built, operated, and used by sixteen countries. It is the largest and most involved international space cooperation project in human history.

At present, the ISS consists of fifteen compartments. There are five experimental sections in various countries, namely, Destiny (U.S. laboratory module), Columbus (European Research Laboratory), Kibo (means hope, Japanese experience module), Poisk (means explore, Russian Mini research module), and Rassvet (means dawn, Russian Mini research module). There are three Nodes named Unity, Harmony, and Tranquility. Three Pressurized Mating Adapters attach to the Nodes' berthing mechanisms. The other sides of the adapters allow for docking vehicles. There also is a joint airlock named Quest, an observation cabin named Cupola, a functional cargo block named Zarya (means sunrise), a docking compartment named Pirs (means pier), a service module named Zvezda (means star), and a permanent multipurpose module. Figure 3.31 shows the ISS in space.

The ISS is a complicated system that consists of more than ten subsystems. Figure 3.32 indicates some of the subsystems.

3.5.2.4　Space probe system

The components of the space probe system may vary depending on its mission. Basically, it is

Chapter 3 Flight vehicle system composition

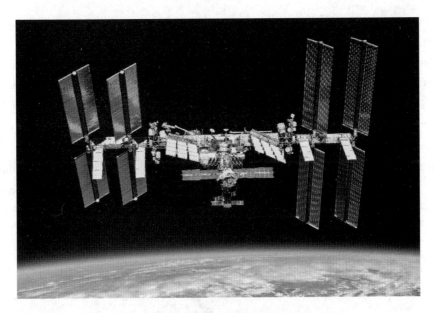

Figure 3.31 ISS in space

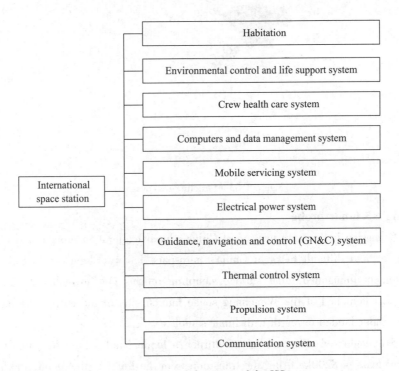

Figure 3.32 Subsystems of the ISS

similar to that of a satellite system. For a planetary landing exploration mission, the space probe may need a lander to perform a hard landing or soft landing. For a recoverable space probe, it may need a returner to bring it back to the Earth. The basic components of the space probe system could refer to Chapter 3.5.2.1.

1) Voyager 1 space probe

Voyager 1 is a U.S. space probe launched in 1977 to fly by Jupiter and Saturn. It crossed into interstellar space in August 2012 and continues to collect data. Voyager 1 and its sister ship Voyager 2 have been flying longer and farther than any other spacecraft in history.

The payloads of Voyager 1 include the imaging science system (ISS), the ultraviolet spectrometer (UVS), the infrared interferometer spectrometer (IRIS), the planetary radio astronomy experiment (PRA), the photopolarimeter (PPS), the triaxial fluxgate magnetometer (MAG), the plasma spectrometer (PLS), the ow-energy charged particles experiment (LECP), the plasma waves experiment (PWS), the cosmic ray telescope (CRS), the radio science system (RSS). Besides, it also carries a copy of the Golden Record—a message from humanity to the cosmos that includes greetings in 55 languages, pictures of people and places on Earth and music ranging from Beethoven to Chuck Berry's "Johnny B. Goode." Figure 3.33 shows the structure of Voyager 1.

Figure 3.33 Structure of Voyager 1

2) Chang'e 5 lunar probe

Chang'e 5 mission is the sixth mission of China's lunar exploration project, and it is also one of the most complex and difficult tasks in China's aerospace history. Chang'e-5 is China's first lunar probe to implement unmanned lunar surface sampling returns. On November 24, 2020, the CZ-5 carrier rocket was launched at the Wenchang space launch site. In the early morning of December 17, Chang'e 5 lander landed on earth with lunar samples.

Chang'e 5 is composed of an orbiter, a returner, a lander, and a lifter in series from bottom to top, with a total mass of 8200kg. After the transition from the Earth to the Moon, braking in the near Moon and flying around the Moon, the combination of lander and lifter is separated from the combination of orbiter and returner, the orbiter carries the returner to stay in orbit, and the lander carries the lifter to perform soft landing in the pre-selected area of the front of the Moon to carry out the automatic lunar surface sampling. After the completion of the lunar surface mission, the lifter takes off the lunar surface with lunar samples, and after entering the lunar orbit, it docks with the orbiter and returner assembly. After docking with the orbiter and returner assembly, the lunar sample is transferred to the returner. Then, the lifter separated again, the orbiter and the returner returned to

Earth orbit, and finally, the returner returned to Earth alone with the sample. Figure 3.34 shows the structure of Chang'e 5.

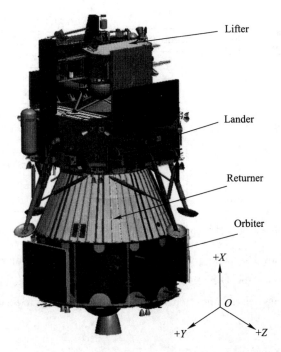

Figure 3.34　Structure of Chang'e 5

3.5.2.5　Space shuttle system

As a typical spacecraft system, the space shuttle system also can be divided into payload and the platform although the platform is even more complex. Till now, only the U.S. and USSR have successfully built and launched the space shuttle, each space shuttle can be roughly divided into a launch vehicle (or called booster, depending on its design) and a shuttle itself. The components of the shuttle are mainly like the manned spacecraft but have some particular systems to perform its mission, for example, the manipulator system to release the satellite. Figure 3.35 indicates the main two parts of a space shuttle system.

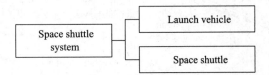

Figure 3.35　Main parts of the space shuttle system

1) U.S. space shuttle

The U.S. space shuttle system is made up of three main components. The two solid rocket boosters (SRBs) provide 80 percent of launch thrust. The huge rust-colored external tank (ET), feeds fuel to three space shuttle main engines (SSMEs) during launch. The Orbiter serves as the crew's home in space, and it is the only reusable part of the system. Figure 3.36 indicates the main components of the U.S. space shuttle system.

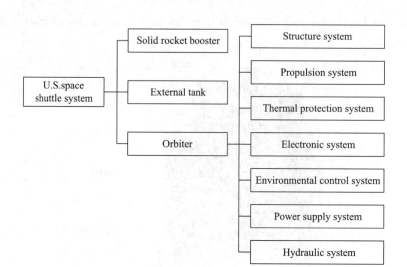

Figure 3.36 Main components of the U. S. space shuttle system

The Orbiters, including Discovery, Atlantis, and Endeavour, are about the same size and weight as a DC-9 aircraft. Generally, the Orbiter contains the pressurized crew compartment, the huge cargo bay, and the three main engines mounted on its aft end. Figure 3.37 indicates the main parts of an Orbiter.

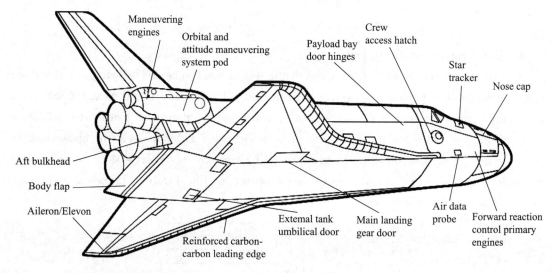

Figure 3.37 Main parts of an Orbiter

2) USSR space shuttle

The USSR space shuttle system consists of the space shuttle named Buran and the heavy launch vehicle named Energiya. Energiya is a two-stage carrier rocket that transports Buran to the specified altitude. Then Buran starts its engine, breaks away from the carrier rocket, and enters the orbit alone. Figure 3.38 indicates the main components of the USSR space shuttle system.

Buran was first launched from Baikonur Space Center on November 15, 1988. Its shape is similar to that of the US space shuttle. Figure 3.39 indicates the Buran's main structure.

Chapter 3 Flight vehicle system composition

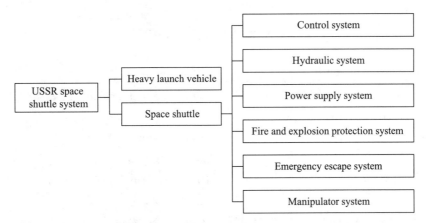

Figure 3.38 Main components of the USSR space shuttle system

Figure 3.39 Buran's main structure

Questions

1. What is CLU of the Javelin system? How does it work?
2. Try to describe the composition of a ground-to-air missile system and analyze its working principles.
3. What are the basic compositions of the carrier rocket system?
4. Describe the differences in system compositions between missile and aircraft.
5. Describe the similarities and differences of various spacecraft systems.
6. What are the differences between the compositions of manned spacecraft system and unmanned spacecraft system (e. g. satellite)?
7. What is the relation between the spacecraft system and the space engineering system?
8. Describe the functions of the main compositions of UAV system.

Words and phrases

aileron	副翼
astronomical observation	天文观测
ATK (Alliant Techsystems) Launch Tube Assembly	艾莲特技术公司的发射筒组件
berth	停泊
bunker	碉堡,掩体
Cadmium	镉
CLU (Command Launch Unit)	指令发射装置
docking module	对接舱
elevator	升降舵
external tank	外燃料箱
flap	襟翼
fuselage	机身
hydraulic system	液压系统
imaging infrared	红外成像
inter-stage section	级间段
inter-tank section	箱间段
lethal	致命的
LOX/ kerosene engine	液氧煤油发动机
Mercury	汞
orbital module	轨道舱
probe	探测器
PTZ (Pan/Tilt/Zoom)	云台
recoil	后坐力
rudder	舵,方向舵
seeker	导引头
service module	服务舱
space shuttle	航天飞机
tandem	一前一后
Telluride	碲
transition section	后过渡段
TT&C station	测控站
versatility	通用性,多功能性

References

[1] Barnard R H, Philpott D R. Aircraft Flight[M]. 3rd ed. UK: Pearson Education, 2014.
[2] 赵育善,吴斌. 导弹引论[M]. 西安:西北工业大学出版社,2002.
[3] 文仲辉. 导弹系统分析与设计[M]. 北京:北京理工大学出版社,1989.
[4] 于剑桥,文仲辉,梅跃松,等. 战术导弹总体设计[M]. 北京:北京航空航天大学出版社,2010.
[5] 罗格. 世界航天器与运载火箭集锦[M]. 北京:宇航出版社,2000.
[6] 龙江,等. 飞机系统[M]. 成都:西南交通大学出版社,2017.
[7] 张庆伟,林左鸣. 世界民用飞机手册[M]. 北京:航空工业出版社,2009.
[8] 李发致,等. 无人机构造[M]. 北京:高等教育出版社,2018.
[9] Austin R. Unmanned Aircraft Systems: UAVs Design, Development and Deployment[M]. Wiley,2010.
[10] 《世界无人系统大全》编写组. 世界无人系统大全[M]. 北京:航空工业出版社,2015.
[11] 张庆军,刘杰,等. 航天器系统设计[M]. 北京:北京理工大学出版社,2018.
[12] 戚发轫,李颐黎. 巡天神舟:揭秘载人航天器[M]. 北京:宇航出版社,2011.
[13] Kitmacher G H. Reference Guide to the International Space Station[M]. NASA,2015.
[14] 季晓光. 苏联"暴风雪"号航天飞机[M]. 北京:航空工业出版社,2015.

Chapter 4　Flight vehicle propulsion system

A flight vehicle propulsion system provides power for a flight vehicle and propels it forward. So it is also called a power unit, or propulsion system. It is one of the most important parts of a flight vehicle system. A propulsion system generally includes an engine and its auxiliary system.

According to the working principles, there are two types of propulsion systems. One is the direct reaction propulsion system, which pushes a flight vehicle forward by itself. That is to say, the thrust is produced by exhausting the working fluid outside. It is also called a jet-propelled system, such as turbojet, turbofan, ramjet, rocket engines, and combination engines, and so on.

The other is the indirect reaction propulsion system, which is composed of engines and propellers. Engines mainly output the mechanical energy and the propellers generate the reaction force. Such as piston engines, turboprop, turboshaft, aviation electromotor, and so on.

4.1　Aeroengine system

Commonly aero-engines are referred to operate within the atmosphere, which needs to obtain the oxidizer of the fuel by the ingestion of atmospheric oxygen, so they are also called air-breathing engines. According to the definition, the aeroengines generally include piston engines, gas turbines, ramjets, and so on.

Gas turbines are rotary engines that extract energy from a flow of combustion gas. They have an upstream compressor coupled to a downstream turbine with a combustion chamber in-between. There are many different variations of gas turbines, such as turbojet engines, turbofan engines, turboprop engines, and turboshaft engines.

4.1.1　Aviation piston engines

4.1.1.1　Piston engine

Piston engine, also called reciprocating engine, which uses one or more pistons to convert pressure into rotational kinetic energy. It is mainly composed of cylinder, piston, connecting rod, crankshaft, valve mechanism, propeller reducer, casing and so on.

Piston engines do not produce aerodynamics directly, but they can drive a propeller to generate the aerodynamics. In brief, the piston engine converts a certain amount of energy from a fuel and oxygen mixture into the kinetic energy of a piston. The energy of the piston is then used to turn a shaft. Finally, the shaft turns a propeller. The engines have been used since the first motor airplane succeeded in 1903.

The characteristics of the piston engines are that the structure is relatively complex. Furthermore, they can only produce limited power, in general, 70 ~ 2000kW. But they have a wide application for low-speed flights. So far, piston engines have been utilized for helicopters and light low-speed airplanes.

The principles for piston engines are illustrated in Figure 4.1. Generally, they are four-stroke reciprocating piston engines. That is to say, the piston passes through four strokes in the cylinder, namely the intake stroke, the compression stroke, the expansion stroke and the exhaust stroke.

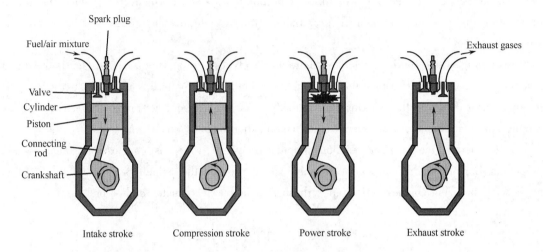

Figure 4.1 A single-cylinder reciprocating piston engine power cycle

The power produced by a piston engine is converted into thrust by the propeller. This device is composed of two or moreblades (really just small wings) attached to a central shaft. As the propeller is rotated by the engine, the blades move through the air like wings and create lift in a direction perpendicular to their motion (parallel to the shaft). The component of this lift created by the propeller which is directed along the propeller shaft is thrust. The concept is illustrated in Figure 4.2 and Figure 4.3.

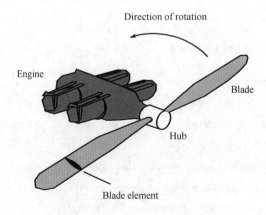

Figure 4.2 Engine and propeller

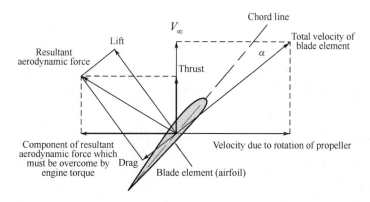

Figure 4.3 Blade element geometry and aerodynamic forces

4.1.1.2 Auxiliary systems of piston engine

An aviation piston engine is developed on the basis of piston engine. To be as the aviation piston engine, besides piston engine, some necessary auxiliary systems are needed to guarantee the aircraft piston engine work normally. They are composed of several parts as follows.

1) Air intake system

The higher the altitude is, the lower the air density becomes, and the engine will produce little power. One way to overcome the power loss with altitude is adding a pump to the air intake to increase the amount of air into the cylinders. There are two common methods for this. The first is called turbocharging. A turbocharger makes use of the energy expelled in the exhaust to run a small pump in the air intake. A "supercharger" is another method to pump additional air into the cylinders. A supercharger can be powered mechanically through a belt on the engine shaft, or with an electric motor. The purposes of both are the same, to increase the amount of air (oxygen) into the cylinders at higher altitudes where the air is less dense. The result is that a turbocharged or supercharged engine can maintain constant power up to a higher altitude.

2) Fuel system

The fuel system is composed of a fuel pump, gasifier, or fuel-injection equipment. The fuel pump presses the gasoline into the gasifier where the gasoline is atomizing and mixed with the air. Then the mixed gas goes into the cylinder.

3) Ignition system

The mixed gas in the cylinder is ignited by the electric spark which is created by the high voltage.

4) Cooling system

Some of the heat generated by burning the fuel is translated into kinetic energy and carried away by the exhausted gas. The rest is given to the wall of the cylinder and other related assemblies. To guarantee the piston engine to work normally, the heat must be brought away by the cooling system. There are two kinds of the cooling system, gas cooling and liquid cooling which are shown in Figure 4.4.

5) Start-up system

The extra forces are needed when the engine starts up. Usually, the crankshaft is turned by the

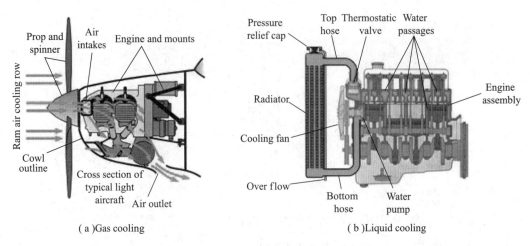

Figure 4.4　The cooling type of the aviation piston engine

electromotor.

6) Timing system

The timing system will turn on and close the intake valve and air-evacuation valve in time.

4.1.2　Turbojet engines & turbofan engines

4.1.2.1　Turbojet engines

Propellers become very inefficient at a high subsonic speed, and no practical supersonic propellers have ever been developed. Propulsion for the high subsonic, transonic, and supersonic flight regimes is usually provided by either turbojet or turbofan engines. These power plants produce thrust without using a propeller.

Figure 4.5 illustrates a schematic diagram of a typical turbojet engine. As shown in Figure 4.5, a turbojet engine takes air in through an inlet diffuser. The diffuser is designed so that its cross-sectional area is greater at its downstream end. This causes the velocity of the air flowing into the inlet to decrease and its static pressure to increase. A compressor then increases the static pressure further as it delivers the air to the combustor or burner. Fuel is mixed with the air and burned in the combustor. The hot gas is exhausted through a turbine which acts like a windmill to extract power to turn the compressor through a shaft. Then the gas flows out of the engine through a nozzle which causes it to accelerate until the static pressure of the exhaust approximately equals the ambient pressure.

Turbojet is an engine that uses a turbine to drive a compressor for increasing air pressure. There are two types of compressors in turbojet. One is the axis-flow type, shown in Figure 4.6, which is composed of a fixed part and rotating part, called the rotor. The other is centrifugal type, using centrifugal forces, the requirements are met.

There are two fundamental problems with the turbojet. First, the turbojet produces thrust with a very high exhaust-gas velocity. We have shown that this requires more power and is thus inefficient. Another problem is that the higher the exhaust gas velocity is, the more noise the engine produces. This noise is unacceptable today and FAA (Federal Aviation Administration) noise

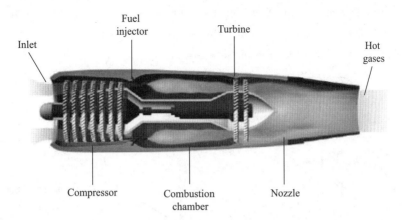

Figure 4.5 Schematic diagram of a turbojet engine

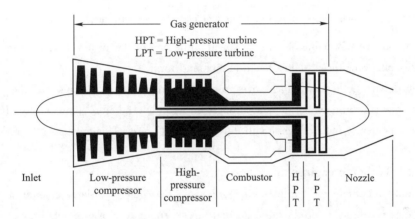

Figure 4.6 Axis-flow type sketch

standards do not permit turbojets to operate at many airports.

Older airplanes, such as the Boeing 707, that used turbojets, are now a rare sight at most airports. The Boeing 727 and 737 originally had turbojets, and many of the earlier versions of these airplanes cannot fly into many airports because of noise restrictions. In the case of the Boeing 737, the airplane has gone through two major redesigns to improve efficiency and lower noise. Because of its unique requirements, the Concorde uses turbojets. It is noisy and is considered a "gas-guzzler" by any standards.

The engine power is developed by a turbojet, thus the amount of air drawn into the engine depends on the amount of fuel injected into the burner, not on the speed of the airplane. If more fuel is added, more force is put on the turbine, causing more air to be brought in by the compressor. It is a characteristic of jet engines that the engine power and the thrust are approximately only dependent on the throttle setting. Thus an engine can develop full engine power and thrust sitting on the runway or at cruise speed.

Now here comes the interesting part. The propulsive power is the thrust times the speed of the airplane. Hence although the engine runs full power and thrust when it starts to roll for takeoff, it produces almost no propulsive power. This is shown in Figure 4.7. Since wasted power is the difference between engine power and propulsive power, almost all of the power is wasted! The propulsive efficiency is almost zero at takeoff and increases with speed, which is the ratio of the propulsive power divided by the engine power. One sometimes hears that jet efficiency increases with speed. This is the source of that increase inefficiently.

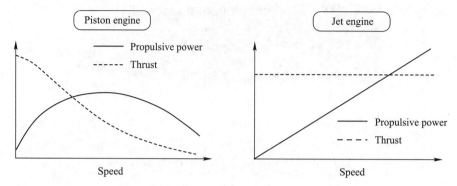

Figure 4.7 Propulsive power and thrust as functions of speed for a propeller and a jet engine

A jet's propulsive power increases with speed while one of the piston-driven engines is roughly constant influencing how these planes make climbs and turns.

Turbojets have another advantage over piston-driven airplanes. Because of the diffuser, the amount of air they take in does not depend strongly on altitude. Therefore, they can fly high where the parasitic power is greatly reduced while still developing full power. This enhances the efficiency of jet airplanes.

4.1.2.2 Turbofan engines

To optimize the efficiency of a jet engine, one wants to accelerate a large amount of gas but to as low a velocity as necessary to produce the needed thrust. The nature of turbojets limits the amount of air that can be processed. The solution is the turbofan engine, as shown schematically in Figure 4.8.

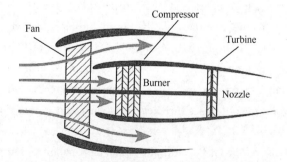

Figure 4.8 In a turbofan, much of the air bypasses the core

The turbofan engine is designed around a turbine engine, but with much more energy produced in the burner being converted into mechanical energy by the turbines. Most of the energy is used to turn a large fan in front of the engine. The fan is very much like a propeller, but with 30 to 40 blades instead of just 2 to 4. The large fan accelerates a large amount of air, at a much slower speed than that of the exhaust of a turbojet producing the same thrust or power. Thus it is much more efficient.

It is important to understand that the air that goes through the fan does not go through the core of the engine but goes around the outside of the core. This can be seen in the photograph of a turbofan in Figure 4.9. Most of the air that goes through the fan bypasses the core. The ratio of the air that goes around the core to that which goes through the core is called the *bypass ratio*. Typical engines today have bypass ratios of about 8:1, meaning that eight times as much air goes around the core as goes through it. Ideally, the velocity of the exhaust gas and that of the air from the fan would be the same. In such a situation, with a bypass ratio of 8, about 90 percent of the thrust would come from the fan and 10 percent from the exhaust of the turbojet powering it.

Figure 4.9 A turbofan engine

A turbofan is more efficient and therefore has a lower TSFC (thrust specific fuel consumption) because it accelerates more air (bypass air in addition to core air) for the same amount of fuel burned. Turbofan efficiency increases with increasing bypass ratio, but so does engine size and weight. Figure 4.10 and Figure 4.11 illustrate two types of turbofans. The one on the left has a relatively low bypass ratio, and all of its bypass air flows with the core air into an afterburner. The turbofan on the right has a much higher bypass ratio and no afterburner.

An additional benefit, and a necessary one, is that lower exhaust velocity produces less noise. Jet engines today are much quieter than they were 30 years ago. The fanjet also gives engine designers a way to increase thrust with growing efficiency. They can increase the mass flow through the engine

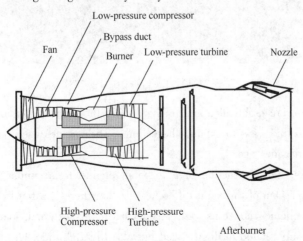

Figure 4.10 Turbofan with a low bypass ratio (0.2 ~ 1.0)

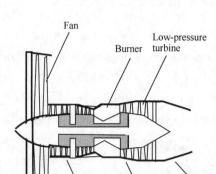

Figure 4.11 Turbofan with a high bypass ratio (2.0~8.0)

while decreasing the exhaust velocity. The result of this is that the 45,359.23kg thrust engines on the Boeing 777 have a fan so large that the engine's diameter is within inches of the fuselage diameter of a Boeing 737. One can fit six seats and an aisle in one of these engines, though these would be very uncomfortable seats. Figure 4.12 shows a photograph of an adult standing in front of a Boeing 777 engine. This gives a feeling of their size.

Figure 4.12 Engine of Boeing 777

Unfortunately, these are practical limits to the size of the engines. But there are always clever engineers out there who find ways to expand these limits. The engines used on the Boeing 777 were unimaginable just two decades ago.

Figure 4.13 shows the construction of a turbofan engine. It is similar to turbojet. Or it is a derivative of the turbojet. One of the main differences is that there is a fan in it.

The thrust of jet engines concerns with airflow mass, gas external speed, and flight velocity V. Airflow mass is not only related to flight speed but also flight altitude. Jet engines are mainly used in aircraft, ground-to-ground missiles, and cruise missiles.

With turbo-engines, there is a limitation for speed, usually for low speed. Turbofans are the

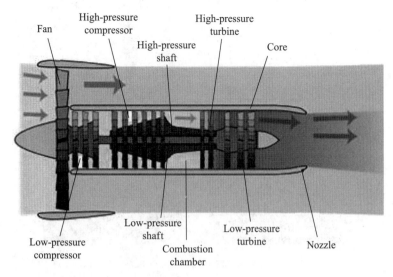

Figure 4.13 Turbofan engine sketch

engine of choice for commercial transports and business jets. Military jets also use turbofans but must compromises efficiency if the airplane is capable of supersonic flight. The huge inlets of the fanjets are a disadvantage when it comes to supersonic drag. Military fighters typically use engines with bypass radios on the order of 3.

4.1.3 Turboprop engines & turboshaft engines

A turboprop power plant replaces the fan of the high-bypass-ratio turbofan with a propeller. The excess power in the turbine is used to turn a propeller, rather than a fan. Figure 4.14 shows an example of a corporate turboprop airplane. A propeller cannot be directly attached to this shaft from the turbines because it turns at a very high speed. So a gearbox or reduction gear is used to reduce the rotation speed of the propeller. At lower flight speed the propeller is more efficient than the fanjet. Since it has a larger diameter, it gets thrust from accelerating more air at a low speed. This, as we knew, is the most efficient way to get propulsion. Propellers become less efficient at high speeds because of the effects of compressibility. Its operating characteristics at full power are similar to high-bypass-ratio turbofans and it typically has a lower TSFC. However, the turboprop loses thrust at high speeds more like a piston engine/propeller power plant. The thrust-to-weight ratio for a turboprop is generally higher than the one for a piston engine, but TSFC is usually higher also. Turboprops are usually designed so that the high-energy air from the burners expands almost completely to ambient static pressure in the turbines, so almost all of the energy is converted into shaft power.

Turboprops are used on smaller commuter aircraft and have found a growing market in smaller aircraft. They also have found growing use in the general aviation and corporate airplane market. Their biggest advantage is that they can produce greater power than the equivalent piston-driven engine with much less noise and maintenance. Turboprop engines are much lighter than piston engines of the same power but are also more expensive.

A turboshaft engine is a form of gas turbine that is optimized to produce shaft power, rather than jet thrust, which are used primarily for helicopters and auxiliary power units (APU). In concept,

Figure 4.14 Turboprop fan

turboshaft engines are very similar to turbojets, difference is that nearly all energy in the exhaust is extracted to spin the rotating shaft.

A turboshaft engine is made up of two major parts assemblies: the gas generator and the power section, as shown in Figure 4.15. The gas generator consists of a compressor, combustion chambers with igniters and fuel nozzles, and one or more stages of the turbine. The power section consists of additional stages of turbines, a gear reduction system, and the shaft output. The gas generator creates the hot expanding gases to drive the power section. Depending on the design, the engine accessories may be driven either by the gas generator or by the power section.

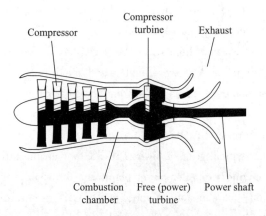

Figure 4.15 Schematic of turboshaft engines

In most designs, the gas generator and power section are mechanically separated so that they may each rotate at different speeds appropriate for the conditions. This is referred to as a free power turbine. A free power turbine can be an extremely useful design feature for vehicles, as it allows the design to forego the weight and cost of complex multi-ratio transmissions and clutches.

The general layout of a turboshaft is similar to that of a turboprop. The main difference is that a turboprop is structurally designed to support the loads created by a rotating propeller, while the propeller is not attached to anything but the engine itself, shown in Figure 4.16. In contrast, turboshaft engines usually drive a transmission that is not structurally attached to the engine. The transmission is attached to the vehicle structure and supports the loads created instead of the engine. However, in practice, many of the same engines are built in both turboprop and turboshaft versions, with only minor differences.

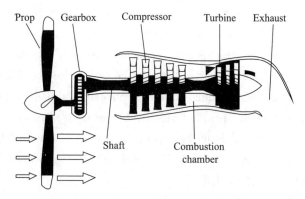

Figure 4.16 Schematic of turboprop engines

4.1.4 Ram-engines

At very high Mach numbers, the air which enters a jet engine inlet is slowed down and compressed so much that the turbomachinery (compressor and turbine) is not needed and may be eliminated. The engine is a little more than an afterburner connected to the inlet. This device is called a ramjet or a flying stovepipe because the air is compressed by the ram effect. Ram effect is the increased static pressure that results when the air is slowed by the inlet. A ramjet cannot function at low speeds because the compression of the air in its inlet is not sufficient. This requires ramjet-powered aircraft to be accelerated to operating speed by some other propulsion system. At Mach number above about 3.0, however, ramjets are more efficient than afterburning turbojets, and they have a much higher thrust-to-weight ratio.

Figure 4.17 shows a typical ramjet. Same principles as turbojet, but the construction are different which are simple, no turbine, compressor, and other rotating parts. As well it is high efficiency. The main drawback is that it cannot start up by itself. It is suitable for high supersonic flights.

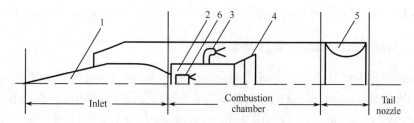

Figure 4.17 Ramjet engine sketch

1—Inlet cone; 2—Precombustion chamber; 3-Injecting; 4—Flamestabilizer; 5—Tail nozzle; 6—Igniter.

By the flow state in the combustion chamber, two types are involved in ramjets which are subsonic combustion ramjet and scramjet. Scramjet is an acronym for Supersonic Combustion

Ramjet. As in conventional ramjets, a scramjet relies on high vehicle speed to forcefully compress and decelerate the incoming air before combustion, but whereas a conventional ramjet always decelerates the air to subsonic velocities using a multishot intake system before combustion. Airflow in a scramjet can remain supersonic. This allows the scramjet to efficiently operate at extremely high speeds.

All scramjet engines have fuel injectors, a combustion chamber, a divergent thrust nozzle, and an intake, which compresses the incoming air. Figure 4.18 illustrates a schematic diagram of a typical scramjet. Due to the nature of their design, scramjet operation is limited to near-hypersonic velocities. As they lack mechanical compressors, scramjets require the high kinetic energy of a hypersonic flow to compress the incoming air to operational conditions. Thus, a scramjet-powered vehicle must be accelerated to the required velocity by some other means of propulsion, such as turbojet, or rocket engines. In the flight of the experimental scramjet-powered Boeing X-51A, the test craft was lifted to flight altitude by a Boeing B-52 Stratofortress before being released and accelerated by a detachable rocket to near Mach 4.5.

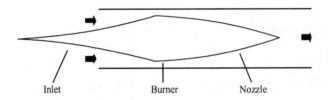

Figure 4.18　Schematic of scramjet engines

While scramjet technology has been under development since the 1950s, only very recently have scramjets successfully achieved powered flight. In May 2010, this engine has been tested to reach Mach 5 for 200s on the X-51A Wave rider.

4.2　Rocket engine system

Oxygen in the atmosphere is very important to the foregoing engines. It is used as the oxidizer of the fuel, so these engines can only be used as the power on the aircraft. However, the rocket engine carries the combustion chamber and oxidizer by itself, so it can work not only in the atmosphere but also in space. The rocket engines can be used as the power of rockets, missiles, and space flight vehicles.

4.2.1　Main performance parameters

Rocketengine technology can combine extremely high thrust (meganewtons), exhaust speeds (around 10 times the speed of sound in air at sea level), and thrust-to-weight ratios (> 100) simultaneously as well as can operate outside the atmosphere while permitting the use of low pressure and hence lightweight tanks and structure. Rocket engines can be further optimized to even more extreme performance along with one or more of these axes at the expense of the others.

4.2.1.1　Thrust

Rocket engines are reaction engines and obtain thrustby Newton's third law. They produce thrust by the expulsion of a high-speed fluid exhaust. This fluid is nearly always gas which is created by high-pressure combustion of solid or liquid propellants, consisting of fuel and oxidizer components, within a combustion chamber. The fluid exhaust is then passed through a supersonic propelling nozzle which uses heat energy of the gas to accelerate the exhaust to high speed, and the reaction to it pushes the engine in the opposite direction.

As mentioned in Chapter 2 (Figure 2.1), incendiary agent burns in the combustion chamber of a rocket, and the gas fluid with high speed, which is u_e, spreads afterward outside. On the other hand, the gas fluid reacts on the rocket to form a reacting force. It produces a thrust P to push the rocket forward with a velocity of V.

Also how to produce thrust can be explained in terms of pressure distribution, which indicates the pressure exerts on the surface of an engine. As shown in Figure 4.19, it is divided into internal pressure acting on the internal surface of an engine and external pressure distribution that is a free airstream. Thrust occurs when the internal pressure and external pressure are not in equilibrium.

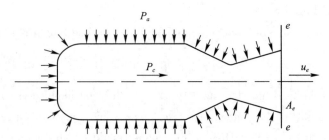

Figure 4.19 Pressure distribution of a rocket engine

The thrust of the rocket engine can be expressed as follows:

$$P = \dot{m} u_e + A_e (p_e - p_a) \tag{4-1}$$

where, $\dot{m}$ is exhaust gas mass flow, u_e is exhaust velocity, A_e is cross-section area at nozzle exit plane, p_e is gas pressure at nozzle exit plane, p_a is ambient (or atmospheric) pressure.

The thrust of the rocket engine consists of two parts, momentum thrust by reacting force and pressure thrust. The first term $\dot{m} u_e$ represents the momentum thrust, which is the principal part and contributes above 90% of the gross thrust; the remaining $A_e (p_e - p_a)$ term represents the pressure thrust, also called static thrust, which improves slightly with increasing altitude. Note that, there is only one effect from the altitude which affects the atmospheric pressure, but no effect from the speed of a rocket.

However, in jet engines, P is affected not only by flight velocity but also by flight altitude.

$$P = \dot{m}_a (u_e - V) + A_e (p_e - p_a) \tag{4-2}$$

Here, V means flight velocity,

$\dot{m}_a$ stands for airflow quantity per second.

In the vacuum environment, the thrust produced by the rocket engine is defined as vacuum thrust, where, the atmospheric pressure $p_a = 0 \text{MPa}$. It is expressed by.

$$P_V = \dot{m} u_e + A_e p_e \tag{4-3}$$

Maximum thrust for a rocket engine is achieved by maximizing the momentum contribution of the equation without incurring penalties from over-expanding the exhaust. This occurs when $p_e = p_a$. It is called a design state or completely expanded state in such conditions. The corresponding thrust is called characteristic thrust or optimal thrust, which is expressed as follows.

$$P_{max} = \dot{m} u_e \quad (4-4)$$

To characterize the performance of the nozzle, thrust coefficient c_p is used so that the thrust formula can be rearranged slightly as follows.

$$P = c_p p_c A_t \quad (4-5)$$

Where, p_c is combustion pressure, A_t is the area of the nozzle throat.

$$c_p = f(A_e/A_t, p_e/p_c, p_a/p_c, k) \quad (4-6)$$

Here, k is the ratio of specific heat of constant pressure to that of constant volume, e is a subscript of the external nozzle, A_e/A_t is area ratio, p_e/p_c, p_a/p_c are pressure ratios.

4.2.1.2 Total impulse

Total impulse is defined as the integration of the thrust to time, which is illustrated by.

$$I_0 = \int_0^{t_k} P dt \ (N \cdot s) \quad (4-7)$$

Here, P is thrust, t_k is the total time for engine work, I_0 is the total impulse of an engine.

It expresses the ability of the engine. The flight range which a missile can reach depends on I_0.

How to determine a P-t curve? Three main factors should be concentrated in engine design, including total weight of propellant, flight speed, and control requirements. Figure 4.20 shows the variation of thrust to time in different temperature.

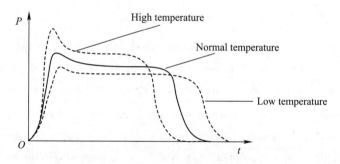

Figure 4.20 Variation of thrust to time in different temperature

For certain engines, the thrust formula can be modified to

$$P = \frac{\dot{G}}{g} u_e + A_e (p_e - p_a)$$

$$= \frac{\dot{G}}{g} \left[u_e + \frac{A_e (p_e - p_a)}{\dot{G}/g} \right] = \frac{\dot{G}}{g} u_{ef} \quad (4-8)$$

Effective exhaust velocity u_{ef} mainly depends on the performance of the propellant.

Based on the Eq. (4-7) and Eq. (4-8), the total impulse can be defined as:

$$I_0 = \int_0^{t_k} \frac{\dot{G}}{g} u_{ef} \mathrm{d}t \qquad (4-9)$$

If the thrust of engine is constant all the time, then the total impulse is:

$$I_0 = \frac{G_p}{g} u_{ef} \qquad (4-10)$$

G_p is the total weight of expended propellant.

As a result, the total impulse is related to the u_{ef} and G_p. If the propellant is chosen, it mainly depends on G_p.

4.2.1.3 Specific impulse

Specific impulse is defined as the impulse per unit weight of propellant. It is expressed by.

$$I_s = \frac{I_0}{G_p} \qquad (4-11)$$

Here, G_p is the total amount of propellant,

I_s is the specific impulse.

It is an important parameter of the rocket propulsion system performance, similar in concept to the miles per gallon parameter used in automobiles. A higher number means better performance. It affects flight velocity.

Fora solid rocket engine, the specific impulse of double-base powder is 180 ~ 200s, of modified double-base (MDB) powder is 240s and compounded powder (added AL powder) is 240 ~ 250s.

For a liquid rocket engine, specific thrust is used instead of specific impulse, which indicates thrust produced by propellant quantity combusted per unit time. The specific impulse of liquid hydrogen (LH) is about 250 ~ 270s, and the combination of liquid oxygen (LOX) and liquid hydrogen (LH) is about 340 ~ 360s.

4.2.1.4 The ratio of thrust to weight

It is the ratio of the thrust to the weight of an engine, which is defined as

$$N = \frac{P}{mg} \qquad (4-12)$$

N indicates the quality of an engine. Therefore, decreasing a weight of an engine is very important for engine design.

4.2.1.5 Aftereffect impulse

Aftereffect impulse is produced when the remainder propulsion agent continues burning after the engine stop command was sent by the control system. It is the integration of the residual thrust to time, which is defined as follows. The magnitude of the aftereffect impulse affects the missile hit precision and spacecraft injection accuracy.

$$I_c = \int_{t_b}^{t_m} P \mathrm{d}t \qquad (4-13)$$

Here, t_b is the beginning time when the engine stop command was sent,

t_m is the ending time when the thrust disappeared.

4.2.2 Solid rocket propulsion system

A solid rocket propulsion system mainly consists of a solid rocket engine, igniter, and safety unit, thrust vector control unit, and thrust-stop unit. Compared with the liquid rocket propulsion system, the solid rocket propulsion system has the characteristics of simple construction, convenience for utilization and maintenance, high reliability, and lower cost. It is widely used in rockets and missiles, especially in tactical missiles.

As seen in Figure 4.21, it is composed of solid propellant, combustion chamber, nozzle, general-conduit, and igniter.

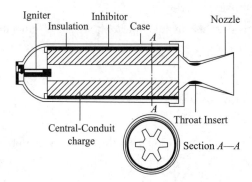

Figure 4.21 Sketch of solid rocket engine

4.2.2.1 Solid propellant

It is also called gunpowder or powder. Typical kinds of gunpowder are as follows.

1) Double-base powder

It is homogeneous. And it is mainly composed of two parts, nitrocellulose (nitrocotton), which is about 50% ~ 60%, and nitroglycerin, about 25% ~ 30%, and others.

2) Compounded powder

It is inhomogeneous which means the oxidizer contains a high percent active oxygen, and the incendiary agent is such as Al powder which has the advantage of high energy. Under the conditions of normal combustion, it has lower pressure and is beneficial to decrease the weight of the engine unit. On the other hand, environmental effects on combustion performance are relatively slight, as well as combustion speed can be easily adjusted.

3) Modified double-base (MDB) powder

It is composed of double-base powder and oxidizer with Al powder. Then a new type of gunpowder is formed.

4.2.2.2 Combustion

The solid propellant is ignited in the combustion chamber, forming the gas with high temperature and high pressure. Then the gas spurts out with high velocity through expanding. As a result, the reaction force is formed, i.e. thrust.

In the chamber, the combustion speed of the solid propellant connects with the combustion area. Of course, it depends on the requirements for the engine, for example, the magnitude of total impulse, and thrust.

As shown in Figure 4.22, the relations between the thrust and combustion area are illustrated. There are different geometric shapes of general conduit charge, such as circle, star shape, and so on.

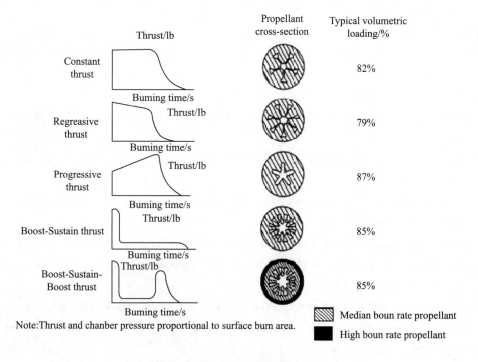

Figure 4.22 Thrust vs. combustion area

If the combustion area is constant, the thrust will be constant too. In other words, if the combustion area is variable, the thrust will be changeable. At the same time, if a large thrust is needed, a large combustion area is required. Figure 4.23 shows some typical combustion manners.

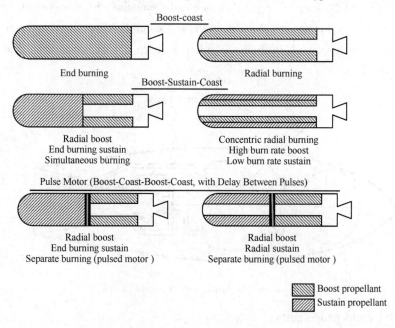

Figure 4.23 Typical combustion manners

The velocity of combustion is affected by pressure and initial temperature. Usually, the bigger the pressure is, the faster the rate of combustion is. At the fixed total impulse, if the combustion time is longer, the rate is lower.

4.2.2.3 Nozzle

The nozzle works in high temperature, high pressure, and high rate gas flowing. So the requirements for it are relatively strict, which means the materials, structure, and heat-protection of the nozzle must be considered.

Usually, a Laval nozzle is utilized whose three sections are convergent section, throat section, and expanded section.

Sometimes, an adjustable nozzle is used to control the thrust such as thrust vector control.

4.2.2.4 Igniter

An igniter burns the propellant in a short time to make an engine work stably. It is not only a dangerous part but also easy to have obstacles.

In general, an igniter is composed of a detonated tube, amorce, and linkers. In small and medium solid rocket engines, the igniters are used. In large solid rocket engines, ignition engines are utilized.

4.2.2.5 Thrust-stop unit

The thrust-stop unit stops the thrust before the propellants burnt up. It is the main content of the thrust control. It is especially used for the ballistic missile which closes the engine at the end of the active stage of trajectory.

4.2.3 Liquid rocket propulsion system

The liquid rocket propulsion system is more complex. Figure 4.24 shows the typical sketch of the liquid rocket engine. The propellant is one or several kinds of chemical liquids. Liquid rocket engines have the characteristics of much higher specific impulse, thrust-to-weight ratio, adjustable thrust, long time working, higher cost and efficiency, and no effects from the velocity on thrust and so on. Therefore, it is widely used in ballistic missiles, carrier rockets, shuttle, and space flight vehicles. There are many kinds of liquid rocket engines.

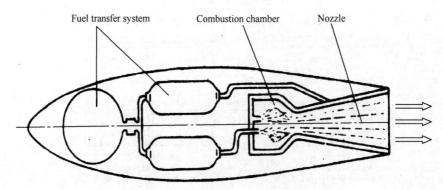

Figure 4.24 Liquid rocket engine sketch

4.2.3.1 Liquid propellants

The liquid propellant is composed of an incendiary agent and oxidizer in the liquid state. According to the propellant types, the mono-combination-unit liquid rocket engine has the mixture state of incendiary agent and oxidizer. The dual-combination-unit liquid rocket engine has the incendiary agent and oxidizer stored separately before they enter the combustion chamber. Furthermore, there is also a multi-combination-unit liquid rocket engine.

There are several propellant agents as follows. For incendiary agents, liquid hydrogen is the best and kerosene is the second. For oxidizers, liquid oxygen is often used and liquid fluorine is the second used whose boiling point is too low to store. As known, the hydrogen-oxygen rocket engine is used in the carrier rocket.

The requirements for liquid propellants are denoted by their energy, utilization characteristics, and economy.

4.2.3.2 Composition of liquid rocket engine

The liquid rocket engine is composed of a thrust barrel, turbo pump, gasifier, and valve adjusters.

1) Thrust chamber

The thrust chamber is the component that can generate thrust. It is composed of an injector, combustion chamber, and nozzle. The injector is the head of the thrust chamber; the combustion chamber and nozzle are considered as the body of the thrust chamber. Figure 4.25 shows the construction of a typical thrust chamber.

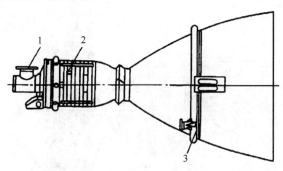

Figure 4.25 Construction of a typical thrust chamber
1—head; 2—body; 3—liquid trap.

(1) Injector

The propellant is injected into the combustion chamber by the injector. The injector mainly includes the coping and the injecting tray.

(2) Combustion chamber

The liquid propellant is atomized, mixed, and burned in the combustion chamber. Usually, the shape of the combustion chamber is column, sphericity, or spheroidicity. The shape and cubage of the combustion chamber are important to the combustion efficiency of the propellant. The shape of the combustion chamber which provides big thrust is sphericity or spheroidicity, while the shape of the combustion chamber which provides small thrust is cylindrical form.

(3) Nozzle

It is a Laval nozzle that has been introduced in Chapter 2. It is used to convert thermal energy into kinetic energy. The airflow accelerates to the throat of the pipe, where it reaches a velocity of Mach 1. After the throat, the air continues to accelerate to supersonic, and then the thrust is formed.

2) Turbo pump

Turbo pump is an important part of pumping liquid rocket engine. It is composed of turbine and propellant pump.

(1) Functions and characters

Turbo pump is used to increase the pressure of the propellants and transport them into the thrust chamber at the required pressure and flux. By burning the liquid propellants, HTHP fire gases are generated to impulse the turbine. For the pump-inlet pressure is low, it does not need the storage box to increase the pressure. This work is done by the pump. So the structural quality of the storage box can be diminished.

(2) Classification and compositions

Based on the types of transmission or disposition of turbine and pump, turbo pump of liquid rocket engine can be divided into coaxial turbo pump, gearing turbo pump, and double turbo type turbo pump.

A turbo pump consists of a propellant pump, turbine, measuring devices, auxiliary devices, and so on. Figure 4.26 shows the construction of a typical turbo pump.

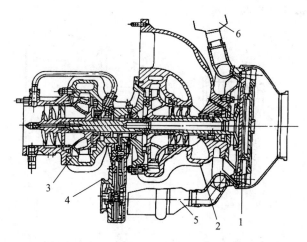

Figure 4.26 Construction of a typical turbo pump

1—Turbine; 2—Fuel pump; 3—Oxidant pump; 4—Gear-type pump; 5—Gasifier; 6—Powder starter.

3) Gasifier

A gasifier is used to generate the working substance which is used to drive the turbo pump.

(1) Compositions

A gasifier is a device that can provide high-temperature fire gases. The structure is similar to the thrust chamber of the liquid rocket engine. It is also composed of an injector, a combustion chamber, and a set of nozzles. Figure 4.27 gives the construction of a typical gasifier.

There are two kinds of gasifiers. They are single-unit gasifiers and dual-unit gasifiers. At present, dual-unit gasifiers are widely used. The propellants and principles of the gasifers are the same as those of the main thrust chamber. The difference between them is the mixing ratio of the

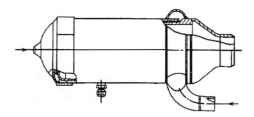

Figure 4.27 Construction of a typical gasifier

propellants. So the temperature of the gasifiers' working substances is not too high. Besides, the volume of it is small.

(2) Principles

When the dual-unit gasifier starts to work, a small amount of oxidizer and fuels are sent into the gasifier through the main pipe behind the pump. They are mixed and burned together there and some fire gas at certain temperature and pressure is generated to drive the turbine and turn it into the normal working stage.

4) Valve and adjuster

Valves and adjusters are the indispensable components of the liquid rocket engine. No matter what the propellant supply system is, the whole working process of a liquid rocket engine includes three stages such as startup, turn levels, and close down. The medium in the propellant supply system and liquid or gas control system should be cut down or turned on in time; at the same time, working parameters need to be adjusted and controlled during the whole process to guarantee the normal work of the engine. These works such as controlling, regulating and examining, which are all done by the valves and adjusters whose functions are introduced as follows:

(1) Regulating and changing the mixing ratio of the propellants;

(2) Controlling the procedure of startup or closedown;

(3) Controlling and regulating the working conditions of the pressurizing system in the storage box;

(4) Controlling the working conditions of TVCs;

(5) Controlling and detecting the working conditions of the whole engine.

4.2.3.3 Integral system composition

The liquid rocket engine is only one of the subsystems of the liquid rocket propulsion system which includes several subsystems. In detail, the propellant supply subsystem is used to supply the propellant with a certain proportion to the thrust chamber. Pre-cold and eliminating subsystem is used for pre-cold of the pipes and combinations, as well as preserve of propellant after engine shutdown. Also, the pressure-increase subsystem for the propellant store box is one of the subsystems. Next is the propellant-adjusted subsystem, with which the mixture proportion could be assured. Another important subsystem is for adjusting the thrust and thrust vector. Finally, the pipes and frames are the important parts of a liquid rocket propulsion system. These subsystems are introduced as follows.

1) Propellant supply system

To supply the liquid propellant with a certain proportion and flux to the thrust chamber, propellant supply systems are needed. They can be divided into extruding supply systems and pumping supply systems.

(1) Extruding supply system

The system uses the high-pressure gas or liquid to press the store box so that the liquid propellants are sent into the thrust chamber. According to the working modes, extruding supply systems can be divided into cold-air extruding supply systems and hot-gas extruding supply systems. The cold-air extruding supply system adopts the high-pressure bottle while the hot-gas extruding supply system utilizes the products of decomposition and combustion in the liquid gasifier. The combustion product in the solid gasifier can also be utilized in the hot-gas extruding supply system. At present, cold-air extruding supply systems are widely used. Figure 4.28 shows a typical extruding supply system.

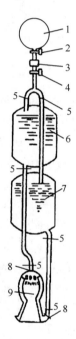

Figure 4.28 Typical extruding supply system

1—High-pressure gas-bottle; 2—High-pressure valve; 3—Pressure reducer; 4—Low-pressure valve;
5—Septa; 6—Incendiary agent storage; 7—Oxidant storage; 8—Board for controlling the flux; 9—Combustion chamber.

The working process of extruding supply system is introduced approximately as follows:

The pressure of inert gas (nitrogen, helium, or air) stored in the high-pressure bottle is within $25 \sim 30$ MPa. When the system starts to work, the pressure of the gases drops down to the required pressure within $3.5 \sim 4.5$ MPa bypassing the high-pressure demolition valve, pressure reducer, and low-pressure demolition valve. If the propellant is bipropellant, following the respective pipeline, each unit will be pressed into the head of the thrust chamber. The flux of each unit is controlled by the flow controller.

For the hot-gas extruding supply system, the high-pressure bottle is replaced by a gasifier. It has its start-up system and combustion chamber in it. The working process of the system is the same as that stated before.

(2) Pumping supply system

Pumping supply systems are widely adopted in the liquid rocket engine. Propellants stored in the storage box are pressed by the pump and sent into the thrust chamber.

Compared with the extruding supply system, the structure of it is more complex because the high-pressure bottle is replaced by the turbo pump. However, since the propellant before the pump is at low pressure, forces acting on the storage box are relatively small and the quality of the storage box is lighter. Figure 4.29 illustrates a typical pumping supply system.

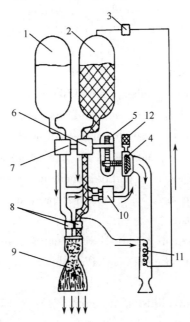

Figure 4.29 Typical pumping supply system

1—Incendiary agent storage; 2—Oxidant storage; 3—Pressurization valve; 4—Turbine; 5—Gear box;
6—Oxidant pump; 7—Incendiary agent pump; 8—Basic valve; 9—Thrust chamber; 10—Gasifier;
11—Evaporator; 12—Powder starter.

2) Pre-cold and eliminating system

To guarantee the liquid rocket engines such as liquid oxygen/hydrogen rocket engines work normally, pre-cold and eliminating systems are needed.

(1) Eliminating system

The liquid oxygen/hydrogen propellants are both cryogenic, especially the liquid hydrogen whose boiling point is only about 20K. Before filling, if the air exists in the pipelines and channels, there may be chemical reactions between the liquid hydrogen and oxygen in the air even to trigger the explosion. Besides, cryogenic hydrogen may freeze the oxygen and vapor in the air. It results in the malfunctions of valves and other components. So to the oxygen/hydrogen engines, an inert gas such as nitrogen, helium and so on must be used to eliminate and replace the gas left in the engine system before filling the liquid oxygen/hydrogen propellants.

(2) Pre-cold system

A pre-cold system must be set in the liquid rocket engine which uses cryogenic propellants. The purpose of pre-cold is to drop the temperatures of components such as pipelines and channels to the temperatures of liquid hydrogen and liquid oxygen. This can avoid the aerification of cryogenic propellant which may result in a lot of kickbacks such as fluctuations of propellant flow and pressure, the postponement of start-up, oxygen enrichment in the thrust chamber, and ablation.

3) Pressure-augmentation system for propellant storage box

When the propellants are sent into the thrust chamber, certain pressure and flux must be required. The function of the pressure augmentation for the storage box is to guarantee the pressure. The pressure-augmentation system for extruding supply system and pumping supply system are different from each other. A high-pressure bottle or gasifier is used in the store box of extruding supply system while a turbo pump which is driven by the gasifier is used in the storage box for the pumping supply system.

4) Propellant adjusted system

The system is used to regulate the mixing ratio of the propellants in the liquid rocket engine to guarantee the propellants (incendiary agents and oxidizers) are all exhausted at the same time.

During the flight, carrier rocket may be subjected to the impacts such as flight g-load, aerodynamic heating, mass deviation, and so on. The actual mixing ratio of the propellants may deviate from the rated one so that a certain unit of propellants is used up beforehand. Engine-running time is shortened so that performances of it are all depressed. A propellant-adjusted system can reduce the surplus of the propellants.

The propellant adjusted system is a closed-cycle control system that is composed of a liquid level sensor, convertor, controlling machine, power amplifier, propellant control valve, battery, and so on.

5) Thrust and thrust vector adjusted system

During the flight, the orbit of the carrier rocket may be changed by random disturbances, so that sometimes it requires to alter the thrust. So controlling-force system and the controlling-moment system are both needed to realize the control in the flight process. Regulating the size of the thrust chamber and changing the direction of the thrust are two methods adopted by the system.

(1) Thrust-adjusted system

Thrust can be changed by regulating the flux of the propellants. Thrust is related to the flux of propellants and pressure in the combustion chamber. So the purpose of changing the thrust can be achieved by regulating the flux or the pressure.

(2) Thrust vector adjusted system

According to the preliminary orbit, the flight poses of a carrier rocket must be regulated and amended during the flight. Thrust and thrust vector adjusted systems are required for the carrier rocket to complete the maneuvering flight including pitching, yawing and rolling. The actuating mechanisms of thrust vector control system on the liquid rocket engine include gas rudder, secondary injection, movable nozzle, and so on.

6) Pipes and frames

(1) Pipes

Pipes are the connecting pieces between different components of the liquid rocket engine. Limited by the positions and other causes, many components cannot be connected directly, then the pipes must be adopted. A certain intensity and pressurizers must be required for the pipes.

(2) Frames

Engine frames connect the engine to rockets or missiles. They can transfer the thrust generated by the engine. They belong to the propulsion system and can be considered as parts of the body as well.

4.2.4 Combination engines

Combination engines are the combinations of different kinds of engines. Different types of engines have different performances and characteristics. To improve the performance of the engines, different kinds of engines can be combined. Learning from others' strong points to offset one's weakness, different flight purposes can be satisfied by the combination engines. For example, the ramjet engine is not good at low-speed flight but superior at high-speed performance. If it can be combined with other engines, not only the merits of it will be exerted but also its range of use will be expanded.

At present, ramjet engines, turbojet engines, and rocket engines are designed to be combination engines.

Figure 4.30 shows the diagram of the rocket-ramjet engine which is combined with a rocket engine and a ramjet. Ramjet engine cannot work alone and is not good at low-speed flight. So the rocket engine is used as its roll booster. It has the advantages of both the rocket engine and ramjet. And the disadvantages of ramjet are got rid of.

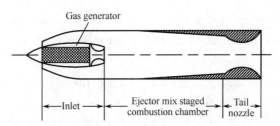

Figure 4.30 Sketch of rocket-ramjet engine

Figure 4.31 illustrates the diagram of the combination of the solid and liquid rocket engines. It derives from solid and liquid rockets. Dating back to the 1960s, the research was started. The purposes were to acquire the advantages of both solid and liquid rocket engine. But up to now, due to the complex construction and difficulties in techniques, the development has not been made.

A combination of solid and liquid rocket engines uses a combination propellant. As shown in Figure 4.31, the combination propellant includes a solid incendiary agent and liquid oxidizer. It can improve the mean specific impulse of the propellant for the density of liquid oxidant is smaller than that of the solid incendiary agent. The solid propellant is filled in the combustion chamber, while the solid oxidant is stocked in the storage.

The working process is illustrated as follows: the high-pressure gas is stored in the high-pressure gas bottle. The pressure drops to the required bypassing the pressure reducer. Then it goes into the liquid propellant storage.

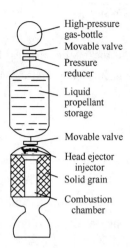

Figure 4.31 Combination of solid and liquid rocket

Pressed by the gas, the liquid oxidant enters into the ejector injector of the combustion chamber. The atomized liquid oxidant is injected into the combustion chamber. When the engine works, a little liquid fuel that can burn with the liquid oxidizer is injected from the ejector. The solid incendiary agent is warmed and its surface is atomized by the burning gases, the atomized solid incendiary agent is mixed with the liquid steam, then burn together. The high-temperature and high-pressure burning gases are accelerated in the nozzle to produce an exhaust of high-speed gases with a great deal of thrust.

Figure 4.32 shows the construction of a combination of the turbojet engine and ramjet engine. The ramjet engine cannot work when the velocity is zero, so the turbojet engine starts to work first. When the Mach number comes up to $3 \sim 3.5$, the turbojet engine stops while the ramjet engine begins to work. They use the same combustion chamber. The high-temperature and high-pressure gases are accelerated in the nozzle to produce an exhaust of hypersonic gas with a great deal of thrust.

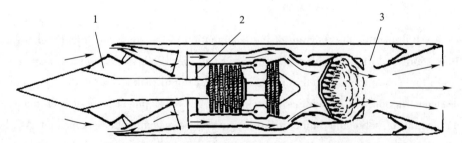

Figure 4.32 Construction of combination of the turbojet engine and ramjet engine
1—Variable inlet duct; 2—Inlet guide vanes; 3—Adjustable nozzle.

4.3 Thrust vector control system

With the development of flight vehicle control technologies, conventional aerodynamic controls can not meet the demand, even are incompetent under certain conditions, such as exo-atmospheric flight. Furthermore, to keep pace with the increasingly severe demands of high maneuverability and fast response, we should break traditional concepts. So unconventional flight controls emerge as the times require which is Thrust Vector Controls (TVC). TVC technologies have been widely used in missiles, Space Shuttle boosters, fighters, Kinetic Kill Vehicles (KKV), satellites, etc. In what follows, particular emphasis is placed upon thrust vector control system (TVCs) appropriate to missiles.

4.3.1 Classification of thrust vector control system

According to different implementation mechanisms, four principal groups are involved in thrust vector control system, which are movable nozzle, secondary fluid injection, mechanical obstructions, and reaction jet control.

4.3.1.1 Movable nozzle

Members of this group are flexible nozzle, ball, and socket swivel nozzle. Figure 4.33 shows the flexible nozzle. It consists of a nozzle mounted directly to the rocket motor aft closure using a laminated flexible mounting. The laminations consist of concentric spherical layers of elastomeric and thin metallic shells bonded to form a flexible sandwich. The mounting is stiff in the axial direction but can be easily deflected sideways. This mounting may interface between a conventional motor closure and an optimized nozzle.

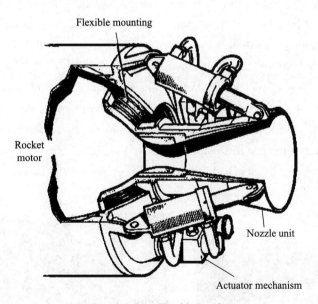

Figure 4.33 Flexible nozzle

4.3.1.2 Secondary fluid injection

In this system, fluid is injected through the nozzle expansion cone into the motor gases. The penetrating fluid causes an oblique shock to occur in the supersonic nozzle gas flow, which gives rise to an uneven pressure distribution and causes the efflux to be deflected. Members of this group are liquid injection and hot gas injection, shown in Figure 4.34 (a) and Figure 4.34 (b).

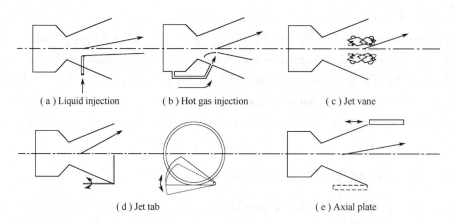

Figure 4.34　Secondary fluid injections and mechanical obstructions

4.3.1.3　Mechanical obstructions

Systems that employ obstructions in the efflux come into this category. Members of this group are jet vanes, jet tabs, axial plates, and so on. Jet vanes are control surfaces inside the nozzle which have an advantage in differential deflection and roll control. Jet vanes can also share common actuators with aerodynamic control. However, jet vanes are continually exposed to the rocket exhaust plume, which results in about 3% ~ 5% loss in specific impulse even if they are not deflected. Jet tabs are simple devices that switch in and out of the plume. The jet tab sets a bow shock inside the nozzle, similar to that of liquid injection TVC. The axial plate consists of a plate that extends back and forth. When the plate is fully extended, the plume attaches to the plate because of the Coanda effect.

4.3.1.4　Reaction jet control

Reaction jet control can be based on either short-duration impulse, "shotgun shell-type" jets similar to PAC-3 (as shown in Figure 4.35), or longer duration controlled jets similar to THAAD (as shown in Figure 4.36) or SM-3. During the endo-atmospheric flight, reaction jet control interacts with the free stream air. The total force is a combination of the thrust force and the jet interaction force. The jet interaction force results from the pressure distribution on the airframe surface. Although the jet interaction force usually augments the thrust force, there are relatively high uncertainty and risk in the magnitude of the interaction force.

Figure 4.35　PAC-3

Figure 4.36　THAAD

4.3.2 Performance and selection of thrust vector control system

The performance of the thrust vector control system can be divided into four aspects:

(1) Jet deflection angle—that is the amount by which the efflux may be deflected.

(2) Side thrust coefficient—defined as the side thrust divided by the undisturbed axial thrust.

(3) Axial thrust loss—the thrust loss incurred by operating the device.

(4) Actuation effort—the total force characteristics required to be applied to the device to achieve a given response.

Jet deflection angle and side force coefficient describe the lateral force capability of any thrust vector control system. For thrust vector control system using shock wave formation, it is normal to describe the lateral force capability in terms of side force coefficient and an equivalent jet deflection angle. Actuation effort is a necessary parameter when sizing the driving mechanism.

The selection of thrust vector control system depends on the missile system mission definition and requirements, such as required maneuvers, accuracy, reliability, ease of handling, and other factors. It is not possible to state categorically that one particular thrust vector control system is better than another. The system selected will always depend upon the specified vehicle mission and the best system may change with time as advancements are made in materials and new inventions appear on the scene.

4.4 New-type propulsion system

Unlike the stated before, new-type propulsion systems have been discovered in recent 20 years. Such as nuclear rocket propulsion system, solar energy rocket propulsion system, and electric rocket propulsion system. And some new concepts for propulsion systems have been developed.

4.4.1 Solar energy propulsion system

Figure 4.37 shows the concepts of solar energy rocket propulsion systems. The high-pressure gas stored in the bottle is expelled through the pressure regulator along the pipe to the outside. First, it needs to mix with argon in the argon storage to avoid oxidation of the working materials. Then the solar energy propulsion system uses the parabolic rejector to gather solar energy to complete the heat exchange and gets mixing gas with high temperature and high pressure. Finally, the thrust can be produced by the expansion of mixing gas in the nozzle. According to this concept, solar energy will be used to produce the thrust as the power to push the flight vehicle into space. Solar energy propulsion systems have advantages of simple structure, strong reliability, no pollution, high specific impulse. They are suitable for orbital transportation of medium and large spacecraft and interstellar voyages. However, they make heat exchange indirectly by intermediate materials, such as rhenium and tungsten. That is to say, solar energy propulsion systems have serious heat loss and low efficiency. And the direct heat absorption solar energy propulsion systems are technically complex to implement. Therefore, both solutions are still in the conceptual stage, and there is still a long way to go before they can be put into practical use.

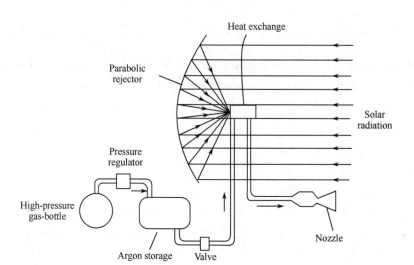

Figure 4.37 Solar rocket engine concept

Besides, solar sail propulsion systems are another type of propulsion system using solar energy as power. The solar sail propulsion systems are composed of a central drum and a sail plate, which use solar radiation pressure as a power to push the vehicle in space. They have unique advantages of continuous acceleration and fuel savings, which can be applied to long-duration, long-range, ultra-long-range space flight and have broad application prospects. Figure 4.38 shows the shape of solar sail propulsion systems.

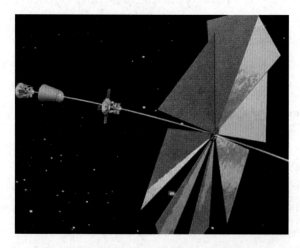

Figure 4.38 Solar sail propulsion systems

4.4.2 Nuclear energy propulsion system

As shown in Figure 4.39, nuclear energy is the energy source. By using nuclear energy to heat working materials, the thrust is produced through spurting out the working materials at high speed. The principle of a nuclear rocket propulsion system is in terms of nuclear-releasing ways, such as nuclear fission, nuclear fusion, and radioactive isotope disintegration.

The working principle of nuclear energy rocket propulsion systems is similar to that of chemical

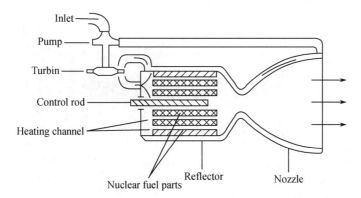

Figure 4.39 Nuclear rocket engine principle diagram

rocket propulsion systems, except that the energy of the working materials is provided by the thermal energy of nuclear reaction instead of the chemical energy generated by the combustion of the propellant. Compared with chemical rocket propulsion systems, nuclear energy rocket propulsion systems have the advantages of high specific impulse, large thrust, long life, and multiple starts. However, nuclear energy rocket propulsion systems still have shortcomings in terms of excessive reactor mass and radiation protection difficulties, while their testing and development face political disputes.

Nuclear energy propulsion systems can be based on in-situ resource utilization (ISRU) technology using resources (ice, water, and methane, and so on) on planets, their satellites, and asteroids to create working materials, and use the planetary atmosphere as the infinite thermal energy of the reactor and working materials to achieve unlimited orbital flight and detailed measurements of the state of planetary atmospheric components, density, and temperature. In conclusion, Nuclear energy propulsion systems have a broad application prospect in the field of space science research and deep space exploration.

4.4.3 Electric propulsion system

The electric propulsion systems can be divided into electrothermal propulsion systems, electrostatic propulsion systems, and electromagnetic systems, in which electrothermal propulsion systems use electrical energy to increase the thermal energy of propellant that is then ejected. According to the heating method, electrothermal propulsion systems can be divided into resistojet and arcjet. Figure 4.40 shows the principle of arcjet. The cathode and anode discharge generates an electric arc that heats the gas and converts it into hot plasma, which expands in the nozzle to produce thrust.

The electrothermal propulsion systems have the advantages of high specific impulse, high efficiency, and long life of electric propulsion system, and also have the characteristics of simple structure, low cost, safety and reliability, easy operation and maintenance, low pollution, and so on. They are suitable for small and low-cost satellite orbit adjustment, altitude control and position keeping, and so on.

In electrostatic propulsion systems, the gas is first ionized, and then the charged ions are

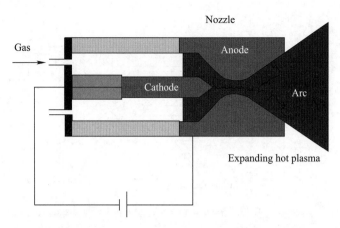

Figure 4.40 Electrothermal rocket engine principle diagram

accelerated by the electric field force (Coulomb force) and ejected to push the flight vehicle forward. Figure 4.41 shows the principles of electrostatic propulsion systems.

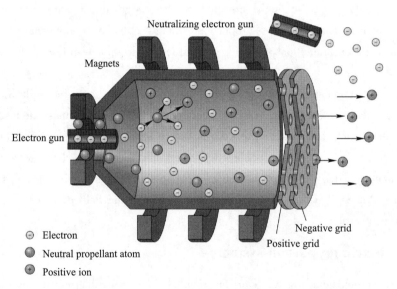

Figure 4.41 Electrostatic rocket engine principle diagram

Electrostatic propulsion systems are the most economical, and the thrust can be adjusted by simply adjusting the electric field strength. Besides, because its specific impulse is much larger than other existing propulsion systems, only a small amount of propellant is required to achieve high final velocities. However, they have the disadvantages of small thrust, large input power, large size, complex structure. Meanwhile, there is mutual repulsion between positive charges, and the space charge saturation effect may harm efficiency. Also, they can only be applied in a vacuum environment. Therefore, Electrostatic propulsion systems are promising for a wide range of applications in long-distance space flight.

The electromagnetic propulsion systems use the interaction between electric current and magnetic field (Lorentz force) to ionize and accelerate the propellant and produce thrust. Their working materials are heated into a plasma state of propellant gas. The electromagnetic propulsion systems can be divided into magneto-plasma-dynamic thruster (MPD), pulsed plasma thruster (PPT), and Hall Effect thruster.

Figure 4.42 shows the working principles of electromagnetic propulsion systems, working materials will be ionized at high temperature, to form plasma, electrons escaped, the neutral plasma is formed. The neutral plasma behaves electrical conductivity. And it acts on magnetic fields, producing the force with acceleration by electromagnetic induction.

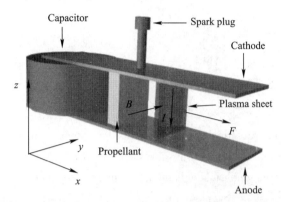

Figure 4.42 Electromagnetic rocket engine principle diagram

Electromagnetic propulsion systems have high specific impulse capability at low power, while the structure of them is simple, and they are currently used in some space vehicles. The main problems limiting their development are low propulsion efficiency and plume pollution. To achieve widespread application on space vehicles, key technology problems of electromagnetic propulsion systems need to be broken.

4.4.4 Micro propulsion system

With the emergence of small satellites, microsatellites, nano-satellites, pico-satellites, and formation flight technology, higher requirements for space propulsion technology have been put forward, and micro propulsion systems have received a lot of attention in recent years, which use modern micro electro mechanical system (MEMS) technology and micro-manufacturing technology to improve the maneuvering and control performance of space vehicles through miniaturized structure, integrated functions, and cost reduction measures.

Micro propulsion systems include cold gas propulsion, chemical propulsion, and electric propulsion systems. The cold gas propulsion system produces thrust by ejecting a gas working materials with a certain pressure from the nozzle, and this propulsion system is extremely simple in structure, but its specific impulse is also low; the principles and working characteristics of chemical and electric propulsion are similar as those of conventional rocket propulsion systems and electric propulsion systems. Figure 4.43 shows the schematic structure of a single solid-propellant micro thruster. We can see its small size.

Note that, regardless of the method used to produce thrust, some phenomena specific to small scales must be taken into account in the design of micro and small space propulsion systems, such as fluid surface tension and surface layer effects, and so on.

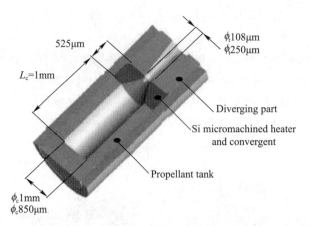

Figure 4.43 Schematic structure of a single solid-propellant micro thruster (SPM)

The earliest application direction of micro propulsion systems is mainly for attitude control and position keeping of spacecraft. With the development of space technology, the size and mass of spacecraft are getting smaller and smaller, and many micro propulsion systems have started to be applied as the main propulsion system of tiny spacecraft. The micro propulsion system allows for precise orbital maneuvers, gravitational compensation, and attitude control. Also, compared with conventional propulsion systems, micro propulsion systems have a very high specific impulse. This property makes the micro propulsion systems have a great advantage for increasing spacecraft lifetime, interplanetary navigation, and deep space exploration. In practical applications, MEMS are usually deployed on spacecraft in arrays. Figure 4.44 shows a typical 7×1 MEMS-based solid propellant micro thruster array

Figure 4.44 7×1 MEMS-based solid propellant micro thruster array

The micro propulsion systems mainly include micro-pumps, micro-valves, micro-thrusters, and related control circuits. The manufacturing of these micro precision components requires MEMS as technical support. However, MEMS technology has not yet matured, and there are still problems of mechanical processing difficulties, high cost, and poor sealing based on MEMS manufacturing micro propulsion systems. Despite all the problems, MEMS micro propulsion systems represent the development direction of micro propulsion technology, which can miniaturize the propulsion system to an unprecedented level, and have a broad application prospect in micro spacecraft applications, and spacecraft precision attitude control.

Questions

1. Please describe the main parameters of a flight vehicle engine system.
2. Please describe the compositions of a rocket engine thrust.
3. Try to analyze the factors affecting the engine thrust.
4. If a missile flies with acceleration at first, then with a constant speed, try to describe the thrust curve, and the combustion mode of the engine.
5. Try to analyze the main differences between rocket engines and jet engines.
6. If a rocket engine works at the height of 10km, mass flow quantity $\dot{m} = 5230$kg/s, pressure at the exit of nozzle $p_e = 0.074$ MPa, gas speed $u_e = 2000$m/s, cross area at the exit of nozzle, $A_e = 11.70$m^2.

    Try to get
    (1) Thrust at the height of 10km,
    (2) Vacuum thrust of the engine,
    (3) Thrust at sea level.
7. Give the advantages and disadvantages of solid rocket propulsion systems and liquid rocket propulsion systems.
8. Describe the classification and characteristics of thrust vector control systems.
9. List three kinds of new propulsion systems and summarize their advantages and disadvantages.
10. Introduce the classification of the electric propulsion system and its working principle briefly.

Words and phrases

ablation　冲蚀,消融,烧蚀
aerification　气化
aft closure　后盖
afterburner　加力燃烧室
aftereffect impulse　后效冲量
air intake system　进气系统
air-evacuation valve　排气活门
aisle　走廊,过道
ambient　周围的
amorce　点火药
anode　阳极
assembly　部件,组件
atomizing　雾化(作用)
axial plate　轴向挡板

axis-flow type　轴流式
ball and socket swivel nozzle　球窝式旋转喷管
blade　叶片,桨叶
board for controlling the flux　流量控制板
bypass ratio　双涵道空气流量比
cathode　阴极
centrifugal type　离心式
characteristic thrust　特征推力
Coanda effect　康达效应
coaxial turbo pump　同轴式涡轮泵
combustion chamber　燃烧室
commuter aircraft　定期短途班机
compounded powder　复合推进剂
compressor　压缩器

Coulomb force　库仑力
Concord　"协和"号飞机
conduit　管道
coping　顶盖
crankshaft　曲柄轴
cryogenic　低温的
cylinder　汽缸
detonated tube　发火管
double base powder　（硝化甘油与硝化纤维）双基火药
double turbotype turbo pump　双涡轮式涡轮泵
dual-combination-unit　双组元
effective jet velocity　有效喷气速度
elastomeric　弹性体的,人造/合成橡胶的
evaporator　蒸发器
exo-atmospheric flight　大气层外飞行
extract　提炼出,分离出,榨出,蒸馏出
extruding supply system　挤压式供应系统
FAA　联邦航空局
flexible nozzle　柔性喷管
flight regime　飞行状态
fluctuation　波动,动摇,变动
fluorine　氟
four-stroke reciprocating piston engine　四冲程的活塞式发动机
frame　机架
fuel pump　燃料泵
fuel-injection equipment　燃料喷射装置
gas-guzzler　油老虎
gasifier　气化器
gasoline　汽油
gearbox　变速箱
gearing turbo pump　齿轮传动式涡轮泵
general conduit charge　普通管形装药
Helium　氦气
high-pressure demolition valve　高压爆破活门
homogeneous　均质的
HTHP　高温高压
incendiary agent　燃烧剂
inert gas　惰性气体

inhomogeneous　异质的
injecting tray　注射盘
inlet diffuser　进口扩压器
inlet diffuser　进气道
inlet guide vane　进口导流片
ISRU　原位资源利用
jet tab　扰流片
jet vane　燃气舵
jet-propel system　喷射推进式系统
kerosene　煤油
KKV　动能杀伤拦截器
liquid fluorine　液态氟
liquid hydrogen　液氢
Lorentz force　洛伦兹力
low-pressure demolition valve　低压爆破活门
malfunction　故障
mass deviation　质量偏差
mechanical obstruction　机械阻流
MEMS　现代微机电系统
methane　甲烷
MDB　改性双基推进剂
mono-combination-unit　单组元
MPD　磁等离子流体动力推力器
nitrocellulose　硝化纤维
nitrocotton　硝化棉
Nitrogen　氮气
nitroglycerin　硝化甘油
nuclear fission　核裂变
nuclear fusion　核聚变
oblique shock　斜激波
oxygen enrichment in the thrust chamber　推力室富氧
parasitic power　寄生功率
piston engine　活塞式发动机
postponement　延迟,延期
PPT　脉冲等离子体推力器
pre-cold and eliminating system　吹除与预冷系统
pressure-increase system for propellant store box　推进剂储箱增压系统

pressurization valve　增压活门
propellant adjusted system　推进剂利用系统
propeller　螺旋桨
pumping supply system　泵压式供应系统
radioactive isotope disintegration　放射性同位素衰变
ram engine　冲压发动机
ramjet　冲压式喷气发动机
rated　额定的,标称的
reduction gear　减速齿轮
scramjet　超声速燃烧冲压发动机
septa　隔膜
shaft　轴,杆状物
shotgun shell-type　猎枪弹壳式
specific impulse　比冲
sphericity　球形
spheroidicity　椭球形
subsonic combustion ramjet　亚声速燃烧冲压发动机

surface layer effects　附面层效应
surfacetension　表面张力
surplus　剩余,盈余,过剩
throttle setting　油门调节
thrust barrel　推力室
total impulse　总冲
TSFC　单位推力耗油率
turbine　涡轮
turbocharger　涡轮增压器
turbofan　涡轮风扇
turbojet　涡轮喷气
turbomachinery　涡轮机组
turboprop　涡轮螺旋桨
turbo pump　涡轮泵
TVCS　推力矢量控制系统
vacuum thrust　真空推力
vapor　水蒸气
windmill　风车

References

[1] 廉筱纯,吴虎. 航空发动机原理[M]. 西安:西北工业大学出版社,2005.
[2] 唐金兰. 固体火箭发动机原理[M]. 北京:国防工业出版社,2013.
[3] 王春利. 航空航天推进系统[M]. 北京:北京理工大学出版社,2004.
[4] 谭永华. 航天推进技术[M]. 北京:中国宇航出版社,2016.
[5] 荣思远,刘家夫,崔乃刚. 太阳帆航天器研究及其关键技术综述[J]. 上海航天,2011,(2):53-62.
[6] 何伟锋,向红军,蔡国飙. 核火箭原理、发展及应用[J]. 火箭推进,2005,31(2):37-43.
[7] 毛根旺. 航天器推进系统及其应用[M]. 西安:西北工业大学出版社,2009.
[8] 于达仁,刘辉,丁永杰,等. 空间电推进原理[M]. 哈尔滨:哈尔滨工业大学出版社,2014.
[9] Dan M G, Ir A K. Fundamentals of electric propulsion: Ion and hall thrusters[M]. New York: John Wiley & Sons,2008.
[10] Mazouffre S. Electric propulsion for satellites and spacecraft: established technologies and novel approaches [J]. Plasma Sources Science and Technology,2016,25(6):033002.
[11] Garrigues L, Coche P. Electric propulsion: comparisons between different concepts[J]. Plasma Physics and Controlled Fusion,2011,53(12):124011.
[12] 徐泰然. MEMS 和微细统一设计与制造[M]. 北京:机械工业出版社,2004.
[13] Liu B, Li X, Yang J, et al. Recent Advances in MEMS-Based Micro thrusters[J]. Micromachines,2019, 10(12):818.

Chapter 5 Flight vehicle aerodynamic configuration

The shape and layout of the large parts of the flight vehicle are closely related to the aerodynamic forces experienced by the flight vehicle, and fundamentally affect the flight characteristics and performance of the flight vehicle, especially the air vehicle. The layout and position of the external form of the vehicle are often referred to as the aerodynamic configuration. The design tasks of the flight vehicle are different, their maneuverability requirements are also different, which leads to different forms of aerodynamic configuration of the flight vehicle.

5.1 Design requirements for configuration

Aerodynamic configuration design is to determine the theoretical configuration and to give the aerodynamic characteristics. In detail, it is to design the shape, dimension and relative position of the body, lifting surface, stabilizing surface, control surface and the other components which exposed in the air, and then to give the pressure distribution, temperature distribution and the change law of aerodynamic coefficient with the flying height, velocity, angular velocity, attitude angle and so on. The design of aerodynamic configuration plays a vital role in the overall design of flight vehicles. A flight vehicle's flying path, flying height, maneuverability and stability depend largely on its configuration. In other words, once the configuration is determined, its main flight performances are established.

There are some requirements forconfiguration as follows:

(1) Have good aerodynamic characteristics such as high lift-drag ratio;
(2) Meet control and stability requirements; meet dynamic response requirements;
(3) Have effective aerodynamic control surfaces, and ensure the required servo power is as small as possible;
(4) The configuration is simple and easy to implement;
(5) Meet the range, speed and other performance requirements;
(6) The arrangement of parts is reasonable and convenient;
(7) Have small size, light weight, higher reliability and maintainability.

A configuration is a synthetic research result according to design requirements of aerodynamic characteristics and flight performances. In other words, it is a multi-surface balance. The excellent configuration is achieved through many cycles, so that it will ensure good performances and normal work of flight vehicles.

5.2 Flight vehicle configuration & parameters

Generally speaking, a flight vehicle usually has wings, body or fuselage and tails. The airfoil and wing platform are two important characteristics of the wing. The body can be divided into nose, central body and boat-tail. The relative position of wings and body is another important feature for a flight vehicle. The position of the tail is usually at the rear of the body in which the function of the fixed tail is as wing. If the tail is movable, it is called rudders. All of the above, and the related terms and parameters will be introduced as follows.

5.2.1 Airfoil

The airfoil is the wing profile of the flight vehicles, whose parameters are descriloed the airfoil shape. As shown in Figure 5.1, for an airfoil, all of the central points between the upper and lower surface can be connected into a line which is called mean camber line. The start-point and end-point of mean camber are leading edge and trailing edge respectively. The straight line between leading edge and trailing edge is called chord line. The longest line perpendicular to the chord line and connecting the upper and lower surface of airfoil is called max thickness. The longest line perpendicular to the chord line and connecting the mean camber line and chord line is max camber.

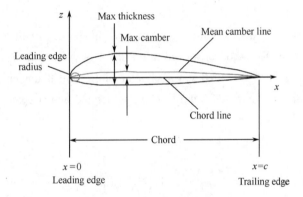

Figure 5.1 Airfoil of NACA 2412

The airfoil series are shown some tgpical airfoils with parameters and wind tunnel test data by NACA. NACA, which is the abbreviation for the National Advisory Committee for Aeronautics, is the term that should be mentioned when talking about airfoil. It is a U.S. Government agency, forerunner of NASA, which performed many wind tunnel tests of airfoils and other shapes in the 1930's and 1940's. NACA used a series of codes with 4, 5, and 6, and more digits to systematically classify the airfoils they tested.

NACA 2412 means that the airfoil has 2% camber, its point of maximum chamber is located at its 40% chord line, and its maximum thickness is 12% of its chord length. If an airfoil with a NACA 2412 section has a chord length of 4m, its maximum thickness would be 48cm.

Thereis a little difference between 4-digit and 5-digit NACA airfoils. Taking NACA 23015 for an example, the first number "2" implies its designed lift coefficient is 0.15 multiplied by "2", the second number "3" means its point of the maximum chamber is located at its 3/20 (15%) chord line, the third number "0" shows the type of mean camber line, the last two numbers "15" means its relative thickness is 15%. See some references for more details of the various NACA airfoil series and designs. Besides NACA, the Gottingen series airfoil from Germany and RAF series airfoil from

Britain are also very famous.

Some airfoils as shown in Figure 5.2 are usually used in missiles which are double-wedge airfoil, hexagonal airfoil, double-curve airfoil and blunt trailing edge airfoil.

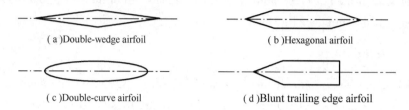

(a) Double-wedge airfoil (b) Hexagonal airfoil

(c) Double-curve airfoil (d) Blunt trailing edge airfoil

Figure 5.2 Some airfoils for missile

5.2.2 Wing form & tail configuration

1) Wing form

There are several wing forms, just like trianglar, trapezoidal, rectangular, elliptied, sweepback wing and so on.

Figure 5.3 illustrates a view of a wing form with some of the important dimensions, angles and parameters used to describe the shape of an aircraft wing. The wing span b, is measured from left wing tip to right wing tip, which is called gross span including the diameter of fuselage. The span from left wing tip to right wing tip without the fuselage is defined as net span.

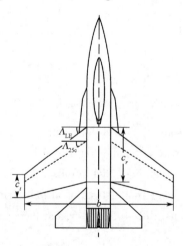

Figure 5.3 Wing geometric parameter definitions

The symbol c is used for the chord length of an airfoil at any point along the wing span. The subscript r indicates the chord length at the wing root or the aircraft central line. The subscript t denotes the wing tip chord. So c_t is tip chord, and c_r is root chord.

The symbol Λ_{LE} is used for wing sweep angle with the subscript LE denoting the wing leading edge. The subscript 25 denotes the line connecting the 25% chord positions on each airfoil of the wing, quarter sweep angle $\Lambda_{0.25c}$.

The symbol λ is used for the wing taper ratio, or ratio of tip chord to root chord. Here, $\bar{c}$ is mean aerodynamic chord (MAC), and aerodynamic center (a. c.) is the position of $0.25\bar{c}$. The ratio of wing area S and span b is called standard mean chord (SMC).

The symbol AR is used for a parameter called aspect ratio. It is the ratio of wing span and MAC. Aspect ratio indicates how short and stubby or long and skinny the wing is.

During the computation of aerodynamics of aircraft, S is as the reference area and $\bar{c}$ as the reference length.

2) Tail configuration

Tails are usually composed of moved and unmoved parts. The unmoved part is called stabilizer or fin. There are vertical fin and horizontal fin in tails. The moving part is control surface including rudder and elevator. Three basic types of tails in aircraft are classified by the relative position of vertical fin and horizontal fin as shown in Figure 5.4.

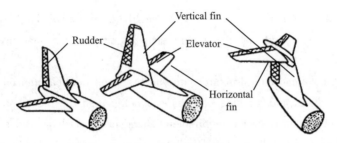

Figure 5.4 Three types of tails

5.2.3 Fuselage

The fuselage of a flight vehicle is used to carry the payload and connect other parts. The volume, flight drag, and radar cross-section (RCS) should be considered in the design of fuselage. The important parameters of fuselage are the shape and dimensions of cross-section. The fuselage with a round shape is easy to manufacture, with rectangular shape has smaller RCS, with elliptic shape has better aerodynamic performance. The flight drag increases with the growth of diameter, so the cross-section diameter should be minimized in design.

Figure 5.5 illustrates key parameters of fuselage. The symbol L_B is used for the total length of fuselage and the symbol D is used for the diameter of fuselage. Generally, there are nose, central fuselage and boat-tail in the fuselage. The symbol L_n and L_t are respectively used for the length of nose and boat-tail. The symbol D_b is used for diameter of boat-tail.

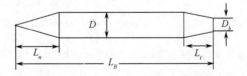

Figure 5.5 Fuselage parameters

The ratio of L_B and D is called fuselage fineness whose symbol is λ_B. The ratio of L_n and D is called nose fineness whose symbol is λ_n. The ratio of L_t and D is called boat-tail fineness whose symbol is λ_t. The ratio of D_b and D is called boat-tail contraction ratio whose symbol is η_t. The nose and boat-tail configuration will be discussed in detail separately in the next.

1) Nose

There are sharp cone shape, hemispherical shape, ellipsoidal shape and parabolic shape as shown in Figure 5.6.

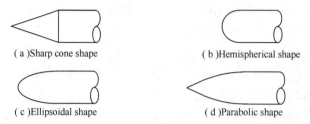

Figure 5.6 Nose shapes in conventional flight vehicle

From the view of aerodynamic performance, the aerodynamic drag of sharp cone shape is smaller than that of parabolic shape, and that of parabolic shape is smaller than that of ellipsoidal shape in the situation of fixed nose fineness.

From the view of volume and structure requirements, hemispherical and ellipsoidal shapes are better than parabolic shape, and the parabolic shape is better than sharp cone shape.

From the view of guidance system requirements, hemispherical shape and ellipsoidal shape are suitable for infrared seeker and TV seeker. Parabolic shape is suitable for radar seeker. Some of the missile head's curve equations are derived directly from the radar beam requirements.

For the aircraft, the vision requirements should be considered except for aerodynamic drag and guidance system requirements.

2) Boat-tail

There are three common types of boat-tail which are cylindrical boat-tail, shown in (a) of Figure 5.7, contractive boat-tail, shown in (b) and (c) of Figure 5.7, and expensive boat-tail which is called skirt tail, shown in (d) of Figure 5.7.

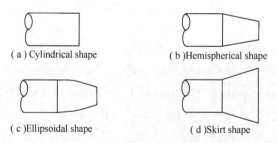

Figure 5.7 Boat-ail shapes in conventional flight vehicle

The boat-tail contraction ratio is mainly determined by the layout of flight vehicle. The cross-section diameter of boat-tail should be enough to place the nozzle of engine, and sometimes the space of servos and radio installations around the nozzle also should be considered. Generally, the dip angle of generatrix should not exceed 12° ~ 15° in determining the boat-tail fineness, for airflow separation may occur if the angle is larger. Skirt tail is usually used in ballistic missiles, and to increase the stability of the missiles.

5.3 Typical flight vehicle configuration

The typical configurations are much different between aircrafts and missiles. An aircraft generally has only two wings, but a missile may have four, six or even eight wings; compared with the control surfaces of aircraft including aileron, rudder, and elevator, there are all-moving wings, grid fin and so on for missile. The angle of incidence, angle of twist and dihedral angle should be considered when designing the wing of aircraft, but it is not required for a missile. The typical configuration of missiles and aircraft will be introduced separately.

There are various types of control surfaces which are important parts in controlling flight vehicles, so the control surfaces will be analyzed in detail. Here control surfaces of missiles will be focused on at first. Besides, high-lift devices in aircraft is very important, such as flaps, slats, leading edge extension and so on.

5.3.1 Typical missile configuration

Missile configuration is relatively simple, and is commonly composed of head, body, wings and tails. It does not mean that there are only a few types of missile configuration. In fact, the types of missiles are various.

1) Normal configuration

Most tactical missiles use normal configuration, whose wings are at the front of tails which are the control surfaces. It can be controlled by means of the movement of the control surfaces. The normal configuration of a missile is shown in Figure 5.8. The exocet missile shown in Figure 5.9 uses normal configuration.

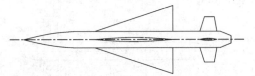

Figure 5.8　Normal configuration

Figure 5.9　Exocet missile

2) No-wing configuration

It is a special form, in which wing area is zero. So it has relatively small maneuvering. The no-wing configuration of a missile is shown in Figure 5.10. In general, it is used for ground to ground missiles, such as "Patriot" of USA (Figure 5.11). And rockets also have the configuration without wing.

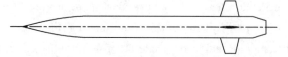

Figure 5.10 No-wing configuration flight vehicle

Figure 5.11 Patriot

3) No-tail configuration

Figure 5.12 shows no-tail configuration, in which wings and tails are close to each other, so that there are no tails. The efficiency for rudder of this configuration is very low. The MIM-23 shown in Figure 5.13 use no-tail configuration.

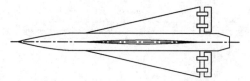

Figure 5.12 No-tail configuration

Figure 5.13 MIM-23

4) All-moving wing configuration

As shown in Figure 5.14, the wing is the main lifting surface. But in this configuration, the wing is also the control surface. That is to say, by operating wing surface, the pitching, yawing and rolling movement of a flight vehicle is realized. The advantages are as follows. It reacts very fast because of the large control surface. And the flight angle of attack is small. At the same time, there are some disadvantages, such as large hinge moment for wing operation, and relatively small ability for rolling control.

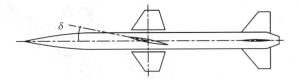

Figure 5.14 All-moving wing configuration

Figure 5.15 illustrates strong vortices from the wing interact with tail at all-moving wing configuration. The AIM-7 shown in Figure 5.16 uses all-moving wing configuration.

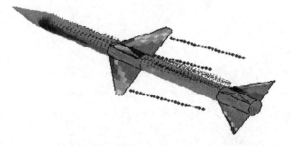

Figure 5.15 All-moving wing interaction with tail

Figure 5.16 AIM-7

5) Canard configuration

The control surface is in the front of wing, so it is called front wing control, or canard configuration (Figure 5.17). It is easy to adjust the static margin in relative to the variation of the missile mass center, and the control efficiency is high. But it is difficult to make rolling-control. The KH-25 shown in Figure 5.18 uses canard configuration.

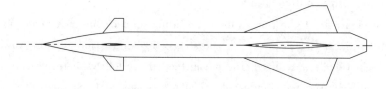

Figure 5.17 Canard configuration

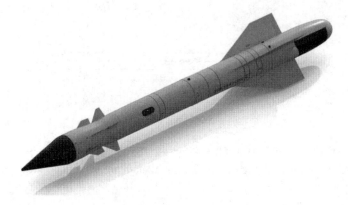

Figure 5.18 KH-25

6) " – " form configuration

The configuration is just like airplane form. As shown in Figure 5.19, it is plane-symmetrical. Most tactical missiles for attacking the targets on the sea, such as anti-ship missiles and cruise missiles, have this configuration. The YJ-18 shown in Figure 5.20 uses " – " form configuration.

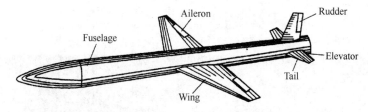

Figure 5.19 " – " form configuration

Figure 5.20 YJ-18

For the selection of missile configuration, there are a lot of types can be chosen which depend on several aspects, such as the characteristics of attacking targets, the aerodynamics, the structure arrangements, the requirements of the maneuvering performances, and so on. One point must be pointed out that most missile shapes are normal configuration.

The concepts of missile configuration are from other flight vehicles, for example, " - " form configuration dates from an airplane, and the cross form configuration is from a rocket. The configuration of a missile is the combination of wings, body and tails. The longitudinal combinations of wings and tails are " - - + ", " - - × ", " + - + ", " × - + ", " + - × " or " × - × ". Figure 5.21 shows the different ways of classification.

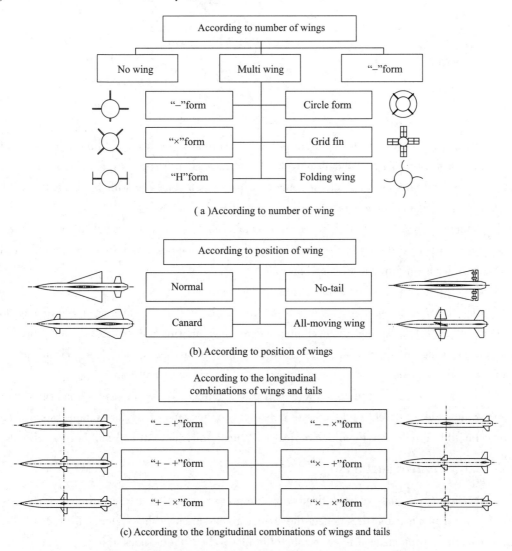

Figure 5.21 Different missile configuration

5.3.2 Typical aircraft configuration

As shown in Figure 5.22, for an aircraft, from the view of configuration, it is composed of anose, afuselage, wings and tails which consist of fixed and moved parts. From the view of compositions, there are pilot cabins, engines, electronic communications units, hydraulic units, and executing structures, and so on.

There are many types of aircraft. Generally, several kinds of airplanes such as transport aircraft

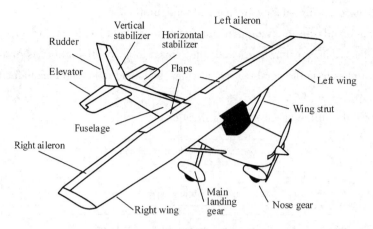

Figure 5.22 Aircraft configuration

and military airplanes are built for different purposes. There are three kinds of transport aircraft. The first is with passengers. The second is only with payloads, no passengers, i.e., cargo aircraft. The third is a combined airplane which is not only with passengers, but also with payloads. For military aircraft, generally, fighters, bombers and others are designed.

From the wing types, fixed wing and rotating wing aircraft are divided. A helicopter is an aircraft with a rotating wing. According to the power, airplanes with engines and no power are defined. For example, a glider is an airplane with no power.

A primary trainer is another kind of airplane for training pilots.

Typical aircraft configuration is shown in Figure 5.23. On the one side, according to wing form of the aircraft, sweep-back wing, triangle wing and rectangular wing are divided. On the other side, according to wing number and wing position, the dual-wing plane and monoplane are manufactured. Monoplanes are given with upper wing, medium wing and lower wing.

At the same time, according to tail form and position, three types, such as horizontal tail, V-type tail and vertical tail are given. And there are three kinds of horizontal tails, for example, normal form, canard form and no-tail form. From the vertical tail, single tail, dual tail and multi-tail are divided.

5.3.3 Control surfaces

The main control surfaces of a fixed-wing aircraft are attached to the airframe on hinges or tracks so they may move and thus deflect the air stream passing over them. This redirection of the air stream generates an unbalanced force to rotate the plane about the associated axis.

Usually, three main control surfaces are elevators, rudders and ailerons. Elevators, the moving part of the horizontal tail, make aircraft or missiles move in pitching direction. Rudders, the moving parts of the vertical tail, make aircraft or missiles move in yawing direction. Ailerons, the moving parts of the wing, move in differential ways to make the flight vehicles in rolling movement.

Aocording to the location of the control surfaces, the rear control surfaces, forward control surfaces and moving wings are given.

1) Rear control surfaces

The majority of tactical missiles have fixed lifting surfaces (often called wings) with their center

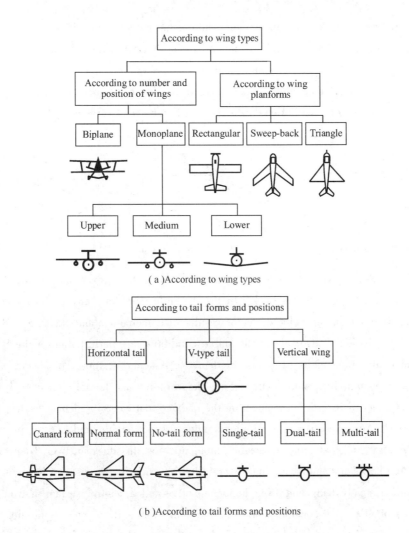

Figure 5.23 Configuration with wing and tail types

of pressure somewhere near the missile center of gravity, and rear control surfaces. In subsonic missiles, it may be more efficient to use the controls as flaps immediately behind the wings as the flap controls the circulation over the whole surface. With supersonic flow, the control surface cannot affect the flow ahead of itself and therefore it is placed as far as possible in order to exert the maximum moment on the missile. Rear control surfaces often make for a convenient arrangement of components. Usually, it is desirable to have the propulsion system placed centrally in the missile so that the center of gravity movement due to propellant usage is minimized. It is convenient and sometimes essential to have the warhead and fuse at the front together with any associated electronics including the guidance receiver. This leaves the control system to occupy the rear end with the propulsion bleat pipe passing through its center. If there are four servos, it is impossible to design a neat servo package around this pipe.

When considering the lateral forces and moments on missiles, it is convenient first of all to consider the combined normal forces due to incidence on the body, wings and control surfaces as acting through a point on the body called the center of pressure (c. p.), and to regard the control surfaces as permanently locked in the center position. As stated before, if the c. p. is ahead of the center of gravity (c. g.), then the missile is said to be statically unstable. If it coincides with the c. g., then it is said to be neutrally stable; and if it is behind the c. g., it is said to be statically stable. This is the reason that feathers are placed at the rear end of an arrow to move the c. p. aft. These three possible conditions are shown in Figure 5. 24 ~ Figure 5. 26.

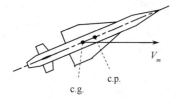
Figure 5. 24 Unstable

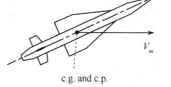
Figure 5. 25 Neutrally stable

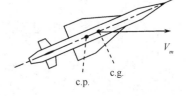
Figure 5. 26 Stable

The missiles are shown with a small incidence, which the body does not point in the same direction as the missile velocity vector V_m. In the unstable condition, any perturbation of the body away from the direction of the velocity vector results in a moment about the c. g., which tends to increase this perturbation. Conversely in the stable condition, any perturbation of the body direction results in a moment which tends to oppose or decrease this perturbation. The distance between the c. p. and the c. g. is called the static margin. Since lateral force and lateral maneuver by aerodynamic means being obtained by exerting a moment on the body so that some incidence results, it follows that if the static margin is excessive, the missile is unnecessarily stable and control moments will be relatively ineffective in producing a sizeable incidence for missile maneuver. There has to be a compromise here between stability and maneuverability. Now consider a missile whose forward speed is a constant, with a steady body and wing incidence of α and a control surface movement from the central position of δ. We will consider motion in one plane only and assume the missile is not rolling; the effects of gravity are neglected. Figure 5. 27 shows the normal force N due to the body, which acts through the c. p. when wings and rear control surfaces are assumed to be in the central position. But there will be an additional force N_c due to the control surfaces being deflected by an amount δ. Let this force act at a distance ℓ_c from the c. g. Neglecting the small damping moment due to the fact that the missile is executing a steady turn, this picture can represent dynamic equilibrium if the control surface moment $N_c \ell_c$ is numerically equal to $N \cdot x^*$ where x^* is the static margin. If $\ell_c/x^* = 10$, say then $N = 10 N_c$, and the total lateral force is $9N_c$. Note that this force is in the opposite direction of N_c. Since x^* is typically 5% or less of the body length, it is easily seen that a small change in the static margin can significantly affect the maneuverability of the missile. Thus the standard method of obtaining a large lateral force on a missile is to have a large moment arm by placing the control surfaces as far from the c. g. as possible, and by designing in a small static margin. If a missile has no autopilot (i. e. no instrument feedback), a sizeable static margin has to be allowed to ensure stability in flight, say 5% or more of the overall length. With instrument feedback zero or even negative static margins used, maneuverability improves. In conclusion, it should be noted that the overall c. p. can never be regarded as in a fixed position. The c. p. of the body in particular will vary with incidence and Mach number.

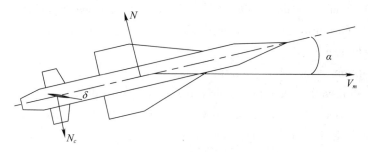

Figure 5.27 Rear control surfaces

2) Forward control surfaces

Since the main object of setting a control surface is to place it as far from the c. g. as possible, a position as far forward as practicable appears to be a logical choice. Forward control surfaces are often called "canards" named after ducks who apparently steer themselves by moving their head. Figure 5.28 shows another possible case of dynamic equilibrium. In this case, it is seen that the lateral force due to the missile as a whole now adds to the force due to the deflection of the control surface, and therefore if $\ell_c/x^* = 10$ as before, then the total normal force is $11N_c$ compared with $9N_c$ with rear controls. Also, the final sense of the total normal force is in the same sense as the control force. Therefore, canards are slightly more efficient in the use of lateral control forces. If the reader thinks that canards will automatically render the missile unstable, it will note that the canard-controlled missile has been drawn with the main lifting surfaces rather further aft to make the overall c. p. aft the c. g. The position of the overall c. p. relating to the c. g. with the control surfaces centralized is the stability criterion.

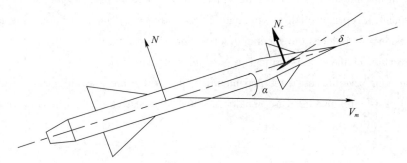

Figure 5.28 Canard control

Since canards appear to be superior to rear controls, why do we find so many missiles with rear controls? Firstly, we shall see that with feedback instruments in a control system the static margin can be zero or even negative whilst maintaining overall stability, so the difference in total normal force available can often be negligible. Secondly, there is the question of convenience in packaging as already mentioned which usually favors rear controls. Finally, and this is probably the chief reason, the downwash from canards onto the main lifting surfaces can, in many configurations, nullify any attempt to control the missile in roll. A long thin missile suffers less in this respect than a short one. There are two ways around this problem. If the missile is a twisting and steered one, the small front portion of the missile can be attached to the main body by means of a bearing which allows the rear body to rotate freely thus uncoupling the head from rolling moments induced in the wings. An alternative is to mount the wings on a collar which allows them to rotate comparatively freely around the body as already mentioned.

3) Moved wings

Using servos to move the main lifting surfaces and employ small fixed rear stabilizing surfaces is unusual. The main reason for adopting this configuration would be, for a given lateral acceleration, to minimize the body incidence. If the propulsion system is a ramjet, the air intake is likely to choke if the body incidence is large, say 15° or more. Alternatively, if the missile is a sea-skimmer with height control governed by a radio altimeter, it may be necessary to specify a maximum body incidence. In Figure 5.29 the normal force from the wings for a given incidence is shown as half of that due to wings, body and stabilizing surfaces combined, the situation is possible for fairly large fixed stabilizing surfaces. If the distance between c. p. and the wings forward of the c. g. is twice the static margin, then the steady state body incidence is equal to the wing deflection, giving a total normal force three times the original normal force from the wings. In other words, if a wing deflection of 10° produces a body incidence of 10° it will result in a final wing incidence of 20°. However, there are some distinct penalties involved in the use of moving wings. Clearly, the servos will be appreciably larger to cope with the increased inertia of the load and the larger aerodynamic hinge moments. Also, it has already been demonstrated that moving wings are an inefficient way of producing a large normal force due to the small moment arm available. Owing to the fact that the whole bending moment at the wing root has to be taken by a shaft, the wing will have to be designed much thicker around the mid chord. This not only increases the structure mass, but also it will increase the drag at supersonic speed, and pressure drag varying with the thickness-to-chord ratio squared. It is desirable to make the center section of the missile body square in cross section to eliminate a large wing-body gap when the wing is deflected, such a gap considerably reduces the generated normal force. And finally, since the control moment arm is small, the position of the c. g. is critical as a small shift will make an appreciable change in the control moment arm. Nevertheless, if the maximum g requirements are low and speed is subsonic, such as an anti-ship missile, the overall mass penalty may not be excessive if small moving wings are used.

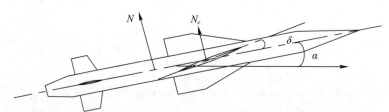

Figure 5.29 Moved wings

5.3.4 High-lift devices

A larger wing will provide more lift and reduce takeoff and landing distance, but will increase drag during cruising flight and thereby lead to lower optimum fuel economy. High-lift devices are used to smooth out the differences between the two goals, allowing the use of an efficient cruising wing, and adding lift for takeoff and landing. Figure 5.30 illustrates high-lift devices in aircraft.

Chapter 5 Flight vehicle aerodynamic configuration

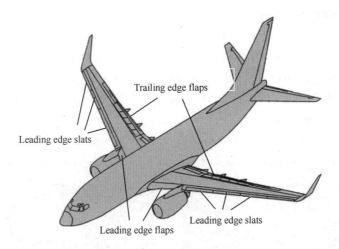

Figure 5.30 High-lift devices in aircraft

1) Flaps

The most common high-lift device is the flap, a movable portion of the wing that can be lowered into the airflow to produce extra lift. The purpose is to re-shape the wing section into more camber. Flaps are usually located on the trailing edge of a wing, while leading edge flaps are occasionally used as well. Some flap designs also increase the wing chord when deployed, and the wing area to produce more lift. Such complex flap arrangements are found on many modern aircraft. The first "travelling flaps" that moved rearward were starting to be used just before World War II, and have been followed by increasingly complex systems made up of several parts with gaps in between, known as slotted flaps. Large modern airliners make use of triple-slotted flaps to produce the massive lift required during takeoff.

2) Slats

A leading edge slat is wing aerodynamic feature of some aircraft to reduce the stall speed and promote good low-speed handling qualities. The slat can be divided into fixed type, which is also called slot, auto-moved type and manual-moved type. Air below the wing can accelerate through the slat towards the low pressure region above the wing, and exit from the slat moving parallel to the upper wing surface. This high-speed flow then mixes with the boundary layer attached to the upper surface and delays boundary layer separation from the upper surface. In this manner they allow to fly at higher angles of attack and thus reduce the stall speed. Figure 5.31 illustrates flaps and slats.

(a) Leading edge slat

(b) Leading edge flap

(c) Trailing edge flap

Figure 5.31 Flaps and slats

3) Leading edge extension

Another high-lift device is the leading edge extension (LEX) which is also sometimes referred to strake including body strake and wing strake as shown in Figure 5.32. A LEX typically consists of a small triangular fillet between the wing leading edge root and fuselage. In normal flight the LEX generates little lift. At higher angles of attack, however, it generates a vortex that is positioned to lie on the upper surface of the main wing. The swirling action of the vortex increases the speed of airflow over the wing, so that it reduces the pressure and provides greater lift. LEX systems are notable for the potentially large angles in which they are effective, and are commonly found on modern fighter aircraft.

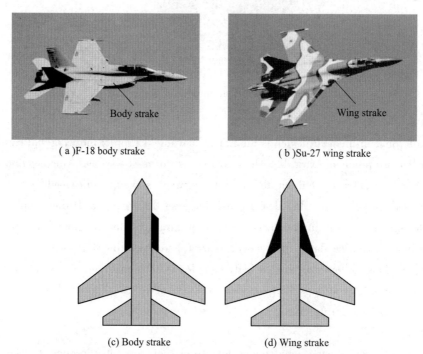

(a) F-18 body strake   (b) Su-27 wing strake

(c) Body strake   (d) Wing strake

Figure 5.32   Leading edge extension of fuselage and wing

5.4   Configuration of hyper-velocity vehicles

5.4.1   Lifting body configuration

A lifting body is a configuration in which the body itself produces lift. In contrast to a flying wing without a conventional fuselage, a lifting body is a fuselage that generates lift without the shape of a typical thin and flat wing structure. This configuration is mainly used in space-shuttle and other reusable re-entry vehicles.

The lifting body research arose from the idea of spacecraft re-entering the earth atmosphere and landing much like a regular aircraft. Following atmospheric re-entry, the traditional capsule-like spacecraft have very little control over where they land. A steerable spacecraft with wings as an aircraft could significantly extend its landing envelope. However, the vehicle's wings would have to be designed to withstand the dynamic and thermal stresses of both re-entry and hypersonic flight. A proposed solution eliminated wings altogether, the designed fuselage body itself is to produce lift. Thus the lifting body becomes reality gradually.

Space shuttle is not a pure lifting body as seen. Considering the maneuverability, the landing distance and other flight envelope reasons, the Shuttle's delta wing design was given. The design of X-33 and X-37 also adopted lifting body idea, as shown in Figure 5.33. The plan of X-33 has been cancelled by NASA in 2001. The X-37 (also known as the Orbital Test Vehicle) is an American unmanned vertical-takeoff, horizontal-landing (VTHL) space plane. The space plane's first orbital mission, USA-212, was launched on April 22, 2010 using an AtlasV rocket. Its return to earth on December 3, 2010 was the first test of the vehicle's heat shield and hypersonic aerodynamic handling. Subsequent flights gradually extended the mission duration, reaching 780 days in orbit for the fifth mission, the first to launch on a Falcon 9 rocket. The latest mission, the sixth, launched on an Atlas V on May 17, 2020.

(a)X-33　　　　　　　　　　　(b)X-37

Figure 5.33　X-33 and X-37

Some aircraft with wings also employ bodies that generate lift. The Short SC.7 Skyvan (an aircraft of America) produces 30% of the total lift from the fuselage, almost as much as the 35% each of the wings produces. Fighters like the F-15 Eagle also produce substantial lift from the wide fuselage between the wings. Because the F-15 Eagle's wide fuselage is so efficient at lift, an F-15 was able to land successfully with only one wing.

5.4.2　Blended-wing-body

A Blended-wing-body (BWB) configuration has a combination of wings and the body. The wings are smoothly blended with the body, so there is no clear boundary between those two parts. The advantages of the BWB approach are high-lift, low-drag, light-mass and large-volume. Due to the elimination of the right angles between wings and the body, BWB configuration will also help to reduce RCS, and improve stealth performance. A typical aircraft with BWB configuration is shown in Figure 5.34.

In conventional aircraft, because the wing and body are two separate parts, the infiltration area and the aerodynamic drag are large. In order to reduce the drag, designers added foreskin in the junction of the wing and body. The foreskin was not designed to withstand load, but it was hard not to be affected by aerodynamic force in flight. Thus there were many problems of foreskin such as deformation. Later, according to the advantages and disadvantages of foreskin, the researcher proposed BWB concept which meant designing and manufacturing aircraft by composing the wing and

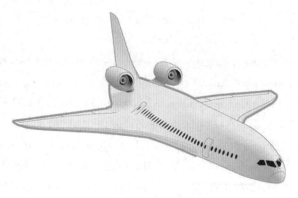

Figure 5.34 BWB configuration

body together. In the past years, the BWB technology was adopted in many high maneuverability fighters (such as the F-16, Su-27, etc.).

Both NASA and Boeing are exploring BWB designs under X-48. Studies suggest that BWB aircraft, configured for passenger flight, could carry from 450 to 800 passengers and achieve fuel savings of over 20 percent. NASA has been developing, since 2000, a remotely controlled model with a 21ft (6.4m) wingspan. X-48 is shown in the second picture of Figure 5.35.

The B-2 Spirit stealth bomber is a design which falls between classic flying wing concepts and the BWB concept. It is usually classified as a flying wing, however, as the protruding body sections are not much larger than the underlying wing shape structure. B-2 is shown in the third picture of Figure 5.35.

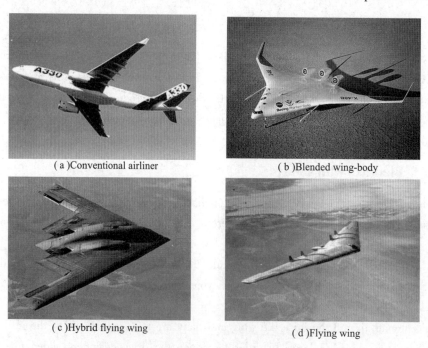

(a) Conventional airliner (b) Blended wing-body

(c) Hybrid flying wing (d) Flying wing

Figure 5.35 Spectrum of aircraft design concepts

5.4.3 Waverider configuration

A waverider is a hypersonic aircraft design that improves its supersonic lift-to-drag ratio by using the shock waves generated by its own flight as a lifting surface.

For conventional flight vehicle, the leading edges are blunt. There will be detachment of the shock wave over the blunt leading edges of the wings in hypersonic flight, and then the detachment would allow the air on the bottom of the craft to flow spanwise and escape to the upper part of the wing through the gap between the leading edge and the detached shock wave. This loss of airflow reduces the amount of lift.

Waveriders generally have sharp noses and sharp leading edges on their wings. The underside shock surface remains attached to this. Air flowing in through the shock surface is trapped between the shock and the fuselage, and can only escape at the rear of the fuselage. With sharp edges, all the lift is retained.

The configuration determines the flow field in the general design of flight vehicles. On the contrary, the configuration of waverider is derived from flow field. According to the generation method and flow field, waveriders can be divided into the following categories: caret wing, conical waverider, star body, viscous optimized waverider and so on. Figure 5.36 illustrates waverider X-43A.

Figure 5.36   X-43A waverider

The waverider remains a well-studied design for high-speed aircraft in the Mach 5 and higher hypersonic regime, although no production design has used the concept to date. X-43A is a hypersonic flight test machine developed by the Dryden Flight Research Center under the NASA. The X-43A is the record holder for the fastest speed among aircraft built by mankind that uses external intake power.

5.5   Flight vehicle configuration variation & development

Dating back to the first flight with motor of Wright Brothers in 1903, the great changes of the flight vehicle configurations have taken place. There are thousands of configurations for both missiles and aircraft.

5.5.1   Missile configuration variations

Compared to airplane configurations, missile configurations are quite simple. Some tactical missile configurations are shown in Figure 5.37.

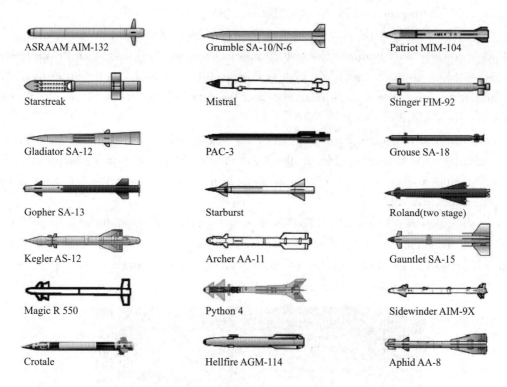

Figure 5.37 Some tactical missile configurations

1) Surface-to-surface missile

A surface-to-surface missile is a missile designed to be launched from the ground or the sea and hit targets on land or sea. Ballistic missiles and aerodynamic missiles are the main two types of missiles in this category. Ballistic missiles only have tail fins or have no wings, while aerodynamic missiles have a pair of large plane wings, which are similar to airplanes. A surface-to-surface missile may be launched from handheld or vehicle-mounted equipment, fixed installations, or a ship. They can be powered by a rocket engine or sometimes launched by explosive charges, and the launch platforms are usually stationary or slow-moving. The V1 flying bomb was the first operational surface-to-surface missile.

Some classic surface-to-surface missiles are as follows. The LGM-30 Minuteman is a U. S. land-based intercontinental ballistic missile (ICBM), in service with the Air Force Global Strike Command. In 2021, the LGM-30G Minuteman III version is the only land-based ICBM in service in the United States. The Iskander missile is the most advanced combat tactical missile-equipped by the Russian army. It can carry many types of warheads and has amazing damage capabilities. The RT-2PM2 is one of the most recent intercontinental ballistic missiles deployed by Russia, which was developed from the RT-2PM Topol mobile intercontinental ballistic missile.

2) Surface-to-air missile

Surface-to-air missiles refer to missiles launched from the ground, ships, or submarines to intercept air targets. This type of missile generally has four wings, arranged in a cross and X-shape, and there are also four rudder surfaces, the arrangement of which is the same as that of the wings. Due to the high speed and mobility of most air targets, most air defense missiles are winged missiles with an axis ymmetric layout. From the beginning of research in Germany in the early 1940s to nearly 60 years in the 21st century, three generations of air defense missiles in the world have been developed, and the fourth generation is still being developed in the 21st century.

Some classic surface-to-air missiles are as follows. The MIM-104 Patriot is a surface-to-air missile (SAM) system, the main air defense system of the U. S. Army. The RIM-161 Standard Missile 3 is a ship-based missile system used by the United States Navy to intercept short-range and intermediate-range ballistic missiles as a part of the Aegis Ballistic Missile Defense System. The Terminal High Altitude Area Defense System (THAAD), formerly the Theater High Altitude Area Defense System, is a ballistic missile defense system of the United States designed to kill short-range, medium-range, and medium-range ballistic missiles through interception and hit interception. S-300 is a series of long-range surface-to-air missile systems developed for defense against aircraft and cruise missiles of Soviet air defense forces. The HQ-2 has long been our main mid-to-high altitude air defense missile, it was an improved version of HQ-1. The HQ-9 is a medium to long-range and active radar homing surface-to-air missile. The HQ-16 is a mid-range and short-range area air defense missile. This type of missiles can quickly intercept multiple targets in all directions and can intercept supersonic anti-ship missiles and subsonic anti-ship missiles. Anti-aircraft missiles have always been improved with the improvement of the performance of the incoming targets. Therefore, the main development trend in the future is to improve the early warning and command, control, and communication systems, further enhance the system's anti-jamming capabilities, and at the same time develop air defense missiles that "can be ignored after launch".

3) Air-to-air missile

Air-to-air missiles are missiles launched from one aircraft to destroy another aircraft. This type of missile also relies on four wings and rudder surfaces to quickly change the direction of the missile for maneuvering. It grew out of the unguided air-to-air rockets used during the World War I . Air-to-air missiles are composed of guidance devices, warheads, fuses, power devices, projectile bodies, and wings. An air-to-air missile weapon system is formed with airborne fire control, launchers, and test equipment. It has the characteristics of fast response, good maneuverability, small size, lightweight, flexible and convenient use.

Some classic air-to-air missiles are as follows. The AIM-9 Sidewinder (for Air Intercept Missile) is a short-range air-to-air missile, which is the most widely used air-to-air missile in the West. The AIM-120 advanced medium-range air-to-air missile, or AMRAAM, is an American beyond-visual-range air-to-air missile (BVRAAM) capable of all-weather day-and-night operations. The PL-12 and PL-15 are China's air-to-air missiles. The missile features an active electronically scanned array radar and has a range exceeding 200km. From the perspective of operational requirements and technological progress, the next development trend and context of air-to-air missiles is relatively clear, that is, the attack range is getting larger, the anti-jamming ability is getting stronger, and the maneuvering ability is getting stronger.

5.5.2　Aircraft configuration variations

The design of a new aircraft is a very structured undertaking following a script pioneered by

Frank Barnwell in 1917. Since then aircraft performance has improved enormously by progress in aerodynamics, materials, and propulsion. The task of the aerodynamic designer is to shape the aircraft to meet the performance required for its mission; shape determines aerodynamic performance.

Figure 5.38 presents a thumbnail history of the airplane's aerodynamic development over the twentieth century, including some of the significant contributors who helped to bring it about.

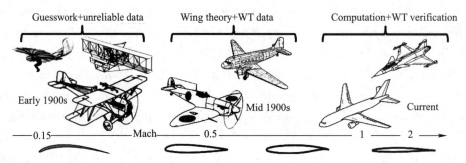

Figure 5.38　Aircraft design development

1) Early exploration era

In 1670, Francesco Lana de Terzi published a work that suggested lighter than air flight would be possible by using copper foil spheres that, containing a vacuum, would be lighter than the displaced air to lift an airship. After that, the research of lighter than air flight has been widely conducted. The typical kinds of aircraft in this stage are balloons and airships. Although airships were used in both World War Ⅰ and World War Ⅱ, and continue on a limited basis to this day, their development has been largely overshadowed by heavier-than-air craft.

By the late 19th century, two problems were identified before the heavier-than-air flight could be realized. The first was the creation of low-drag, high-lift aerodynamic wings. The second problem was how to determine the power needed for sustained flight. During this time, the groundwork was laid down for modern-day fluid dynamics and aerodynamics, with other less scientifically inclined enthusiasts testing various flying machines with little success.

Henson's 1842 design for an aerial steam carriage broke new ground. Although only a design, it was the first in history for a propeller-driven fixed-wing aircraft. After that, a few aircraft using steam engines appeared but none of them achieved controllable and stable flight. Figure 5.39 and Figure 5.40 show some typical aircraft in this era.

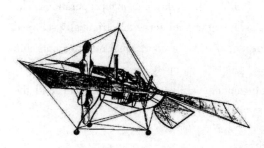

Figure 5.39　Félix du Temple's 1874 Monoplane　　　Figure 5.40　Jean-Marie Le Bris and his flying machine, Albatros Ⅱ, 1868

In the last decade or so of the 19th century, several key figures were refining and defining the modern airplane. Lacking a suitable engine, aircraft work focused on stability and control in gliding flight, as shown in Figure 5.41 and Figure 5.42.

Figure 5.41　The Biot-Massia glider, 1879

Figure 5.42　Otto Lilienthal, 1894

2) Pioneer era

The figure started with the Wright brothers, who built a wind-tunnel to obtain lift and drag data for wing sections. That was needed to estimate the engine power necessary for staying aloft. About the same time, the first mathematical models capable of predicting lift were being worked out by Ludwig Prandtl. His groundbreaking boundary-layer theory enabled engineers to better understand airflow quantitatively. During this period flight had developed as an established technology.

This period saw the development of practical airplanes and their early applications. Although full details of the Wright Brothers' system of flight control had been published in l'Aerophile in January 1906, the importance of this advance was not recognized, and European experimenters generally concentrated on attempting to produce inherently stable machines.

Figure 5.43 and Figure 5.44 show the Wright Flyer and early Voisin biplane.

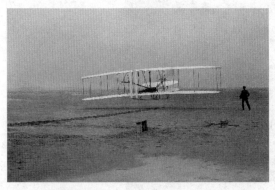

Figure 5.43　The Wright Flyer, 1902

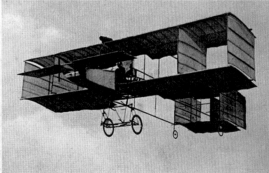

Figure 5.44　Early Voisin biplane, 1908

Almost as soon as they were invented, airplanes were used for military purposes. The first country to use them for military purposes was Italy, whose aircraft made reconnaissance, bombing, and artillery correction flights in Libya. Figure 5.45 shows Nieuport Ⅳ, operated by most of the world's air forces before World War Ⅰ for reconnaissance and bombing.

Figure 5.45　Nieuport Ⅳ

3) The World Wars era

In World War Ⅰ, the combat planes are fixed machine guns acquiring the ability to shoot each other. The years between World War Ⅰ and World War Ⅱ saw great advancements in aircraft technology. Airplanes evolved from low-powered biplanes made from wood and fabric to sleek, high-powered monoplanes made of aluminum. Many American pilots became barnstormers, flying into small towns across the country and showing off their flying abilities. The air races drove engine and airframe development—the Schneider Trophy, for example, led to a series of ever faster and sleeker monoplane designs culminating in the Supermarine S.6B, as shown in Figure 5.46.

Figure 5.46　S.6B seaplane

The theory could be combined with experiments to put more solid data into the designer's hands, thereby eliminating much of the previous guesswork. For example, the cross-sections of World War Ⅰ aircraft wings were all thin with small nose radius. The design was derived from a misinterpretation of the Reynolds number effect in the test data. In 1918 Anton Fokker saw through this error and gave his D-VII aircraft a thick airfoil with a large nose radius. That kept the airflow attached at high angles of attack and gave the plane superior climb and maneuver performance.

The next couple of decades brought us from 100 mph contraptions of wood and fabric held together by bracing wires to the sleek shapes of 400 mph World War II fighters. World War II saw a great increase in the pace of development and production, not only of aircraft but also the associated flight-based weapon delivery systems. Air combat tactics and doctrines took advantage. Large-scale strategic bombing campaigns were launched, fighter escorts introduced and the more flexible aircraft and weapons allowed precise attacks on small targets with dive bombers, fighter-bombers, and ground-attack aircraft.

The first jet aircraft to fly was the Heinkel He 178 (Germany), flown by Erich Warsitz in 1939, followed by the world's first operational jet aircraft, the Me 262, in July 1942 and the world's first jet-powered bomber, the Arado Ar 234, in June 1943. Drag is the force the aerodynamic designer most wants to reduce in every design decision. Melvin Jones' proposal for "the streamlined airplane" showed the gains achievable by cleaning up the shapes to keep the boundary layer attached. The metamorphosis was made possible by technological infusions of materials, structural design, and engine power. Figure 5.47 and Figure 5.48 show Me 262 and Arado Ar 234.

Figure 5.47 Me 262   Figure 5.48 Arado Ar 234

The airplanes that appeared in World War II were similar to contemporary piston engines, with straight, unswept wings and a wood or light alloy construction. They could already fly at supersonic speeds under certain conditions, but their controllability wa often limited at this speed. They generally used swept wings and were equipped with turbojet engines with afterburners.

4) Postwar era

After World War II, commercial aviation grew rapidly, using mostly ex-military aircraft to transport people and cargo. This growth was accelerated by the glut of heavy and super-heavy bomber airframes like the B-29 and Lancaster that could be converted into commercial aircraft.

The Korean War of 1950—1953 forced people to rethink. Facts have proved that the ceiling, acceleration, and climb rate of the first-generation fighters were not high enough. In the late 1950s, countries began to develop second-generation supersonic fighters, emphasizing the so-called "high-altitude high-speed", with a ceiling of more than 20000 meters and a maximum speed exceeding twice the speed of sound. At the same time, the need for multi-task capabilities in battlefield support was rediscovered. Some famous fighters like the Lockheed F-104 Starfighter, Russian MIG-21, English Electric Lightning, and French Dassault Mirage Ⅲ were developed in this period. Figure 5.49 ~ Figure 5.53 show some typical aircraft in this era.

Figure 5.49 D. H. Comet, the world's first jet airliner

Figure 5.50 MIG-21

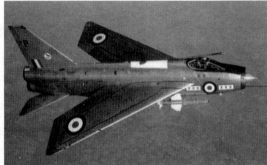

Figure 5.51 F-104

Figure 5.52 Electric Lightning

Figure 5.53 Dassault Mirage Ⅲ

The last quarter of the 20th century saw a change of emphasis. No longer was revolutionary progress made in flight speeds, distances, and materials technology. This part of the century instead saw the spreading of the digital revolution both in flight avionics and in aircraft design and manufacturing techniques. As shown in Figure 5.54 ~ Figure 5.57, types such as the Boeing 747, Airbus A320, Sukhoi Su-27, the General Dynamics F-16 Fighting Falcon are the representatives in this period.

Figure 5.54 Boeing 747

Figure 5.55 Airbus A320

Figure 5.56 Sukhoi Su-27

Figure 5.57 General Dynamics F-16

21st-century commercial aviation has seen increasing interest in fuel savings and fuel diversification, as well as low-cost airlines and facilities. Computational models for transonic flow matured, computer power, and software could support algorithmic shape optimization. Meanwhile, the new generation fighter is developed which has high maneuverability, advanced avionics systems, highly integrated computer networks, and an excellent battlefield. As shown in Figure 5.58 ~ Figure 5.61, types such as the Boeing 787, Airbus A380, the Lockheed Martin F-22, F-35, the Sukhoi Su-57, and the Chengdu J-20 are the representatives in this period.

Figure 5.58 Boeing 787

Figure 5.59 Airbus A380

Figure 5.60 J-20

Figure 5.61 F-35

5) New type of aircraft configuration

Boeing unveiled a concept in 2018 for a hypersonic airliner that could fly across oceans in two hours and have both military and commercial uses. Figure 5.62 shows this new kind of aircraft.

Figure 5.62 Hypersonic passenger plane concept

Figure 5.63 shows the Boeing Company's advanced design concept, which is a variation on the extremely aerodynamic hybrid wing body. Figure 5.64 shows Lockheed Martin's concept that uses a box wing design and other advanced technologies to achieve green aviation goals. Figure 5.65 shows Northrop Grumman's concept, which is based on the extremely aerodynamic "flying wing" design.

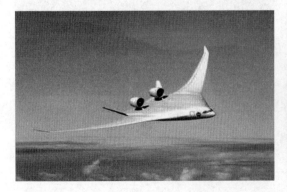

Figure 5.63 Boeing's concept

Figure 5.64 Lockheed Martin's concept

Chapter 5 Flight vehicle aerodynamic configuration

Figure 5.65 Northrop Grumman's concept

Airbus has revealed three concepts for the world's first zero-emission commercial aircraft which could enter service by 2035. These concepts are shown in Figure 5.66 ~ Figure 5.68. They represent a different approach to achieving zero-emission flight, exploring various technology pathways and aerodynamic configurations.

Figure 5.66 ZEROe turbofan design

Figure 5.67 ZEROe turboprop design

Figure 5.68 ZEROe blended-wing body design

5.5.3 Morphing flight vehicle configuration

The concept of morphing flight vehicle can be traced back to 1890, when Clement Ader of France proposed the idea of morphing flight vehicle design. During the development of the morphing flight vehicle, it can be summarized into two lines: rigid morphing flight vehicle and flexible morphing flight vehicle.

1) Rigid morphing flight vehicle

The study of rigid morphing flight vehicle mainly covers variable swept back, retractable wing, folding wing and fuselage deformation. Typical variable swept wing aircraft include the U.S. X-5 variable swept wing aircraft, the F-111 tactical bomber, the F-14 fighter, the B-1 strategic bomber, and the Soviet Union's test flight of the Mig-23 supersonic fighter and Tu-160 supersonic bomber. The variable swept-back aircraft allows for better flight performance at different speeds. In terms of retractable wing aircraft, in 1930 Ivan Makhonine of the Soviet Union designed the K-10, an aircraft with retractable wings, and successfully flew it in 1931. In 1937, the Soviet Union developed the LIG-7, a variable span ratio variant of the aircraft. For folding-wing aircraft, in 1932, Vladimir V. Shevchenko designed the IS-1 single-seat fighter, which was biplane at takeoff, and after takeoff the lower wing folded under aerodynamic force to become a monoplane craft. In 1955, the U.S. proposed the XB-70 Valkyrie bomber research program. The aircraft used a rigid variant technology with hydraulically folded variable design on both wing ends, allowing the XB-70 to have a high lift-to-drag ratio at supersonic range. For fuselage deformation aircraft, the Concorde's variable nose can be adjusted down 5° to 12° during taxiing and takeoff and landing to expand the pilot's field of view, and flat during flight to reduce drag and improve flight efficiency. The V-22 Fish Hawk tilt-rotor aircraft can fly at a high speed like a traditional fixed-wing aircraft by turning its rotor 90° forward in level flight and turning its rotor into a propeller; during maneuvers, its rotor turns upward and becomes a helicopter, capable of vertical takeoff and landing and hovering in the air. Some typical rigid morphing flight vehicles are shown in Figure 5.69 ~ Figure 5.77.

Figure 5.69　X-5 variable sweep aircraft

Figure 5.70　F-14 variable sweep fighter

Figure 5.71　Mig-23 variable sweep aircraft

Figure 5.72　Tu-160 supersonic variable sweep bomber

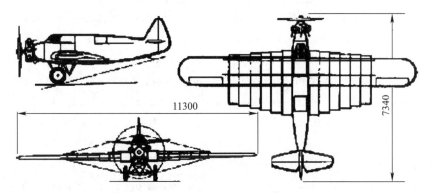

Figure 5.73 LIG-7 variable span aircraft

Figure 5.74 IS-1 folding-wing fighter

Figure 5.75 XB-70 supersonic folding-wingbomber

Figure 5.76 Concorde supersonic airliner

Figure 5.77 V-22 tiltrotor

2) Flexible morphing flight vehicle

Research on flexible morphing flight vehicles addresses the problem of smaller scale variants, including flexible skin materials, drive mechanisms, and local structural design around flexible material applications. As early as the mid-1980s, Rockwell, a U. S. company, first proposed the concept of active flexible wing design to improve the aerodynamic characteristics of aircraft. During the 1980s to 1990s, two large scale adaptive wing research programs were conducted in the U. S., namely MAW (Mission Adaptive Wing) and AFW (Active Flexible Wing) programs. The MAW program investigated variable curvature wings, using a mechanically driven system to cover the wing surface with a layer of flexible material to achieve a smooth transition between the leading and trailing edge control surfaces of the wing. The AFW program conducted experimental wind tunnel studies on a scaled-down model of the F-16 to improve the aircraft's maneuverability by deforming the leading and trailing edge control surfaces. In 2001, NASA proposed the concept of a new flexible variant of the aircraft that could change the wingspan and swept-back angle at any time during flight, and whose wing triplets could freely diverge and merge. From 2004 to 2006, DARPA funded the Morphing Aircraft Structures (MAS) program. NextGen and Lockheed Martin have each proposed two different flexible morphing flight vehicle solutions. A number of universities have also joined the ranks of research on variant vehicles—University of Florida (USA), University of Maryland (USA), University of Bristol (UK), etc. After entering the 21st century, China has paid more attention to the concept of variant vehicles and has carried out specific research on the structural design, material improvement and control analysis of variant vehicles. Relevant universities and research institutes have also launched their own research projects one after another, and have achieved many valuable results. Some typical flexible morphing flight vehicles are shown in Figure 5.78 ~ Figure 5.82.

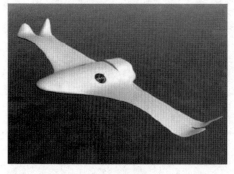

Figure 5.78   NASA flexible morphing aircraft concept

Figure 5.79   NextGen MFX-1 scheme

Figure 5.80   NextGen MFX-2 scheme

Figure 5.81   Lockheed Martinfolding wing scheme

Figure 5.82   Lockheed Martin MPUAV scheme

Questions

1. Please compare various missile configurations by combining the characteristics and usages.
2. What are the principles in selecting the configuration of nose and contraction ratio of boat-tail in missile design?
3. What's the difference between normal configuration and canard configuration on control and stability?
4. Try to describe the three kinds of control surfaces of missiles and the advantages and disadvantages of them respectively.
5. List three kinds of hyper-velocity vehicles and analyze the characteristics of them respectively.
6. Summarize the characteristics of each generation of fighters and give examples.
7. List the main morphing mode of morphing flight vehicles.

Words and phrases

aft　后部,尾部
aileron　副翼
all-moving wing configuration　全动翼布局
aspect ratio　展弦比
blended-wing-body　翼身融合体
blunt trailing edge　钝后缘
boat-tail contraction ratio　尾部收缩比
boat-tail fineness　尾部长细比
body fineness　弹身长径比
boundary layer　边界层
canard configuration　鸭式布局
caret wing　楔形翼
cargo aircraft　运输飞机
chord line　弦,弦线
cockpit　驾驶员座舱
conical　锥形的
deploy　展开
double-wedge　双楔形
double-curve　双弧形
elevator　升降舵
ellipsoidal shape　椭球形
equilibrium　平衡
flight envelope　飞行包线
flying wing　飞翼

generatrix　母线
grid fin　格栅翼
hexagonal　六角形
high-liftdevices　增升装置
hinge　铰链
hydraulic unit　液压组件
hypersonic　高超声速
infrared seeker　红外导引头
leading edge extension　边条
leading edge flap　前缘襟翼
leading edge slat　前缘缝翼
lifting body　升力体
max camber　最大弯度
max thickness　最大厚度
mean aerodynamic chord　平均气动弦长
mean camber line　中弧线
normal configuration　正常式布局
nose fineness　头部长细比
notable　显著的
orbital test vehicle　轨道测试飞行器
parabolic shape　抛物线形
pilot cabin　驾驶舱
reconnaissance-strike　侦察-打击
RCS　雷达散射截面积

re-entry 再入	strake 边条
root chord 根弦	sweep angle 后掠角
rudder 方向舵	swirling 涡流
servo 舵机	taper ratio 梢根比
scramjet propulsion 超燃冲压推进	trailing edge flap 后缘襟翼
sharp cone shape 尖锥形	triple-slotted flap 三开缝襟翼
skirt tail 裙尾	TV seeker 电视导引头
spanwise 沿翼展方向	vertical-takeoff 垂直发射
spectrum 族谱	viscous 黏性
stall speed 失速速度	vortex 涡
standard mean chord 平均几何弦长	waverider 乘波体
star body 星形体	wing span 翼展
static margin 静稳定度	wing tip chord 梢弦
steerable 易操纵的	

References

[1] 苗瑞生,居贤铭,吴甲生. 导弹空气动力学[M]. 北京:国防工业出版社,2006.

[2] 韩品尧. 战术导弹总体设计原理[M]. 哈尔滨:哈尔滨工业大学出版社,2000.

[3] 陈迎春,宋文滨,刘洪. 民用飞机总体设计[M]. 上海:上海交通大学出版社,2010.

[4] 余超志. 导弹概论[M]. 北京:北京工业学院出版社,1986.

[5] 文仲辉. 导弹系统分析与设计[M]. 北京:北京理工大学出版社,1989.

[6] 谢础,贾玉红. 航空航天技术概论[M]. 北京:北京航空航天大学出版社,2005.

[7] 郦正能. 飞行器结构学[M]. 北京:北京航空航天大学出版社,2010.

[8] 樊启发. 世界制导兵器手册[M]. 北京:兵器工业出版社,1996.

[9] 《世界飞机手册》编写组. 世界飞机手册[M]. 北京:航空工业出版社,2011.

[10] 金永德,等. 导弹与航天技术概论[M]. 哈尔滨:哈尔滨工业大学出版社,2002.

Chapter 6 Flight vehicle structures

The configuration and structure of flight vehicles represent two different aspects. Although their configurations are different, missile and aircraft have certain similarities in structures. The missile was invented after aircraft and the manufacture of missile benefited a lot from aircraft. And they have many parts to carry with the same components and similar construction forms. However, due to different payloads and functions, missile and aircraft have different internal arrangements. Following will give a brief introduction about the different arrangements and configurations of missiles and aircraft. Then the whole chapter will discuss the same components and construction forms from structure view.

6.1 Requirements of flight vehicle structure design

After the conceptual and preliminary design of the flight vehicle, comes the detail design and calculation of the structural members. The main assignment of the flight vehicle construction is to integrate all the subsystems, as well as to keep a good aerodynamic configuration so that it can fulfill the performance requirements. Usually, the requirements for flight vehicles are called tactical and technical specification. Construction design is not only to achieve the above assignment, but also to achieve a high efficiency construction which is light weight structure.

The main parts of an aircraft are airframe, control system, power plant, landing gears and airborne equipment. Figure 6.1 shows an internal arrangement or internal layout to give a brief conception about an aircraft construction. The airframe consists of wings, fuselage and tails. The wings are the main part to produce the lift. And the tails are usually located at the end of the aircraft while the fuselage is in the center. The fuselage is mainly applied to integrate the wings, tails, power plant and gears, as well carry the crew, passengers, products and weapons. The landing gears are the support facility for landing, taxing and parking. Maneuver devices include joystick, pedal, draw rod, rocker arm, and pulley and so on. These parts are the most different ones between aircraft and missiles which bring the different configurations. However, the members which construct these parts are the same. In addition to these parts, the power plant is made up engines and other attached systems. The airborne equipment includes instruments, communication, navigation, life insurance and energy supply facilities.

The main parts of missiles are warhead section case, instrument section, liquid rocket propellant storage tank or solid rocket engine case, transition section and tail section. Figure 6.2 shows different separation parts of missiles. Normally, due to the large part of the engine and specific functions of every section, the arrangement is fixed. Warhead is located in front of the missile. The instrument

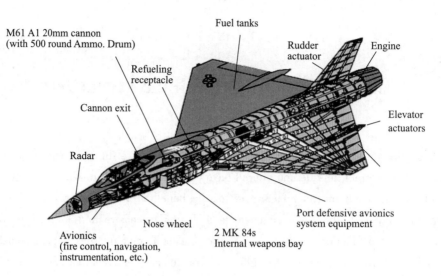

Figure 6.1 Aircraft internal arrangement

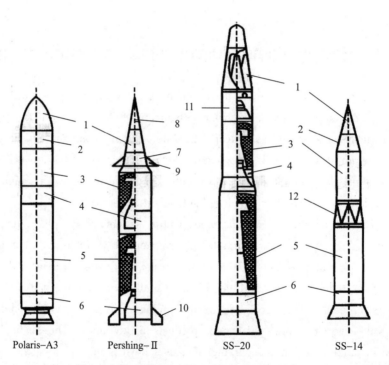

Figure 6.2 Different separation parts of missiles

1—Warhead; 2—Instrument section; 3—Second stage engine; 4—First stage mid section; 5—First stage engine; 6—Tail section; 7—Radar section; 8—Guidance and control section; 9—Air rudder; 10—Stabilizing tail; 11—Boosting section; 12—Trussed mid section.

section is applied to install control system, safety self-destruction system, flight measuring system and other important parts, so it has independent and maintenance case. Usually, the instrument section is located after the warhead. As to multiple carrier rockets, the instrument section is installed at the final stage.

6.1.1 Flight vehicle structure design philosophy

6.1.1.1 Static strength and stiffness design

Design load method is always used in static strength design. The static strength design criteria mean the ultimate load should be equal or greater than the design load.

$$P_d = fP_e \tag{6-1}$$

$$P_u \geqslant P_d \tag{6-2}$$

where, P_d is the design load, P_e is the applied load, P_u is the ultimate load, f is the safety factor.

With the increasing of flight velocity and requirements, the aeroelastic problems extrude. This requires not only enough static strength, but also enough stiffness, which means there cannot be too much deformation. Here is the stiffness design criterion.

$$\delta \leqslant [\delta] \tag{6-3}$$

where, δ is the deformation under the design load, $[\delta]$ is the tolerant deformation of the structure.

The following is the stiffness design criterion for aeroelastic problems.

$$fV_{max} \leqslant V_{cr} \tag{6-4}$$

where, V_{max} is the maximum velocity, V_{cr} is the critical flutter velocity.

6.1.1.2 Strength, stiffness and fatigue design

A large amount of analyses and researches indicate that the safety with static strength and stiffness design criteria is not enough. With the increasing demand of safe life, applications of high strength materials which are always with lower fatigue resistance and augment possibility of structure fatigue damage, in 1960s MIL-A-8866 definite fatigue design criteria, that is to say "safe life" design criteria. Engineers minimize the damage probability within the safe life operation through picking up the right material for the key fatigue parts and detailed designing for the fatigue structures. Here is the design criterion.

$$N_e \leqslant N_{sa} = \frac{N_{ex}}{f_{ex}} \tag{6-5}$$

where, N_{sa} is safe life, N_e is applied life, N_{ex} is experiment life, f_{ex} is dispersion factor, usually is 4.

6.1.1.3 Strength, stiffness, damage allowance and durability design

According to fatigue design, the flight vehicle is ideal without any crack which is impossible in the real design. Thus, damage allowance design makes sure that the construction can still bear the fail-safe load when there are cracks or damages.

As to uncheckable structure, it is ranged to degradation crack propagation structure. The criterion is as follows.

$$N_{a_0 \to a_{cr}} \geqslant 2N_e \tag{6-6}$$

where, a_0 is initial crack length, a_{cr} is critical crack length.

As to checkable structure, it is ranged to fail-safe structure. The criterion is as follows.

$$n_{fa} \geqslant \eta_e = \frac{\eta_d}{f} \quad (f=1.5) \tag{6-7}$$

where, η_e is residual strength coefficient, η_d is design residual strength coefficient, η_{fa} is fail-safe coefficient.

$$\frac{N_{ex,fa}}{2} \geqslant H \tag{6-8}$$

where, H is check interval deadline, $N_{ex,fa}$ is experiment fail-safe life.

Although the damage allowance design criterion solves the safety problems, the maintainability and economic problems stick out. The durability design is to guarantee the structure get the most reliable and economical maintenance during the whole applied life. In 1980s, the USA used "economic life" to set the durability design criterion.

$$N_{ec} = N_e = 2N_{ex,en} \tag{6-9}$$

where, N_{ec} is economic life, $N_{ex,en}$ is experiment economic life.

6.1.1.4 Reliability design

Reliability design regards different kinds of design variables as random variables, and different kinds of design criteria as random criteria.

$$R_s \geqslant R_s^* \tag{6-10}$$

where, R_s is construction reliability, R_s^* is construction reliability index.

6.1.2 Basic flight vehicle structure design requirements

Structure design should synthetically take different factors into consideration. Different functions of different parts need different requirements. However, the final object is to guarantee the best flight quality. So, there are some common basic requirements.

6.1.2.1 Aerodynamic requirement

All kinds of flight vehicles fly in the end atmosphere or through the atmosphere. When related to the aerodynamic configuration, structure design should make sure that the construction can fulfill the aerodynamic configuration accuracy. In addition, to obtain the expected life, drag characteristic, stability and maneuverability (controllability for missiles), there cannot be too much deformation.

6.1.2.2 Quality requirement

As to flight vehicle, reducing of the structure quality brings increased payload, flying velocity and flying distance. When the structure design can meet all the design criteria, the weight should be as light as possible.

6.1.2.3 Operation and maintenance requirement

Every part of the flight vehicle should be checked, maintained and repaired in certain cycle according to the regulations. Good maintenance could improve the safety and support ability. In order to achieve good maintenance, there should be reasonable separation planes, hatches and checking channels to improve the openness and accessibility.

6.1.2.4 Reliability requirement

In order to achieve the reliability index, structure design should make sure that every process takes expected reliability into consideration. Take missile body as an example, the most important to guarantee the reliability is to obtain enough strength and stiffness.

6.1.2.5 Technological requirement

To acquire good technology, engineers should synthetically take machining condition, assembly, production and timing into consideration.

6.1.2.6 Economical requirement

LLC (Life Cycle Cost) requires flight vehicle obtain the best characteristics with the lowest life cycle cost. The most efficient way to decrease the cost is the rationality of the structure design which also has influence on the operation and maintenance cost.

6.1.3 Basic flight vehicle structure design contents

The entire flight vehicle construction can be divided into certain compositions, such as tails, wings, fuselage and gears. These compositions are called components which consist of parts. Parts are the basic units of the structure, because they are not from assembling. When the parts or components have certain functions in the structure system, they are called elements, such as ribs, beams, frames and so on.

According to the structure design conditions and basic requirements, structure design engineers will first come up with reasonable design plans. Then based on them, engineers will design the components and parts in detail. The essential strength calculation and experiment will be done before the final structure drawings and technical documents. After all these, the factories can manufacture the compositions follow the documents.

6.1.4 Flight vehicle structure design methods

6.1.4.1 Qualitative design

Before 1950s, flight vehicle structure design stayed in the qualitative design stage. Engineers did the qualitative analysis based on the experience, design theory and specific requirements. Then they picked up the reasonable design plan and estimated the main section size. If the strength was not satisfied, they would improve their estimation and check again. Once the design plan determined, it cannot be changed during the following design procedure. Strength estimation was mainly based on engineering beam theory while element estimation was on material mechanics.

6.1.4.2 Quantitative design

Delta wing and slender wing are widely applied as the increasing of the velocity. As to the body, the bulkhead size grows with the heavier body, so the influence of elastic deformation must be taken into consideration. Engineering beam theory is no longer suited for the estimation which leads to accurate calculations. After 1960s, FEM (Finite Element Method) has been developed which makes the quantitative design possible. Besides, with the improvement of computers, finite element analysis and structure optimization design which cost a lot in computation become mature.

6.1.4.3 Intelligent design

Because flight structure is a highly static indeterminacy system, finite element analysis is not accurate enough to get the optimization solution. Then structure intelligent design becomes popular recently. The basic theory of intelligent design is topology optimization design. The initial stage of intelligent design automatically determines the element geometry size through finite element methods and structure optimization design, while the highest stage which is still under researches automatically gives the design plan and structure size based on the condition and requirements.

6.2 Wing

The wing is one of the most important components in a flight vehicle. The whole characteristic of the flight vehicle is determined for the most part by wing design. This part will mainly introduce wing stressed members, load transfer analysis and typical wing structure form.

6.2.1 Function and loads

6.2.1.1 Wing function

1) Provide lift

The main function of the wing is to provide lift which balances the gravity of the flight vehicle.

2) Improve flight performance and maneuverability

In addition to the aileron and spoiler for lateral control in the leading edge, there are also variety of lift device such as flap and slat in both leading and trailing edge to improve flight performance and maneuverability.

3) Ensure controllability and stability

Controllability and stability can be ensured through wing geometric parameters and distribution adjustment which could make sure of the center of pressure location and change.

4) Install engines, weapons, and be as the fuel tank

Usually, the engines and some weapons are installed under the wings of an airplane. And the wing is to be as the fuel tank.

6.2.1.2 Wing design requirements

Aside from the whole requirements of the flight vehicle, wing design should meet the following ones: lift-drag ratio must be ensured, the lift coefficient increment of the lift device should be larger, the change of stability, controllability and aerodynamic characteristics from subsonic to supersonic is better to be decreased, the heat transferred into the structure must be avoided, the storage tank should be as large as possible.

6.2.1.3 External load of wing

1) Aerodynamic load

Aerodynamic load is a distributed load directly acting on the wing skin in both suction and pressure ways. This will lead to lift, which is the main external load of wing, and drag. Different wings have different aerodynamic loads situation. But once the total lift is determined, the span wise and chord wise lift distribution can be found, as shown in Figure 6.3.

2) Wing structure mass force

Wing structure mass force, including gravity and inertia force, acts on the whole wing cubage. The magnitude and distribution depend on the structure mass distribution. Due to its small

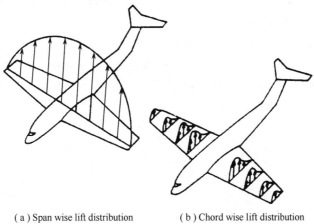

(a) Span wise lift distribution   (b) Chord wise lift distribution

Figure 6.3   Wing lift distribution

value, it is always estimated in proportion to the chord length.

3) Concentrated load from other components

Other components, such as gears, engines, aileron, flap, fuel tank and weapons, are generally connected with wing structure through finite connection points. Therefore, all the loads acted on the components are transferred as concentrated load to the wing. Only the fuel load is a distribution way as an expectation.

The above loads can be classified into two types; one is the distribution load and the other is concentrated load. The simplified external wing loads are shown in Figure 6.4. As seen from the figure, the wing can be regarded as a beam fixed in the body. However, different from engineering beam, it has its own special circumstances.

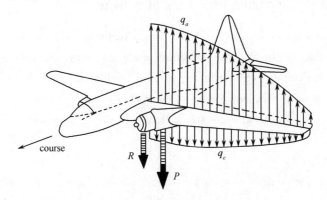

Figure 6.4   Simplified external wing loads

q_a—Distributed aerodynamics; q_c—Distributed wing mass force;

P—Concentrated loads from engine and other parts; R—Support reaction from body.

Generally, engineering beam's length size is much larger than width and height size. But the wing chordwise size has the same order of the spanwise size. So, the chordwise load distribution is also important to analysis.

The fixed ways of the wing are much more complicated than engineering beam. The elastic impact must be taken into consideration during accurate calculation.

6.2.1.4 Internal force of wing

When transferring the load from wing to body, it will cause the internal forces inside the wing structure, such as shear Q, bending moment M and torque M_t, described in Figure 6.5. The direction of axis z is the spanwise while the direction of axis x is the chordwise. The direction of axis y is perpendicular to the $x \circ z$ plane.

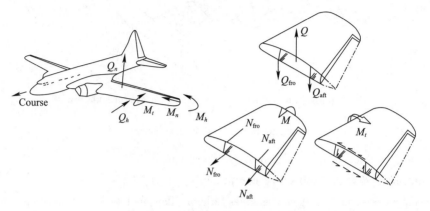

Figure 6.5 Internal forces in the wing

The shear has two components along with axis x and y, which are respectively vertical shear Q_n and horizontal shear Q_h. They also cause the vertical bending moment M_n and horizontal one M_h. Because of the distance between force origin and the center of gravity, there is also torque M_t. Due to the small value of Q_h and M_h, we only analyze the action of Q_n and M_n, and the subscript n is left out.

6.2.2 Main load-bearing members of a wing

Wing structure belongs to thin-wall structure form. As given in Figure 6.6, it mainly consists of skin and skeleton structure. In skeleton structure, there are spar, stringer and longitudinal wall (web) as longitudinal members, while ordinary rib and reinforcing rib as lateral members. They are connected with other wing panels and bodies at the root. The basic function of these members, also mentioned as elements before, is to form and maintain the wing appearance so that the wing can produce aerodynamic and bear the external loads.

6.2.2.1 Skin

The direct function of skin is to form the streamline form appearance. When the aerodynamic force acts vertically on the surface, the skin transfers the loads to the longitudinal and lateral members. Generally, skin bears parts of the loads. The composition of skin and spar or web formed the box-shaped thin-wall structure can bear the torque. The wall panel consisting of skin and stringer can bear the axial force caused by bending moment. The simplest and commonly used is dual skin, while carbon fiber or steel, titanium alloy skin are used in hypersonic flight vehicle.

As seen in Figure 6.7, the wall panel consisting of skin and stringer has two types, composite and integral. Skin in some structure form is very thick, so it is usually constructed as the integral form. In this situation, skin is the only element bearing the bending moment. Integral wall panel

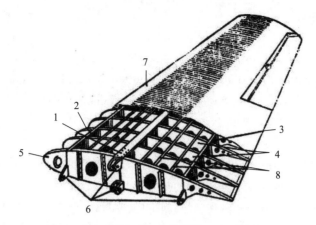

Figure 6.6　Typical wing load-bearing members

1—Spar; 2—Front longitudinal wall; 3—Back longitudinal wall; 4—Ordinary rib;
5—Reinforcing rib; 6—Butt joint; 7—Dural skin; 8—Stringer.

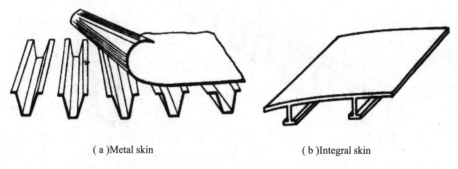

(a) Metal skin　　　　　　　　(b) Integral skin

Figure 6.7　Skin

reduces the number of connections, improve the air tightness of the entire fuel tank and obtain the light-weight wing with smooth surface.

In addition to integral wing panel, sandwich skin is also widely used. Shown in Figure 6.8, sandwich skin is composed of two sheets and foam or honeycomb sandwich. This structure form can reduce the weight, improve the stiffness and surface characteristics, and have good heat insulation, anti-shock performance.

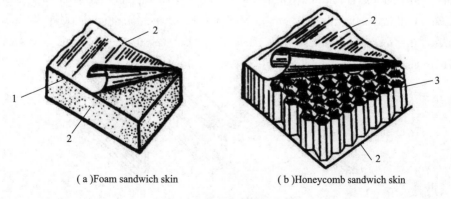

(a) Foam sandwich skin　　　　　　　　(b) Honeycomb sandwich skin

Figure 6.8　Sandwich skin

1—Foam sandwich; 2—Sheet; 3—Honeycomb sandwich.

6.2.2.2 Stringer

Stringers are the longitudinal slender bars in the wing structure. They are usually fixed in the ribs and connected with the skin so they can brace it. The stringer is one of the important members in longitudinal skeleton structure. It mainly bears the axial force caused by bending moment and shear caused by aerodynamics. These forces are determined by the form of the wing structure, and also the section shape and area of the stringer.

6.2.2.3 Spar

As seen in Figure 6.9, a spar is built up by a web and a flange with H shape or trough-type cross section. The spar is simplified bearing member. The flange supports the compression axial force caused by bending moment while web bears the shear and shear flow caused by torque. Spar is the main longitudinal bearing member so that most bending moment acts on it. Most spars are connected to a wing panel or a body at the root.

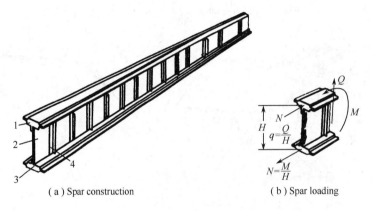

(a) Spar construction (b) Spar loading

Figure 6.9 Spar

1—Upper flange; 2—Web; 3—Down flange; 4—Reinforcing pillar.

6.2.2.4 Longitudinal wall

Longitudinal wall has the similar construction with spar, but has weaker flange. It is always hinged with other parts at the root. Longitudinal wall cannot bear the bending moment, but mainly is used to support and transfer shear. Together with skin and other webs, it can construct the close space which can bear the shear flow caused by torque. It also supports the skin, and improves the anti-buckling of it. Normally, there is no lightening hole in the longitudinal wall. To improve the critical stress, the profile is utilized. The back wall also has the function of blocking out the interior. The construction of longitudinal wall is given in Figure 6.10.

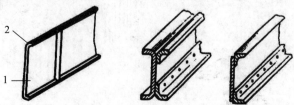

Figure 6.10 Longitudinal wall

1—Web; 2—Weak flange.

6.2.2.5 Rib

As illustrated in Figure 6.11, the main function of ordinary rib is to maintain the cross-section shape of the wing and transfer the local loads from stringers to spars and skin. Rib is usually connected with skin and stringers. When bearing the aerodynamic, it gives the vertical support to skin and stringers through its own stiffness. At the same time, the circum of the rib is connected with skin and webs which can provide the shear flow support.

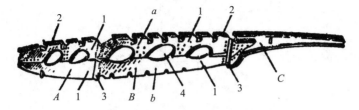

Figure 6.11 Rib

1—Web; 2—Circum crimp; 3—Crimp to web; 4—Lightening hole;
A—Front section; B—Midsection; C—After section;
a—Upper section; b—Down section.

Aside from the functions of ordinary rib, reinforcing rib also bears the concentrated force and moment caused by gear, engine and so on. It will distract the loads to skin, spars and walls. Reinforcing rib is usually used in the nonsequential place and opening of the wing.

6.2.3 Load transfer analysis

Figure 6.12 illustrates a typical load transfer block diagram of a straight wing to give a brief introduction of how aerodynamic is transferred through all the members.

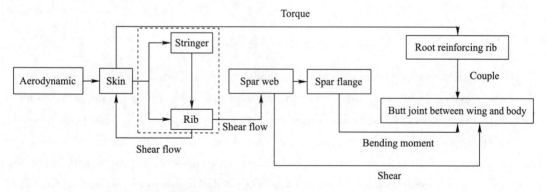

Figure 6.12 Typical load transfer block diagram of straight wing

Aerodynamic acts on skin which is supported by stringers and ribs. As shown in Figure 6.13 (a), the skin directly bears the aerodynamic in both suction and pressure ways. At this time, the skin is tensed (seen in Figure 6.13 (b)). And this tension stress is transferred to stringers and ribs through tensed rivets in lateral distribution form.

A stringer is a spreader beam supported on the rib (seen in Figure 6.14). It bears the distributed loads from skin and transfers them to ribs in smaller concentrated load R_r through the connections between skin, stringers and ribs.

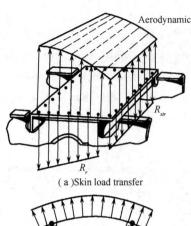

(a) Skin load transfer

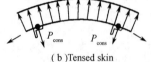

(b) Tensed skin

Figure 6.13　Initial aerodynamic transferred from skin to ribs and stringers

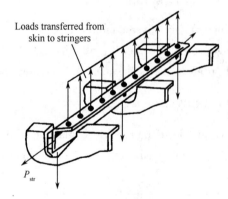

Figure 6.14　Loads transferred from stringer to rib

Rib webs are connected to spar webs in riveting (wed, cementing) ways, while flanges are connected with skin surrounding. These connections limit ribs from moving up and turning relative to the center of gravity. So, it will transfer the load from a stringer and skin to a spar and back to skin in shear flow.

The shear flow from ribs to spar webs causes the accumulative amphitheatre multistep shear along the spar webs. During the accumulating of the shear flow from webs to flanges, wall panels bear the axial force and bending moment. This time, axial force is distributed in proportion flexural rigidity of the longitudinal components.

The shear flow from ribs to skin closed room will cause accumulative amphitheatre multistep torque along ribs acting on the closed room composed of skin and back wall. The closed loop shear flow from No. i rib to skin can be simplified as $\Delta M_{ti} = \Delta q_{ti} \cdot \Delta 2F_{icont}$. It will cause the turning of the cross sections relative to the center of gravity. Therefore, it finally brings torsion deflection.

6.2.4　Wing structure form

The members which bear the forces and moments constitute the main load supporting system. Other components transfer the loads to the main load supporting system, and form the integral load supporting system with it. So-called wing structure form consists of the main load supporting system. It has different force forms, according to the three types: frame-skin construction, integral panel and sandwich construction. As to missiles, there are two more types: grid fin and folding wing.

6.2.4.1 Frame-skin construction

Frame skin construction is the so-called thin-wall construction. At the very beginning, with the increasing velocity and load, flight vehicles use the hard skin to reach higher local and global stiffness. First type is called beam structure, of which the skin is thin and can only bear the torque. With skin getting thicker, the stringers which support the skin reach higher bending bearing. So, the reinforced wall consists of stringers and skin becomes the main bending member which constructs the single block structure. When the skin gets much thicker, the stringers are replaced by longitudinal walls. At this time skin becomes the only bending element which forms the multi-wall structure. According to the bending elements, frame-skin construction can be divided into three types: beam structure, single block structure and multi-wall structure.

1) Beam structure

The main characteristic of beam structure is that the skin is very thin, which is usually made by light alloy or carbon-fiber-reinforced composite. The longitudinal spar is normally strong while the stringer is weak. The cross-section area of spar flange is much larger than the one of stringer, so most bending moments are supported by spar flange. According to the number of spars, beam structure can be divided into single beam, double beam and multi-beam structure.

2) Single beam structure

In single beam structure, spar is usually located in the highest place of the wing profile so that it can reduce the wing weight (seen in Figure 6.15). The bonding joint at the wing root is strong enough to transfer the bending moment and shear to the connecting joint to the body reinforcing frame. There is also one or two longitudinal walls to form the torsion closed room. The root of the longitudinal wall is hinged with the body so that the torque M_t will be transferred to hinge joint in shear ways.

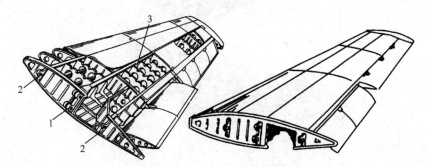

Figure 6.15 Single-beam wing structure with front and back walls
1—Fixed joint; 2—Hinged joint; 3—Spar.

3) Double beam structure

In double beam structure (given in Figure 6.16), the front spar is located at 20%~30% chord length while the back spar is at 60% ~ 70% chord length. There are bonding joints at the wing root. The cross-section area, profile height and inertial moment of the front spar are larger than the ones of the back spar so that it will share most shear and bending moment.

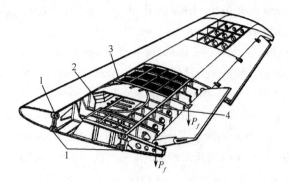

Figure 6.16　Double-beam wing structure
1—Fixed joint; 2—Front spar; 3—Reinforcing rib; 4—Back spar.

4) Multi-beam

In multi-beam structure, the weights of the members which bear the axial force caused by bending moment occupy 50% of the entire weight. To decrease the structure weight, it is better to have more bending moment supporting members and less non-bending bearing members, such as ribs. Multi-beam structure not only solves the light-weight problem, but also improves the torsion stiffness and prevents the aero elastic dangers which are very important to small aspect ratio flight vehicle.

Figure 6.17 gives three types of multi-beam wing structure of military airplane. When the skin has enough stiffness, reinforcing rib can be used in the concentrated forced location instead of

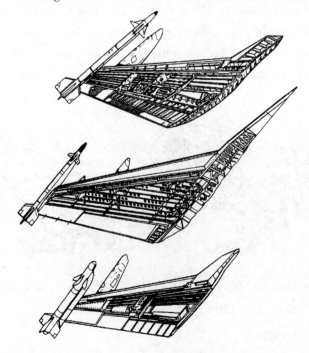

Figure 6.17　Multi-beam wing structure

ordinary ribs. When skin becomes thinner, using serried spars or longitudinal walls can bring high stiffness and light weight. As spars and longitudinal walls have the same function, multi-beam structure can also be regarded as multi-wall structure.

5) Single block structure

Single block structure (shown in Figure 6.18) has thick skin, together with stringers and spar flanges, constituting the wall panel which can bear the axial forces. There are serried stringers in the longitudinal direction, and their cross-section areas are close to the ones of spars. Skin panel together with spars or walls constitute the closed box which strengthen the torsion rigidity.

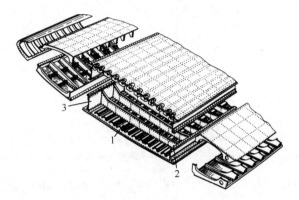

Figure 6.18 Single block wing structure
1—Stringer; 2—Rib; 3—Wall or spar web.

There are several merits about single block structure. The skin deformation under aerodynamic is small, and the flow quality is high. Materials are distracted in the outer edge of the profile which brings high bending, torsion stiffness and strength. The security reliability is higher than the one of beam structure.

On the other hand, single block structure is complicated. The structure around the openings must be strengthened to compensate the bending capability. While connected to the middle wing or body, joints must be distracted in circum.

6.2.4.2 Integral panel

To decrease the drag, thin skin is always used, however, to get higher bearing capability, thick skin is necessary and frame-skin construction is difficult for assembling. So integral panel which is machined by an entire blank is used to construct a wing. The integral panel consists of several huge integral members, such as integral skin wall panels, integral spars and integral ribs. As shown in Figure 6.19, skin and other members constitute the up and down wall panels, and then to assemble an integral wing through riveting.

Integral panel is easy to achieve variable thickness. The skin material is far from the center of the wing profile, so it has good bearing effect and high strength and stiffness. Integral panel has simple construction, smooth surface, good aerodynamic configuration and lower production cost.

6.2.4.3 Sandwich construction

Sandwich panel is the main bearing element which uses the face plate to bear the loads and gets support from sandwich. This structure has higher stiffness, strength and good configuration which can

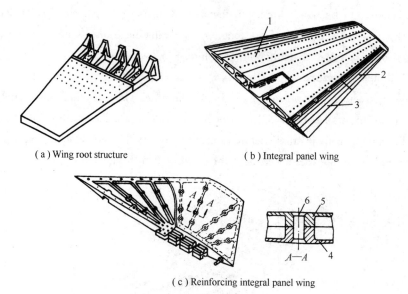

Figure 6.19 Integral panel wing structure
1—Integral panel; 2—Flap; 3—Aileron; 4—Down panel; 5—Upper panel; 6—Rivet.

support bigger local aerodynamic. There is adiabatic material between the two plates which can protect the inner facilities. Although this construction has fewer elements and assembling, the manufacturing is complicated especially at joints. Figure 6.20 shows the honeycomb sandwich wing.

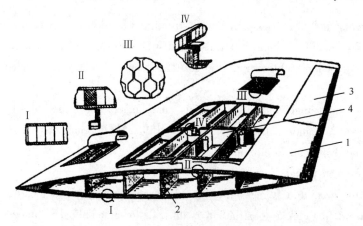

Figure 6.20 Sandwich construction
1—Honeycomb sandwich skin; 2—Longitudinal wall; 3—Aileron; 4—Rib.

6.2.4.4 Grid fin

Modern aircraft has to meet the diverse complex requirements and sometimes contradictory requirements. Traditional flat-wing has become increasingly unable to meet the design requirements, so designers should study and use new type of lifting surface. As an innovative lifting surface, grid fin is one of them.

Grid fin is a spatial system composed of external frame and many internal thin lifting-surfaces. The lift generated by grid fin is usually much larger than the ordinary wing under the same volume.

There are some advantages of grid fin. Because the chord of grid fin is short, the displacement of pressure center is small, the same as the need of hinge moment. The grid fin performs better than ordinary wing in supersonic environment and the ratio of strength and mass is also larger.

The disadvantages of grid fin are that the resistance and radar cross section (RCS) is bigger. Figure 6.21 shows the grid fin, and Figure 6.22 gives the grid fin of a missile.

Figure 6.21　Grid fin

Figure 6.22　Grid fin of a missile

6.2.4.5　Folding wing

In the wing family of missiles, the folding wing is a special type. The folding wing is connected to the body by folding mechanism and can be folded in the body or on the surface of body before missiles launch.

A missile with folding wing has small lateral size which makes it easier to be launched by using box-type. At the same time, with the reduced volume, the carrying capacity of vehicles is greatly increased.

There are two main folding patterns, which are lateral folding pattern and longitudinal folding pattern. The wrap-around fin (Figure 6.23) is a typical one of the former pattern. The famous missile calling BAT (Figure 6.24) adopted both the former and the latter pattern.

Figure 6.23　Wrap-around fin

Figure 6.24　BAT

6.2.5　Connection

6.2.5.1　Butt joint between wings

There are process separation plane and design separation plane in wing structure. The former one is set for easy processing and assembling, so it usually uses non-detachable connections. The latter one is designed for convenient service, maintenance and transportation, so detachable connection is applied. Sometimes, these two separation planes are the same and the joint in the connection is called butt joint.

The butt joint between wings has two types, one is centralized butt joint, and the other is decentralized butt joint. Beam structure transfers the loads in concentrated axial force through spars, so it usually applies the centralized butt joint. Figure 6.25 shows tab butt joint applied in beam structure. Single block structure transfers the bending moment in distributed axial force through up and down wall panels. So decentralized butt joint is applied. There is comb section material joint as surrounding frame, multiple joints as surrounding frame and so on. Figure 6.26 shows comb section material joint as surrounding frame.

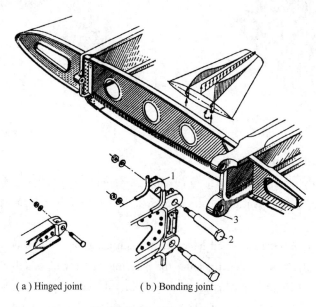

(a) Hinged joint   (b) Bonding joint

Figure 6.25 Tab butt joint
1—Double tabs; 2—Bolt; 3—Single tab.

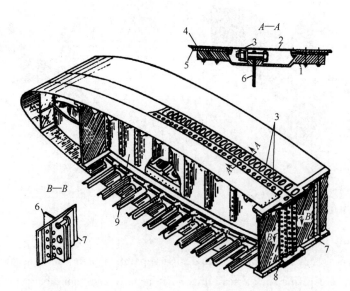

Figure 6.26 Comb section material joint as surrounding frame
1—Butt Joint; 2—Detachable plane; 3—Bolt trough; 4—Skin; 5—Gasket; 6—Rib web; 7—Spar web;
8—Web butt joint angle of section; 9—Reinforcing rib.

6.2.5.2 Connection between wing and body

The connection between the wing and the body depends on the wing vertical location. Figure 6.27 gives certain wing vertical locations. As shown in Figure 6.27 (c), if mid-wing has central wing penetrating the body, the wings are connected to the body in sides. Figure 6.27 (d) shows the mid wing with central wing penetrating the body. As shown in Figure 6.27 (a) and (b), most aircraft use high wing and low wing which could contain enough internal space.

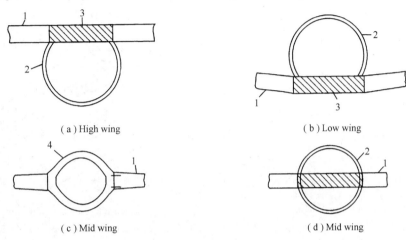

Figure 6.27 Wing vertical location
1—Wing; 2—Reinforcing bulkhead; 3—Penetrating wing; 4—Forging frame.

The above connections are mainly applied in aircraft. As to missile connections between the wing and the body, they have their own characteristics. Normally, the wing and the body are manufactured separately and assembled with detachable joint. According to the bearing type, the joint can be divided into main joint and auxiliary joint. Main joint has three transfer types: centralized, decentralized and trade off. Figure 6.28 ~ Figure 6.33 show the six types of joints, among which tab joint and axial joint are centralized transferring joint, plug-in joint, disk joint and dovetail joint are decentralized joint, and multiple tenon joint is trade off joint.

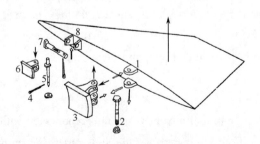

Figure 6.28 Tab joint
1—Main joint; 2—Bolt; 3—Booster joint; 4—Split pin;
5—Axle pin; 6—Booster auxiliary joint;
7—Small axis; 8—Wing auxiliary joint.

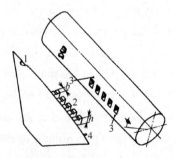

Figure 6.29 Multiple tenon joint
1—Front auxiliary joint; 2—Main joint;
3—Inclined screw hole; 4—Aft auxiliary joint.

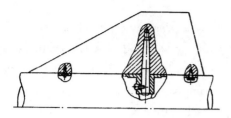

Figure 6.30 Axial joint

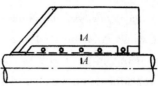

Figure 6.31 Plug-in joint

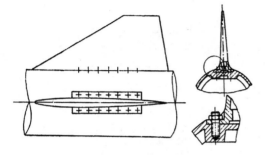

Figure 6.32 Disk joint

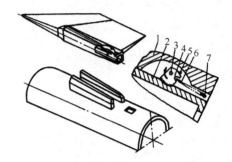

Figure 6.33 Dovetail joint
1—Wing; 2—Body; 3—Location screw; 4—Fixture block;
5—Limiting screw; 6—Spring; 7—Pin.

6.3 Body

6.3.1 Function and loads

6.3.1.1 Body function

Body or fuselage is an important part of the flight vehicle construction. In aircraft, it is normally called fuselage. It has three functions. Firstly, airfoils, engines, solar wings, antenna, gears and any other parts are connected through the body so that an integrated flight vehicle can bear all the loads to form a balance body. Secondly, it is used to carry the effective loads, such as aircrew, passengers, fuel, weapons and facilities. Different types and functions require different inner arrangements. Thirdly, cabins protect the inner environment during launching, flying and reentering.

6.3.1.2 External load of body

1) Concentrated load from any parts

Any parts, such as airfoils, engines, gears, store tank and auxiliary fuel tank, connected with the body will transfer loads to it. Different parts have different transferring form and values, but normally they are all transferred in concentrated load.

2) Mass force

Any inner loading and the body itself have their own mass force which has both concentrated and distributed form. Suppose G_i is the mass force of i loading, $G_i = m_i g \cdot n_i$. Where, m_i is the mass of i loading, g is gravity acceleration, $n_i = n_{y0} \pm (\varepsilon_z x_i / g)$ is the overload coefficient. Take the airplane in

Figure 6.34 as an example, the distributed load q_j of the body mass can be estimated as follows.

$$q_j = \frac{m_j g n H_j}{S_{jc}} \qquad (6-11)$$

where, m_j is the body mass, S_{jc} is the projected area of the body side, H_j is the body height, n is the overload coefficient at the center of mass in every design situation.

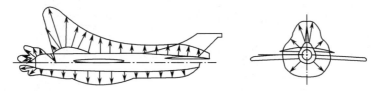

Figure 6.34 Aerodynamic distribution on body

3) Aerodynamic

The air flow around the body is basically symmetrical streamlined body. The aerodynamic on the symmetry plane is much smaller than the mass force. When taking a sudden turn or sideslip, aerodynamic should be considered in the side load of front body, while no consideration in the back body. The local aerodynamic becomes bigger in the head or parts with high camber, so aerodynamic should also be taken into consideration in these parts. But there is no effect to the global load. If the plane is BWB and there is central wing in the middle of the body, the aerodynamic can be considered as wing situation. In fact, aerodynamic can be balanced by itself (shown in Figure 6.34).

4) Pressurize load

Pressurize load can be balanced by itself so that it will not affect the global load. But it is the important design point in pressurize cabin design. During the flight, the pressurize load changes repeatedly, so it is an important part of fatigue load.

5) Vibration load

Engines can cause strong random vibration during the flight. The vibration during launching can lead to deflective launch and aerodynamic shimmy. So, it is necessary to adjust the natural frequency to avoid coupled vibration.

6) Impact load

As to manned spacecraft, there will be strong impact load during landing. In order to keep astronauts safe, buffer measures should be applied.

6.3.1.3 Internal force of body

Under lateral load, all loads acting on the body can be balanced at the joints between body and wings. So, the body can be regarded as an extensional beam supported at the joints. All the external loads under any design situation lead to the shear, bending moment and torque both on the vertical symmetric surface and horizontal plane. Figure 6.35 gives the internal force of one body which indicates the characteristics as follows.

(1) The reaction at the fixed joint between wing and body may be larger than the lift l_w, such as sweep wing $R_1 > R_w$, T-shape sweep tail $R'_z > L_H$. this requires to strength the body structure.

(2) Engines at the back body will increase the bending moment M_z. The dotted line in Figure 6.35

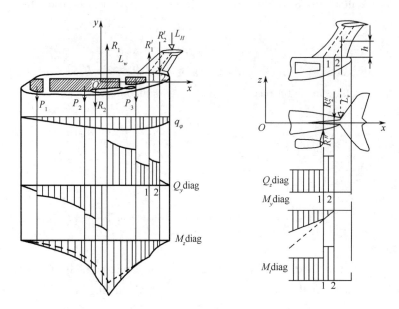

Figure 6.35 Internal force of body

shows bending moment without engine at the back body.

(3) Torque M_t will grow with higher tail. Vertical aerodynamics L_v will cause the shear and bending moment on Oxz plane and torque about the x axis.

(4) When the mass distribution along x axis is dispersive, the body length and M_z will increase.

6.3.2 Main load-bearing members of body

Generally, body construction consists of longitudinal members, lateral members and skin. The function of each member is similar to the member of wing.

6.3.2.1 Skin

Body skin has the same function as wing skin. In global load situation, skin bears the vertical and horizontal shear and torque. At the same time, the reinforced wall made up of skin and stringer bears the axial force. From construction side, skin constitutes the configuration of the cabins and keeps the surface smooth.

6.3.2.2 Stringer and spar

Stringers and spars are both longitudinal members of the body. In stringer construction, stringers and skin constitute reinforced wall to bear the axial force caused by bending. In addition, stringers support the skin to increase the instability critical stress during bending and torsion. In spar construction, spars with larger cross section area bear the axial force caused by bending. Spar can also be the reinforced component at the opening to bear the concentrated load. The stringer in this construction is similar to the one in the wing construction; it bears the aerodynamic acting on the skin and transfers it to the bulkhead.

6.3.2.3 Bulkhead

According to the function of the bulkhead, it can be divided into ordinary bulkhead and reinforced bulkhead. Ordinary bulkhead maintains the cross-section shape and fixes the skin and stringers. It mainly bears the aerodynamic transferred from skin to body and distributed pressure q_1 caused by bending deformation (shown in Figure 6.36). q_1 is the balanced force caused by itself on frame plane. As given in Figure 6.37, when asymmetrical aerodynamic acts on the bulkhead, the rivets in shear around the bulkhead will provide the react shear flow to balance the load. Bulkhead should have higher stiffness to support its own lateral bending. They typical ordinary bulkhead construction is shown in Figure 6.38, it has two flanges and a web on the cross section.

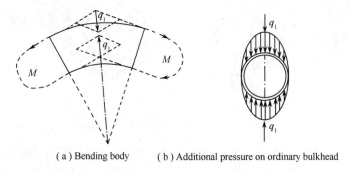

(a) Bending body   (b) Additional pressure on ordinary bulkhead

Figure 6.36   Additional pressure on ordinary bulkhead

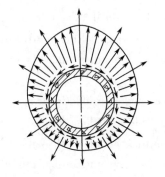

Figure 6.37   Asymmetrical pressure

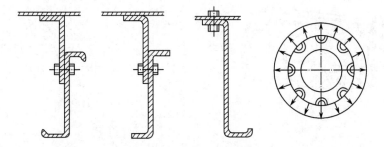

Figure 6.38   Typical ordinary bulkhead construction and cross section

6.3.3   Body structure form

6.3.3.1   Semi monocoque body structure
1) Trussed body structure

Trussed structure is a bearing space truss consisting of two vertical and two horizontal plane trusses (shown in Figure 6.39). The truss is used to be tensed and compressed, while the skin is used to bear the local aerodynamic. The main supporting components of this structure form are spar, pillar, diagonal bracing, transverse bracing, lanyard and skin. Normally, the truss is statically determinate. The internal force can be determined through structural mechanism.

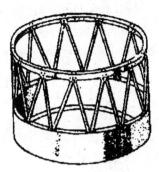
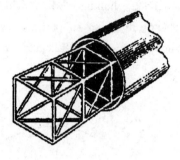

Figure 6.39 Trussed body structure

Trussed body structure was widely used during World War I and World War II, such as Yak-1 and MiG-3. Nowadays, it is still applied in small piston airplanes and some helicopters. Because this structure uses hinged joint, it can work well under heat. Thus, some space vehicles also use trussed structure.

However, with the increasing of the velocity, the vehicle needs an upper fairing to make the configuration more streamlined. Thus, the weight of the vehicle will increase more. And the usage of the inner space of the body is not as good as other frame-skin structure.

2) Stringer body structure

Stringer body structure has serried and strengthened stringers, and thick skin. The thin-wall which is made up of stringers and skin bears the axial force caused by bending, while skin bears the shear flow caused by shear and torque, shown in Figure 6.40. With high bending and torsion stiffness, the structure is light. But there cannot be large opening in the skin otherwise it is difficult to strengthen.

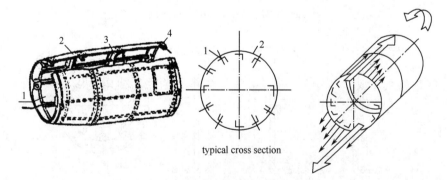

typical cross section

Figure 6.40 Stringer body bearing form
1—Stringer; 2—Skin; 3—Bulkhead; 4—Joint.

3) Braced girder body structure

As shown in Figure 6.41, braced girder has stronger spars with large cross section. But the stringers are very weak, even sometimes they are non-sequence. Besides, the skin is thin. Normally, this structure form has four longitudinal spars. When arranging the spars, not only the best bearing situation should be taken into consideration, but also openings and concentrated forces. The axial force caused by bending is mainly supported by spars. Skin and stringers only bear small part of the axial force. And the entire shear is supported by skin. Braced girder body structure is mainly applied in the front airplane body and main bearing construction of satellites.

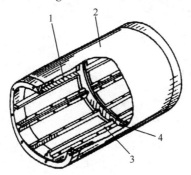

Figure 6.41 Braced girder body structure
1—Stringer; 2—Skin; 3—Braced girder; 4—Bulkhead.

4) Beam body structure

As shown in Figure 6.42, beam body structure consists of skin, frames and longitudinal spars. Most loads act on spar while skin only bears the shear. This structure form is applied in large axial concentrated force situation with big openings. Because the skin only bears the shear, beam structure is heavier than braced girder structure.

6.3.3.2 Monocoque body structure

Monocoque structure is a thick cylindrical shell consisting of skin and a few bulkheads shown in Figure 6.43. There is no longitudinal component and the skin is very thick or uses sandwich structure. The skin is the main bearing member to support bending moment, shear, axial forces and torque. Due to thick skin, this structure has higher torsion stiffness. But it is difficult for openings. Nowadays, this structure is only applied in missiles and space vehicles.

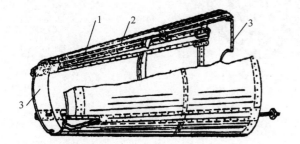

Figure 6.42 Beam body structure
1—Skin; 2—Spar; 3—Bulkhead.

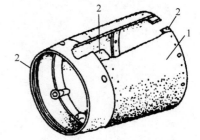

Figure 6.43 Monocoque body structure
1—Skin; 2—Bulkhead.

1) Thick skin integral body structure

This structure consists of thick skin and bulkheads without longitudinal components given in Figure 6.44. The bulkhead only maintains the body shape and connects every part, while the skin bears all the loads. When the cross-section size grows, the critical stress decreases. So only increasing the skin thickness can ensure the bearing capability.

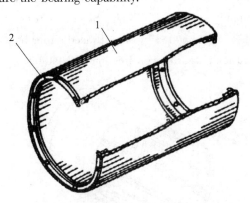

Figure 6.44 Thick skin integral body structure
1—Skin; 2—Bulkhead.

Thick skin integral body structure is simple to assembling and has good aerodynamic configuration and cabin seal. It is usually applied in small-diameter missile body and mini space vehicle. But it is difficult for openings. Usually, hatch over is used if opening is necessary.

2) Reinforced shell body structure

Reinforced shell body structure is made up of integral panel with fillets and frames illustrated in Figure 6.45. The integral panel can be manufactured through forging, casting and chemical etching. The longitudinal and lateral fillets can increase the critical stress. Reinforced fillets can be used at the location where opening is necessary. This structure is usually applied in missile body connection cabins, space shuttle plug-in tank and inlet side walls.

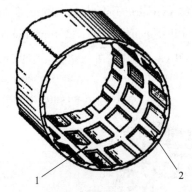

Figure 6.45 Reinforced shell body structure
1—Lateral fillet; 2—Longitudinal fillet.

6.4 Tail

6.4.1 Function and constitution of tail

Tail is the general term of horizontal tail and vertical tail. Horizontal tail is used to ensure the longitudinal stability and maneuverability while vertical tail is used to ensure the course stability and maneuverability. Figure 6.46 (a) shows the conventional tail and Figure 6.46 (b) gives the

T-shape tail. Horizontal tail consists of horizontal stabilizer and elevator while vertical tail is made up of vertical stabilizer and rudder. They are mostly applied in subsonic aircraft. Figure 6.46 (c) shows the all-moving tail plane which is widely used in supersonic aircraft.

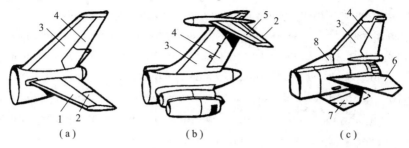

Figure 6.46 Tail constitution
1,5—Horizontal stabilizer; 2—Elevator; 3—Vertical stabilizer; 4—Rudder;
6—All-moving tail plane; 7—Ventral; 8—Dorsal fin.

6.4.2 Basic requirements of tail

The tail efficiency lies on the velocity pressure, tail area, shape and configuration, tail stiffness and supporter stiffness. The basic requirement is to guarantee the stability and maneuverability in the lowest weight. To meet this, first reasonable tail shape, parameters and configuration should be chosen.

Figure 6.47 and Figure 6.48 show the configuration and forces of aircraft. In addition, the lift-drag ratio loss caused by trimming should be as small as possible. Dangerous vibration, such as flutter and buffeting, should be avoided.

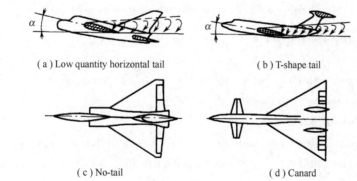

Figure 6.47 Different horizontal tail positions of different configurations

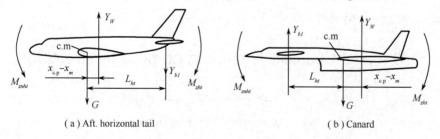

(a) Aft. horizontal tail   (b) Canard

Figure 6.48 Forces under different horizontal tail position

6.4.3 Loads of tail

6.4.3.1 Loads of horizontal tail

There are distributed aerodynamics and mass force acting on horizontal tail. The aerodynamic consists of balance load Y_{ebl}, maneuver load Y_{eman} and lift increment Y_{eg} under gust of disturbed stream. The balance load ensures the balanced state during flight while the maneuver load carries out pitch. The mass force comes from tail structure which is too small to estimate.

The aerodynamic distribution lies on the blow results and strength specification. The span wise load is approximately in proportion to chord length. As to the horizontal tail composed by horizontal stabilizer and elevator, the chord wise distribution is given in Figure 6.49. When $Ma > 1$, the distribution on all-moving tail plane is uniform. The span wise distribution on horizontal stabilizer is in proportion to stabilizer chord length, while the span wise distribution on elevator is in proportion to elevator chord length. The balance condition is $M_{znht} = M_{zht}$ (shown in Figure 6.48).

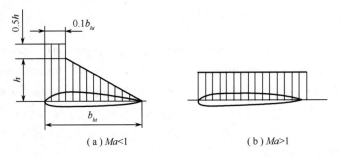

Figure 6.49 Chord wise aerodynamic distribution

6.4.3.2 Loads of vertical tail

The situation on vertical tail is as the same as horizontal tail. But as to multi-engine aircraft, when one of the engines which is not on the symmetric plane stops, horizontal tail will provide a balance load P_{vt} to counteract the moment M_y.

6.4.3.3 Loads of parts

If the horizontal stabilizer is divided into two halves, it can be regarded as cantilever beam. If it is an integral horizontal stabilizer, then it can be estimated like extensional beam. They all bear the distributed aerodynamic and concentrated supported load at suspension points. In this situation, they mainly bear shear Q, bending moment M and torque M_t. The form of stabilizer structure is similar to wing. So, the load transferring is also similar to wing.

6.5 Flight vehicle laminated composite structures

An increasingly large proportion of the structures of many modern flight vehicles are fabricated from composite materials. The use of composites can lead to considerable savings in weight over conventional metallic structures. They also have the advantage that the direction of the filaments in a multi-lamina structure may be aligned with the direction of the major loads at a particular point resulting in a more efficient design. As an extension of the above discussions on conventional flight vehicle structures, in this section we shall discuss some basic characteristics of composite laminates and introduce its applications to airframe structures.

6.5.1 Characteristics of composites

Composite materials consist of any of various fibrous reinforcements coupled with a compatible matrix to achieve superior structural performance. For example, laminate composite material consists of small diameter (around 6 to 10 microns), high-strength, high-modulus (stiffness) fibers embedded in an essentially homogeneous matrix as shown in Figure 6.50. This results in a material that is anisotropic (it has mechanical and physical properties that vary with direction). This remarkable class of materials is cited as a most promising development that has profoundly impacted today's and future technologies of airframe design.

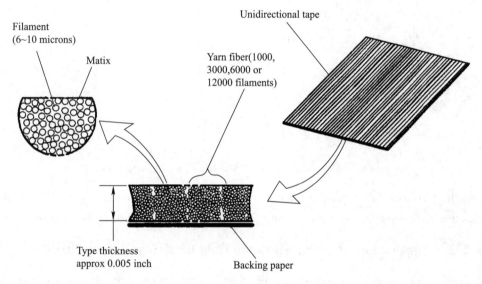

Figure 6.50 Contents of composite materials

The most commonly used advanced composite fibers are carbon and graphite, Kevlar and boron, Carbon fibers are manufactured by pyrolysis of an organic precursor such as rayon or PAN (Polyacrylonitrile), or petroleum pitch. Generally, as the fiber modulus increases, the tensile strength decreases. Among these fibers, carbon fiber is the most versatile of the advanced reinforcements and the most widely used by the aircraft and aerospace industries. Products are available as collimated, preimpregnated (prepreg) unidirectional tapes or woven cloth. The wide range of products makes it possible to selectively tailor materials and configurations to suit almost any application.

The matrix, which may be a polymer, metal, or ceramic, forms the shape of the component and serves the following additional functions: (1) transfers load into and out of the fibers, (2) separates the fibers to prevent failure of adjacent fibers when one fails, and (3) protects the fiber from the environment. Properties of the composite that are significantly affected by the properties of the matrix (matrix-dominated properties) include: (1) temperature and environmental resistance, (2) longitudinal compression strength, (3) transverse tensile strength, and (4) shear strength.

The most important contribution to material strength is that of fiber orientation. Fibers can be unidirectional, crossed ply, or random in their arrangement and, in any one direction, the mechanical properties will be proportional to the amount of fibers oriented in that direction. The most common plies used are unidirectional plies (where all fibers are aligned in one direction) or fabric plies (plain weave, satins, etc.) where fibers are oriented in two mutually perpendicular directions. If each ply in the stacking sequence or layup making up a laminate is denoted by its orientation θ (in degrees) relative to a reference axis ($-90 < \theta < = +90$), as shown for example in Figure 6.51, then a laminate can be denoted by its stacking sequence (or layup). The four standardly used ply orientations are $0°, 45°, -45°$ and $90°$ (although composites are not limited to these orientations).

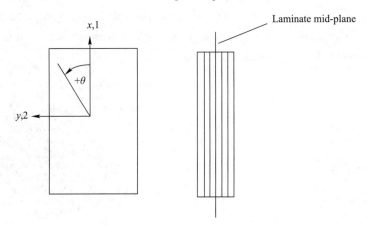

Figure 6.51　Laminate axes and definition of positive θ orientation

Over the past decades, a variety of composite materials have been developed which offer mechanical properties that are competitive with common aluminum and steel but at fractions of their weight.

6.5.2　Application of composite materials to airframe structures

Weight saving through increased specific strength or stiffness is a major driver for the development of materials for airframes. Composite materials are ideal for structural applications where high strength-to-weight and stiffness-to-weight ratios are required. Meanwhile, aircraft and spacecraft are typically weight sensitive structures. Consequently, composite materials are cited as the most promising development that has profoundly impacted today's and future technologies of airframe design. With the weight saving, commercial transports are able to economize on fuel. Weight savings also enable combat aircraft to achieve greater speed and range, extending mission capabilities. In addition, with composite materials it is possible can be arranged in such a way to create structures such as the forward swept wing of the X-29A fighter, as shown in Figure 6.52, which have aerodynamics characteristics that would not be possible if the structures were composed of metal.

Over thirty years ago, airframe manufacturers had very modest experience with the use of continuous graphite/carbon fiber reinforced thermosets, and that only in secondary structures. In recent years, a substantial amount of airframe research has been directed at developing advanced composites for use as heavily-loaded primary structures such as wing, fuselage and empennage components for both commercial and military aircraft. The growth in the use of advanced composites in the airframes in selected aircraft is illustrated in Figure 6.53. However, despite this growth, the reality is, as illustrated.

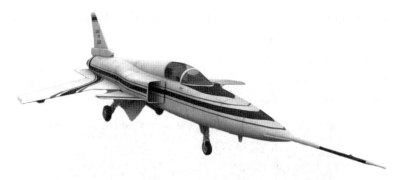

Figure 6.52 X-29A Forward swept composite wing

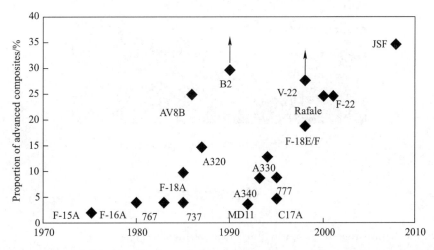

Figure 6.53 Growth of use of advanced composites in airframe structures

In Figure 6.54 for the U. S. Navy F-18 fighter, that airframes (and engines) will continue to be a mix of materials. These will include composites of various types and a range of metal alloys, the balance depending on structural and economic factors. However, the application of advanced composite materials in civil aircraft has generally lagged behind military usage because:

(1) Cost is a more important consideration to commercial aircraft manufacturers,

(2) Safety is a more critical concern, both to the airframe manufacturer and government certifying agencies,

(3) A general conservatism due to financial penalties from equipment downtime.

Figure 6.54 Schematic diagram of fighter F-18 E/F

Since the beginning of the 1980's, an all-or mostly-composite airframe has almost become a must in the developing and manufacturing of business aircraft, as shown in Figure 6.55, as well as general aviation aircraft. Design approaches which differ from those of most transport airframes and used to reduce cost and structural weight. These innovative designs and manufacturing techniques are pioneers in composite airframe structure development.

Figure 6.55 All-or Mostly-composite aircraft

Questions

1. What are the basic structure design requirements of flight vehicle?
2. What are the basic structure design contents of flight vehicle?
3. What are the functions of wing, body and tail?
4. List the external load of wing and their characteristics.
5. What is the main load-bearing members of wing and body?
6. What is the typical structure form of wing, body and tail? Describe the merits and demerits.
7. What are the advantages of composites over traditional material?

Words and phrases

adiabatic material　绝热材料
aero elastic　气动弹性
airborne equipment　机载设备
all-moving tail plane　全动水平尾翼
applied load　使用载荷
asymmetrical　非对称
auxiliary fuel tank　副油箱
balance load　平衡载荷
beam structure　梁式结构
bending moment　弯矩

blank　毛坯
bonding　固接
box-shaped thin-wall structure　盒形薄壁结构
braced girder　桁梁式
buckling　屈曲
buffeting　抖振
butt joint　对接接头
BWB　翼身融合体
camber　曲度
cantilever beam　悬臂梁

carbon fiber 碳纤维
carbon-fiber-reinforced composite 碳纤维增强复合材料
casting 铸造
cementing 胶接
chemical etching 化学腐蚀
chord 弦
comb 梳状
compression axial force 拉压轴力
concentrated load 集中载荷
controllability 操纵性
coupled vibration 耦合振动
course stability 航向稳定性（安定性）
crack propagation 裂纹扩展
damage allowance 损伤容限
decentralized 分散式
deflection 挠度
delta wing 三角形机翼
design criteria 设计准则
design philosophy 设计思想
detachable 可拆卸
diagonal bracing 斜撑杆
disk 盘式
distributed load 分布载荷
disturbed stream 扰动气流
dorsal fin 背鳍
double beam 双梁式
dovetail 燕尾槽
drag characteristic 阻力特性
durability 耐久性
engineering beam theory 工程梁理论
extensional beam 外伸梁
fail-safe 破损安全
fatigue load 疲劳载荷
fatigue resistance 抗疲劳性
fillet 筋条
flexural rigidity 抗弯刚度
flutter 颤振
frame-skin construction 蒙皮骨架式结构
hinge 铰接

horizontal stabilizer 水平安定面
impact load 冲击载荷
interval 间隔
leading edge 前缘
lift coefficient increment 升力系数增量
lift device 增升装置
light alloy 轻质铝合金
lightening hole 减重孔
load supporting system 承力系统
load-bearing member 受力构件
longitudinal stability 纵向稳定性
longitudinal wall 纵墙
maneuver load 机动载荷
maneuverability 操纵性, 机动性
MIL-A-8866 美国空军强度与刚度规范
multi-beam 多梁式
multiple tenon 多榫式
multi-wall 多墙式
non-detachable 不可拆卸
operation and maintenance requirement 使用维护要求
operation and technical requirement 使用技术要求
ordinary rib 普通肋
outer edge 外缘
overload coefficient 过载系数
pillar 支柱
plane truss 平面桁架
plug-in 插入式
preliminary design 初步设计
process 工艺
qualitative 定性的
quality requirement 质量要求
residual strength 剩余强度
reinforced shell 加筋壳
reinforced wall 加筋壁板
reinforcing bulkhead 加强隔
reinforcing rib 加强肋
rivet 铆钉
riveting 铆接

safe life　安全寿命
safety self-destruction system　安全自毁系统
sandwich construction　夹层结构
sandwich panel　夹层板
sandwich　夹芯
semi monocoque　半硬壳式
shear　剪力
shear flow　剪流
sheet　薄板
shimmy　颤振
sideslip　侧滑
single beam　单梁式
single block　单块式
skeleton structure　骨架结构
slender wing　小展弦比机翼
solar wing　太阳能帆板
space truss　空间桁架
spar　翼梁
spoiler　扰流片
static indeterminacy　静不定
statically determinate　静定
static strength　静强度
streamline form　流线形
strength specification　强度规范
stressed member　受力部件
stringer　桁条,桁条式
structure form　结构形式
surrounding frame　围框
suspension　悬挂
tab　耳片,耳片式
tactical and technical specification　战术技术指标
technological requirement　工艺要求
tenon　榫
thick cylindrical shell　厚壁筒壳
thin-wall construction　薄壁结构
titanium alloy　钛合金
torsion closed room　抗扭闭室
trailing edge　后缘
transverse bracing　横撑杆
trough-type　槽型
trussed　桁架式
ultimate load　极限载荷
upper fairing　上整流罩
ventral　腹鳍
vibration　振动
wall panel　壁板
warhead section case　弹头舱壳体

References

[1] 韩品尧. 战术导弹总体设计原理[M]. 哈尔滨:哈尔滨工业大学出版社,2000.
[2] 陈迎春,宋文滨,刘洪. 民用飞机总体设计[M]. 上海:上海交通大学出版社,2010.
[3] Bill R. Secret Projects:Flying Wings and Tailless Aircraft Hinckley[M]. England:Midland Publishing,2010.
[4] 常新龙,胡宽,张永鑫. 导弹总体结构与分析[M]. 北京:国防工业出版社,2010.
[5] 陶梅贞. 现代飞机结构综合设计[M]. 西安:西北工业大学出版社,2001.
[6] 王志瑾,姚卫星. 飞机机构设计[M]. 北京:国防工业出版社,2007.
[7] 刘莉,喻秋利. 导弹结构分析与设计[M]. 北京:北京理工大学出版社,1999.
[8] Kassapoglou C. Design and Analysis of Composite Structures with Applications to Aerospace Structures[M]. New York:John Wiley & Sons,2011.
[9] Baker A,Dutton S,Kelly D. Composite Materials for Aircraft Structures[M]. 2nd ed. AIAA,2004.
[10] 程普强. 先进复合材料飞机结构设计与应用[M]. 北京:航空工业出版社,2019.
[11] Michael Chun-Yu Niu. Composite Airframe Structures[M]. Hong Kong:Hong Kong Conmilit Press limited,2010.
[12] 库玛尔,德雷姆. 英汉航空图解词典[M]. 徐元铭,译. 北京:航空工业出版社,2009.

Chapter 7　Flight vehicle guidance and control system

The guidance system is the most important part of the guided weapon, which is used to detect or measure the flight situation of the missile relative to the target, compute the flight error between the actual position and predetermined position of the missile, and then form guidance commands, control missile to change its flight direction to fly to the target following the predetermined trajectory. Its complexity and advanced nature have direct effect on the operational efficiency, application range and cost of the weapon.

7.1　Basic concepts of guidance and control system

7.1.1　Function and composition

The earliest guidance system was used in the German V1 missile during the World War II. This system consists of a simple gyroscope to maintain heading, an airspeed sensor to estimate flight velocity, an altimeter to maintain altitude, and other synthesized systems. A guidance system has three major sub-sections: inputs section, processing section, and outputs section. The input section includes various sensors, course data, radar and other information sources. The processing section usually uses guidance computers to integrate data, solve guidance equations and finally give guidance command which determines what actions to be taken to maintain or achieve a proper heading. And the outputs section will take actual effect to change flight state by steering actuating parts such as rudders and other devices.

Conventionally, guidance means lead and control, and a guidance system is composed of a lead system and a control system. That is to say, it is divided into two parts: one part is a lead system and another is a control system. While in recent years, some new ideas are adopted that a guidance system is composed of three parts: navigation system which tracks current location, guidance system which leverages navigation data and target information to direct flight control "where to go", and control system which accepts guidance command to affect change in aerodynamic and/or engine controls. In other words, a guidance system is a relatively very large concept. Following the conventional definition, the basic compositions of a guidance system are shown in Figure 7.1.

A lead system means that it can be used to make a missile go onto the target. First, the relative movement between a missile and a target can be detected and measured. Furthermore, the leading command is formed for a missile to attack a target according to the moving deviations of them. A lead system mainly consists of two parts, detecting devices (sensors) and command forming devices. The function of the lead system is measuring the relative motion parameters between missile and target and forming the command used to control the missile according to the predestinate guidance law.

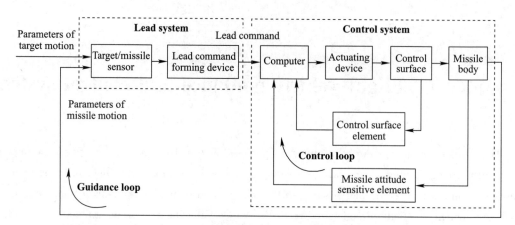

Figure 7.1 Basic composition of a guidance system

A control system is to accomplish some important functions which include, connecting with the lead system, transforming and amplifying the lead command signals, furthermore, driving the control surfaces by the actuating mechanism, in the end, changing the missile flight movement. Therefore, it is composed of several parts, synthesizing units, executing device, control surface, sensitive device, and the controlled object, i. e. missile body. One function of the control system is stabilizing the attitude angle of the missile based on the error information measured by the missile's sensitive device, and the other function is controlling the missile to fly following the trajectory needed according to the control commands transmitted from the lead system.

Generally speaking, the guidance and control system is a large closed loop, and the control system is a small closed loop.

7.1.2 Classification of guidance system

Because different missiles are used to attack different targets, they adopt different principles and devices in the modern guidance system. The categories of the guidance system often depend on the characteristics of the lead system i. e. the source generating the leading information. Table 7.1 shows the main four categories of the guidance system.

Table 7.1 The category of guidance system

Missile guidance system	Autonomous guidance system	Inertial navigation system
		Celestial navigation system
		Doppler navigation system
		Planning guidance system
		Terrain-matching guidance system
		Satellite navigation system
	Remote guidance system	Command guidance system
		Beam guidance system
	Homing guidance system	Active homing
		Semi-active homing
		Passive homing

(Continued)

		Autonomous + Homing guidance system
Missile guidance system	Combination guidance system	Autonomous + Remote guidance system
		Remote + Homing guidance system
		Autonomous + Remote + Homing guidance system

7.1.3 Basic design requirements

To get the missile guided well, there are many requirements for the guidance system. Especially, some fundamental requirements are as follows.

1) Guidance accuracy

The guidance accuracy is the most fundamental and important requirement for the guidance system. Guidance accuracy is often described by miss distance which is defined as the shortest distance between missile and target during the campaign. As usual, the error that causes miss distance can be classified as system error and random error. The system error has a constant value in each missile of the same model. The random error has a random value in each shot of the missile, but the average value of the random error is zero.

The permissible value of miss distance depends on many factors which mainly include hit probability, the mass and the type of the missile's warhead, the type and defense capability of the target. To obtain the miss distance being less than its permissible value, the guidance accuracy must be improved. At present, the miss distance of the tactical missile can be reduced to several meters, and for the strategical missile, the permissible value of miss distance can be relaxed to tens of meters due to its powerful warhead.

2) Combat reaction time

The combat reaction time is defined as the time from the moment when the target is detected to the moment when the first missile is launched, which is decided by performances of the guidance system and defense command, control and communication system. For tactical missiles attacking movable targets, the combat reaction time is mainly decided by the performance of the guidance system. With the development of science and technology, the target can move faster and faster. And the target flying at low altitude is hard to be searched and detected over long distances, so the combat reaction time is required to be as short as possible.

3) Target discriminability

The missile guidance system must have high distance discriminability and angular discriminability if the missile is required to attack the specified target among several targets. The distance discriminability is the discriminability between two targets with the same azimuth but different distances, and it is defined as the shortest distance discriminated between two targets. The definition of angular discriminability is just similar to that of distance discriminability. The target discriminability of the missile guidance system is mainly limited by the measurement accuracy of the correlative transducers. So target transducers with high discriminability are demanded.

4) Antijamming ability

The antijamming ability of the missile guidance system is the capability of maintaining normal operating stages when there are interferences from inside and outside of the missile or missiles are attacked by enemies and meet with electronic countermeasure and anti-missile countermeasure. Tactical missiles are usually required to have strong antijamming ability. Different types of guidance systems have different interference conditions. For the radar remote guidance system, it is easy to counter electronic jamming from the enemy. To improve the antijamming ability of the guidance system, some available ways are as follows: (1) using new technology to reduce the sensitivity of the guidance system to the interference; (2) improving the invisibility and suddenness of the working of the guidance system; (3) using more than one models of guidance system.

5) Reliability

The reliability of the guidance system is the ability of maintaining the parameters of various guidance equipment in their permissible change ranges at the conditions of normal use and maintenance, which can be described by the probability of non-fault during the permitted working hours. This probability is higher, and then the reliability is better.

7.2　Guidance and control components

Most closed loop systems incorporate transducers to measure the system states. The output from the transducer is usually a voltage signal and this is readily compared with the input or reference signal if the input signal is also a voltage. Obvious examples are potentiometers and tachogenerators to monitor the position and speed of a shaft. Considering now any aspect of the motion through space, forces and moments will produce accelerations and hence velocities and displacements with respect to inertial space, or as is often stated approximately with respect to the earth. If we wish to realize a closed-loop system of controlling the motion of a missile, certain instruments will be required to measure accelerations, velocities and displacements in space. Clearly, conventional potentiometers and tachogenerators cannot do this. Instead, such sensitive components as accelerometers, rate gyroscopes and position gyroscopes are used. In order to guidance and control an aircraft, certain lead equipment and actuating devices are also needed to be onboard.

7.2.1　Sensitive components

7.2.1.1　Gyroscopes

A gyroscope (or Gyro) is a device for measuring angular movement or maintaining orientation, based on the mechanical principles of rigid bodies rotating around the fixed point. In general, an object or a structure spinning at a high rate can be called a gyroscope. If a gyroscope rotor is mounted in two mutually perpendicular gimbals, it will have three rotation freedoms and it is called the three degree-of-freedom gyroscope or position gyroscope. When the outer gimbal is fixed, the gyroscope will lose one rotation freedom so is called the two degree-of-freedom gyroscope or rate gyroscope.

1) Elementary theory of gyroscopes

A gyroscope usually has a high moment of momentum resulting from its high angular velocity of spinning, which lets it possesses two important fundamental properties, rigidity and precession.

(1) Rigidity

A gyroscope with the moment of momentum H will maintain its spin axis direction (orientation) when the external torque is zero, which is the rigidity of gyroscopes. Based on the rigidity gyroscopes have replaced the magnetic compass and are used commonly for navigation. Rigidity is the inertial phenomenon of rigid body motion and it complies with the moment of momentum conservation theorem. It barely exists in the real world that no torque acting on one body. So the direction of the spin axis of gyroscope cannot maintain absolutely unchanged, that is, there exists gyroscope drift. But when the gyroscope has large angular momentum and the external torque is small, the change of axis direction is usually small enough to be ignored.

(2) Precession

Considering a gyroscope with a high spinning rate, if a torque acts on it in the direction of perpendicular to its spin axis, the gyroscope will rotate actually around a third axis which will form a right-handed Cartesian coordinate system with the directions of the spin axis and the external torque. This special movement pattern is called the precession of gyroscopes.

The property of gyroscope precession can be illustrated by the following experiment: a gyroscope rotor is spinning around the y-axis at a high rate of Ω and the gyroscope rotor is mounted on two gimbals, as shown in Figure 7.2. Now suspend a mass on the inner gimbal and this will act a torque on the gyroscope in the direction of the x-axis. Then we will find that the gyroscope will not rotate around the x-axis but the z-axis.

Now suppose a torque T is applied to the gyroscope for a short time δt in the right direction to the spin vector as given in Figure 7.3.

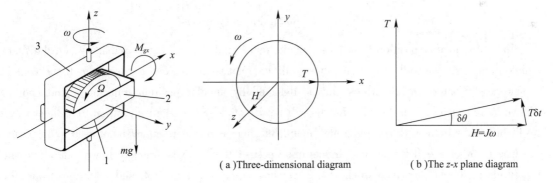

(a) Three-dimensional diagram (b) The z-x plane diagram

Figure 7.2 Gyroscopes characters Figure 7.3 Precession of gyroscopes
1—Rotor; 2—Inner ring; 3—Outer ring.

If the body rotates through $\delta\theta$, the change of angular momentum equals to $T\delta t$. Suppose the original angular momentum is $J\omega = H$, then

$$T\delta t = H\tan\delta\theta \approx H\delta\theta \tag{7-1}$$

And we know that

$$\Omega = \frac{d\theta}{dt} = \frac{T}{H} = \frac{T}{J\omega} \tag{7-2}$$

From Eq. (7-2), we know the precession angular velocity is in direct proportion to the disturbing moment and in inverse proportion to the gyroscope's angular momentum.

2) Free or position gyroscopes

With gyroscope rotors mounted in two mutually perpendicular gimbals, the gyroscope's spin axis has three degree-of-freedom and will hold its direction in inertial space in spite that there has movement in the gimbal's fixed base. This kind of gyroscope has free spin axes with which it is usually used to sense the angular motion of the gyroscope's fixed base. So gyroscopes with three freedoms are also called free gyroscopes or position gyroscopes.

The free gyroscope can be categorized into two kinds, roll and pitch gyroscope vs. yaw and pitch gyroscope, based on the different installation on the flight vehicle.

(1) Roll and pitch gyroscope (vertical gyroscope)

As shown in Figure 7.4, the roll and pitch gyroscope is installed on the flight vehicle with its spinning axis vertical and its rotation axis of inner and outer gimbals coincide with the flight vehicle's pitch and roll directions respectively. The spinning axis will maintain its direction in inertial space even when the flight vehicle framework has angular movement and this is exactly used to measure the angular position of the flight vehicle framework.

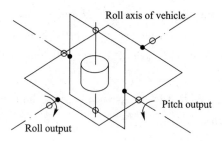

Figure 7.4   Roll and pitch gyroscope

There are angular position transducers in the outer gimbals and inner gimbals. When the flight vehicle framework has roll movement, there will be relative movement between the flight vehicle framework and outer gimbals which can be detected by angular position transducer. And another relative movement between the two gimbals will be detected by the angular position transducer when the flight vehicle framework has pitch movement. But there is no relative movement used to detect the yaw movement of the flight vehicle framework because the gimbals framework will merely rotate around the gyroscope spin axis when yaw movement happens. For the roll and pitch gyroscope, the flight vehicle can have unlimited freedom to rotate about the roll axis without impairing the original orthogonal nature of the three axes. While when the flight vehicle framework has pitch movement, the two sets of gimbals will tend to rotate into the same plane.

When the two gimbals turn into the same plane, the gyroscope will lose one freedom, and this phenomenon is called "toppling". As the angular position will be correctly measured only when the two gimbals remain orthogonal, so for the flight vehicle adopted with roll and pitch gyroscope, the roll movement can be measured always correctly but the pitch movement measurement can be useful only when the pitch rotation angle is in a small range.

(2) Yaw and pitch gyroscope

Asseen in Figure 7.5, yaw and pitch gyroscope are used to measure the yaw and pitch angular movement of the flight vehicle framework. The spin axis is settled along the axis of the flight vehicle framework and the gimbals have yaw and pitch freedoms. Due to the fixed axis character of free gyroscope, the spin axis and inner gimbals will hold their directions in inertial space, but there will be relative movement between the two gimbals and the flight vehicle framework. So with angular position transducers, the relative movement between the two gimbals is measured to indicate the yaw angular position, and the pitch angular position is indicated by the relative between the outer gimbals and the flight vehicle framework.

We know that a free gyroscope can only measure two angular positions of the flight vehicle framework, so at least two free gyroscopes are needed to get all attitude angles. Usually, a free gyroscope may be regarded as an ideal amplifier section and for pitch angle measurement, the transfer function is as follow, taking the measurement of ϑ for example.

$$\frac{u_\vartheta(s)}{\vartheta(s)} = k_\vartheta \qquad (7-3)$$

where, u_ϑ is the output signal, ϑ is the pitching angle of the missile, k_ϑ is the transfer coefficient of the free gyroscope.

3) Two degree-of-freedom gyroscopes

For a two degree-of-freedom gyroscope, the rotor is mounted in only one gimbal, so it only has two degrees of freedom. As given in Figure 7.6, if the framework rotates about the x axis, the rotor together with the gimbal will be forced to rotate around the x axis at the same rate as the framework. As a result, there will be a gyroscopic torque around the y axis due to that the gyroscope will precess about the y axis. The precession angular rate is in direct proportion to the angular rate of the framework about the x axis, and so the framework angular rate or its integral will be obtained if we measure the precession angular rate. Based on different functions, the two degree-of-freedom gyroscope is divided into rate gyroscope and rate integrating gyroscope.

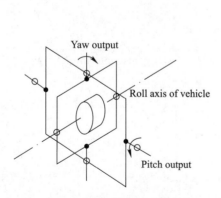

Figure 7.5 Yaw and pitch gyroscope

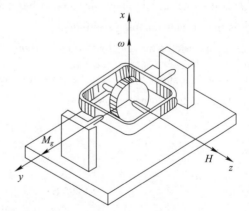

Figure 7.6 Two degree-of-freedom gyroscope

(1) Rate gyroscopes

The function of rate gyroscopes is to measure the angular rate of the "input" axis. A rate gyroscope is a two degree-of-freedom gyroscope plus with torsion spring, damper, potentiometer, and angular position transducer.

As described in Figure 7.7, the rotor is mounted in one gimbal, whose motion about the x axis is constrained by a torsion bar and friction-free spring system. If the framework inputs an angular rate ω_y, a gyroscopic torque will be produced and the gyroscope will precess around the x axis. Then there will be torsion spring moment and damping moment reversing with gyroscopic torque, and there will be a balance when the gyroscope precesses a certain angle. An angular position transducer (potentiometer) is settled to measure the precess angle, which is in direct proportion to the gyroscopic torque and so is in direct proportion to the input angular rate, too.

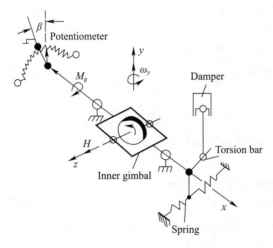

Figure 7.7 Schematic description of the rate gyroscope

Define:

k is the torsion bar stiffness,

β is the angular twist in the torsion bar,

H is the angular momentum of the rotor,

J_g is the total inertia about the torsion bar axis,

f is the viscous damping coefficient,

ω_y is the rate of turn of flight vehicle about the input axis.

Equating torques about the torsion bar axis we have $J_g\ddot{\beta} + f\dot{\beta} + k\beta =$ gyroscopic torque $= H\omega_y$. Taking Laplace transformation, we get

$$J_g s^2 \beta + fs\beta + k\beta = H\omega_y \tag{7-4}$$

Hence

$$\frac{\beta(s)}{\omega_y(s)} = \frac{\dfrac{H}{k}}{\dfrac{s^2}{\omega_n^2} + \dfrac{2\xi s}{\omega_n} + 1} \tag{7-5}$$

where

$$\omega_n^2 = \frac{k}{J_g} \text{ and } \frac{2\xi}{\omega_n} = \frac{f}{k} \tag{7-6}$$

From Eq. (7-5), we know the rate gyroscope is a second-order system and the D.C. gain is

$$\frac{\beta}{\omega_y} = \frac{H}{k} \tag{7-7}$$

Assuming the amplification factor is K_u, the output voltage signal of the angular transducer is given by

$$U = K_u \beta = K_u \omega_y \frac{H}{k} \tag{7-8}$$

(2) Rate integrating gyroscopes

Compared with rate gyroscopes, in the rate integrating gyroscope, the torsion spring is removed and replaced by the torque motor. So the damper plays the major role of balancing gyroscopic torque. The configuration of a typical rate integrating gyroscope is given in Figure 7.8.

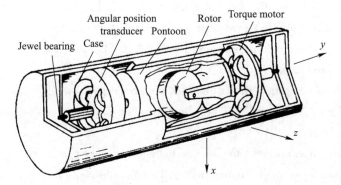

Figure 7.8 Configuration of a rate integrating gyroscope

So we get

$$J_g \ddot{\beta} + f\dot{\beta} = \text{gyroscopic torque} = H\omega_y \tag{7-9}$$

And

$$J_g s^2 \beta + fs\beta = H\omega_y \tag{7-10}$$

Hence

$$\frac{\beta(s)}{\omega_y(s)} = \frac{H}{J_g s^2 + fs} \tag{7-11}$$

If the damping coefficient f is much bigger than rotor moment inertia J_g, the first part of the denominator in Eq. (7-11) can be neglected, and then we have

$$\frac{\beta(s)}{\omega_y(s)} = \frac{H}{fs} = \frac{H/f}{s} \tag{7-12}$$

$$\beta(s) = \frac{H/f}{s} \omega_y(s) \tag{7-13}$$

Then

$$\beta(t) = \frac{H}{f} \int \omega_y(t) \, dt \tag{7-14}$$

This is a pure integral system, and the output of the angular position transducer is in direct proportion to the integral of the input angular rate.

Usually, the cylindrical gimbal of the two degree-of-freedom gyroscope is enclosed in a hermetically sealed outer case and the gap between them is filled with viscous fluid in which the gimbal is floated with neutral buoyancy. The fluid provides viscous shear damping, minimal pivot friction and protection from shock, so the measurement accuracy is improved largely. This kind of gyroscope is called the liquid-floated gyroscope, and the use of the liquid-floated technique is a milestone in the development of inertial instruments.

4) Micro-electro-mechanical gyroscopes

When a mass with a high-frequency vibration is driven by the pedestal to rotate at an angular velocity ω with respect to the inertial frame, a Coriolis acceleration (force) proportional to the angular velocity ω is generated, and this physical phenomenon is called the Coriolis effect. A class of gyroscope that utilizes the Coriolis effect to measure the angular motion of the carrier is called a Micro-electro-mechanical (MEMS) gyroscope. Compared with the traditional mechanical gyroscope, the MEMS gyroscope has the characteristics of small volume, low cost, low power consumption, high reliability, and mass production.

As an autonomous navigation device, the MEMS gyroscope is widely used in aircraft systems such as the F-22 fighter, Predator UAV, Apache helicopter gunship, Sea-Wolf anti-aircraft missile, and Ariane-5 carrier rocket.

The MEMS gyroscope utilizes the Coriolis force phenomenon, and its principle is shown in Figure 7.9. When the object in the figure makes periodic vibrations or other motions along the X axis, and the XY coordinate system rotates along the Z axis with an angular velocity of Ω_Z, a Coriolis force along the Y axis will be generated on the object, and its vector form can be calculated by.

$$F(t) = -2m\Omega_Z \times \dot{x}(t) \tag{7-15}$$

where, $F(t)$ is the Coriolis force, m is the mass of the object, Ω_Z is the angular velocity of the rotation of the coordinate system, $\dot{x}(t)$ is the vector velocity of the object.

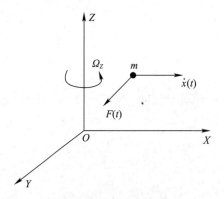

Figure 7.9　Coriolis force

Take a MEMS gyroscope as an example, which adopts a symmetric structure with two mass blocks, as shown in Figure 7.10. Sliders 1 and 1' are detected mass blocks, 2 and 2' are driven mass blocks, and the detected mass block is attached to the driven mass block. Restricted by structural component 3, the detected mass block can be passively moved along the driven axis (X axis) with the driven mass block. While in the direction of the detected axis (Y axis), the detected mass block can move freely under the action of Coriolis force. Therefore, the detected mass block will move along two axes, one is the restricted passive movement along the X axis with the driven mass block, and the other is the free movement along the Y axis with the Coriolis force. 4 (4′) and 5 (5′) are driven electrodes and detected electrodes respectively.

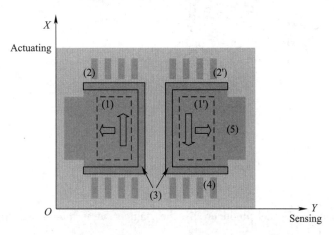

Figure 7.10 Structure of a MEMS gyroscope

According to Eq. (7-15) acceleration generated by Coriolis force is $\alpha(t) = 2\Omega_z \times \dot{x}(t)$. The vibration velocity is a known quantity. If the Coriolis force acceleration $\alpha(t)$ on the detected mass block is obtained, and when combined with the vibration velocity $\dot{x}(t)$ for synchronous demodulation, the rotation angular velocity of the XY coordinate system can be detected. This is the fundamental working process of the MEMS gyroscope.

7.2.1.2 Accelerometers

Accelerometers are important inertial sensors used in the fight guidance and control system. They are used on flight vehicles to measure the acceleration of the input axis, with which the flight range can be calculated through two integrators. The most usual accelerometers include spring-mass accelerometers, liquid-floated pendulous accelerometers, flexure pendulous accelerometers and piezoelectric accelerometers and so on. Accelerometers can also be classified as single-axis accelerometers, two-axis meters and triaxial accelerometers according to the number of axes measured by one single accelerometer.

1) Spring-mass accelerometers

A spring-mass accelerometer is composed of mass, spring, damper, potentiometer and other accessories. A typical spring-mass accelerometer is described in Figure 7.11.

The basic principle of this kind of accelerometer can be described according to Newton's second law. The mass is fixed with the spring, so the linear displacement of mass is relative to the deflection of the spring, and thus, a restoring force is produced by the spring which is in proportion to the displacement. That is

$$F_s = k_s X \tag{7-16}$$

where, k_s is the stiffness coefficient of the spring, F_s is the restoring force, X is the displacement of the

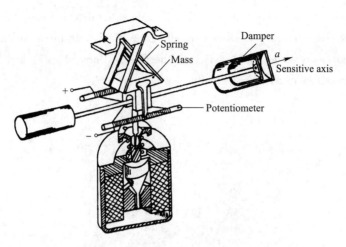

Figure 7.11 Spring-mass accelerometer

spring.

The viscous damping force produced by the damper is in proportion to the relative speed between the mass and the damper. That is

$$F_d = k_d \dot{X} \tag{7-17}$$

where, k_d is the viscous damping coefficient, F_d is the damping force, $\dot{X}$ is the relative speed between the mass and the damper.

Assume the absolute displacement of the flight vehicle framework and the mass are X_d and X_m respectively, and the relative displacement between them is X. The spring and damper are both fixed on the flight vehicle. Based on Newton's second law, we have

$$\sum F = F_s + F_d = m\ddot{X}_m \tag{7-18}$$

As

$$X = X_d - X_m \tag{7-19}$$

we can get

$$m\ddot{X}_d = F_s + F_d + m\ddot{X}_m = k_d X + k_d \dot{X} + m\ddot{X} \tag{7-20}$$

In Eq. (7-20), $\ddot{X}_d$ is the acceleration of flight vehicle framework, which is also defined as a. Taking Laplace transformation for Eq. (7-20), we get

$$ma(s) = X(s)(k_s + k_d s + ms^2) \tag{7-21}$$

$$\frac{X(s)}{a(s)} = \frac{m}{k_s + k_d s + ms^2} = \frac{k_a}{T_a^2 s^2 + 2T_a \zeta_a s + 1} \tag{7-22}$$

where, $k_a = m/k_s$ is the transfer coefficient of the accelerator, $T_a = \sqrt{m/k_s}$ is the time constant of the accelerator, $\zeta_a = \dfrac{k_d}{2\sqrt{mk_s}}$ is damping coefficient of the accelerator.

The transfer relation of spring-mass accelerometers is a second-order system. The potentiometer is used to measure the relative displacement X, and then the acceleration of the flight vehicle framework is gotten.

2) Liquid-floated pendulous accelerometers

The composition of a liquid-floated pendulous accelerometer is similar to the liquid-floated gyroscope. The gap between the pontoon and outer case is filled with viscous fluid in which the pontoon is floated with neutral buoyancy. The neutral buoyancy essentially removes the gimbal bearing friction and makes the gyroscope insensitive to shock and vibration. As shown in Figure 7.12, there is a mass in the pontoon which has a displacement L from the center axis x, and the sensitive axis of the accelerometer is axis z.

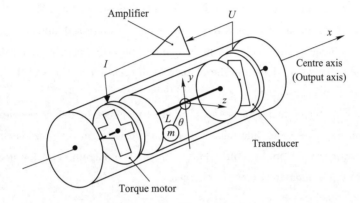

Figure 7.12　Schematic diagram of the liquid-floated pendulous accelerometer

When the acceleration on the sensitive axis, defined as a, is not zero, there will be an inertial moment duo to mass and displacement L.

$$M_a = Lma \tag{7-23}$$

Then, the mass will rotate about axis x and the rotation angle θ is measured by an angular position transducer, whose transfer factor is k_u. So the output voltage U of the transducer is

$$U = k_u \theta \tag{7-24}$$

The output voltage U is transferred to an amplifier whose output is the current signal I. And the current signal is transferred to the torque motor which will produce a moment M_k inverse to the inertial moment and makes the mass balanced. Both the amplifier and torque motor are linear transformation devices and their transfer factors are k_i and k_m respectively. So we have

$$I = k_i U \tag{7-25}$$

$$M_k = k_m I = k_m k_i U = k_m k_i k_u \theta = M_a = Lma \tag{7-26}$$

Hence,

$$k_m I = Lma \tag{7-27}$$

$$I = \frac{Lm}{k_m} a \tag{7-28}$$

we can see that the output of the amplifier, the current signal I, is in direct proportion to the input acceleration a. So the acceleration will be gotten through handling the current signal I.

3) Micro-electro-mechanical Accelerometers

Since the first micro accelerometer in the world came out at Stanford University in 1977, a lot of research and experiments in the structure design, processing technology, processing circuit, simulation modeling, and performance index optimization of MEMS accelerometers have been carried out by various research institutions. According to the detection method, MEMS accelerometer can be divided into piezoresistive accelerometer, piezoelectric accelerometer, tunneling accelerometer, thermal accelerometer, resonant accelerometer, and capacitive accelerometer. The following is a brief description and comparison of detection principles of different MEMS accelerometers.

(1) Piezoresistive accelerometer.

The piezoresistive accelerometer is based on the piezoresistive effect. When the mass block is accelerated, the cantilever beam connected to the mass block is stressed, and the carrier mobility on the cantilever beam changes, thus the resistance of the cantilever beam is changed. The change of acceleration can be detected by detecting the change of the resistance value of the cantilever beam.

(2) Piezoelectric accelerometer.

The piezoelectric accelerometer uses the piezoelectric effect to detect acceleration. When the acceleration acts on the mass block, the mass will produce a force on the piezoelectric crystal, which makes the piezoelectric crystal produce an electric charge. The magnitude of the acceleration can be detected by detecting the number of electric charges.

(3) Tunneling accelerometer.

The tunneling accelerometer is made based on the tunneling effect. The effect of acceleration changes the distance between a pair of tips, thus the magnitude of the tunneling current is altered. The acceleration can be measured by measuring the current passing through.

(4) Thermal accelerometer.

The center of the thermal accelerometer is a heating body, surrounded by a temperature sensor, and inside is a closed air chamber. When working under the action of the heating body, the gas forms a hot gas group inside, and the hot gas group will move under the action of inertial force, forming the change of thermal field. A temperature sensor detects acceleration by detecting changes in the thermal field.

(5) Resonant accelerometer.

The principle of the resonant accelerometer is that the mass block converts the acceleration into inertia force, which is used in the axial direction of the resonator and causes the change of the resonant frequency of the resonator. The change of acceleration can be detected by detecting the change of resonant frequency.

(6) Capacitive accelerometer.

The capacitive accelerometer has a moving mass and a pair of fixed electrodes. The mass block and electrode form a pair of parallel capacitors. When the acceleration acts on the mass block, the mass block moves under the action of the force, which causes the overlapping area or distance of the parallel capacitors to change, thus changing the capacitance. The acceleration can be detected by detecting the capacitance change of the parallel capacitors.

Table 7.2 lists the advantages and disadvantages of several MEMS accelerometers. Among them, the capacitive MEMS accelerometer is the most widely used because of its good repeatability, high resolution, and simple structure.

Table 7.2 Comparison of advantages and disadvantages of various MEMS accelerometers

Type	Advantages	Disadvantages
Piezoresistive accelerometer	Simple structure, simple chip manufacturing process, a simple interface circuit	High-temperature drift coefficient, low sensitivity, high noise, significant spiral deformation, and hysteresis effect
Piezoelectric accelerometer	High dynamic range, high bandwidth, small size, no need for external power supply	Only dynamic signal detection, high output impedance, high requirements for the detection circuit
Tunneling accelerometer	Very high resolution	Very high process requirements, very large low-frequency noise, high power supply voltage
Thermal accelerometer	No adhesion and particle problems, impact resistance, a very low failure rate	Large temperature coefficient, low-frequency response range, high power consumption
Resonant accelerometer	Direct frequency signal output, not easy transmission process distortion, easy to interface with a digital circuit	The micromechanical structure relatively complex
Capacitive accelerometer	Good repeatability, high resolution, good noise characteristics, small temperature sensitivity, low power consumption, simple structure	Easy to be interfered by parasitic effect, the interface circuit relatively complex

7.2.1.3 Altimeters
1) Radar altimeters

A radar altimeter is mounted on the aircraft using electromagnetic waves to measure the height of the aircraft to the ground. The transmitter of the radar altimeter transmits radio signals (usually pulsed waves) to the ground, which are reflected by the ground and received by the receiver. According to the time interval from transmitting to receiving and the speed of wave propagation, the distance between the aircraft and the ground can be calculated.

2) Barometric altimeters

The barometric altimeter is an instrument that uses the relationship between air pressure and altitude to measure the altitude (also known as absolute altitude) of an aircraft. Atmospheric pressure is generated by the gravity of the air at the surface. As the altitude rises, the thickness of the air at the surface decreases and the air pressure falls. So the altitude value can be derived by measuring the atmospheric pressure at the location and comparing it with the standard value, which is the basic working principle of the barometric altimeter.

7.2.1.4 Other sensitive components
1) Static ports

The static port is a sensor to measure the static pressure. Static pressure is the pressure of the air around the aircraft when it is still, that is, the atmospheric pressure at the position of the aircraft.

The static port is generally located on the side below the front section of the fuselage, where is not easy to be disturbed by airflow, as shown in Figure 7.13. The air slowly flows into the static port, and the atmospheric pressure at the position of the aircraft is measured by the pressure sensor. In addition to providing static pressure parameters needed to calculate airspeed, the static port also has the function of measuring the altitude of the aircraft.

Figure 7.13 Static port and its location

2) Airspeed tubes

Figure 7.14 shows the airspeed tube, which is a pressure measurement instrument used to measure fluid flow velocity. The principle is introduced in Section 2.3.4.2. As the plane flies, the air flows into the airspeed tube, and the total pressure of the air is felt at the back of the tube. The total pressure is composed of the dynamic pressure generated by the airflow and the static pressure of the air. It is difficult to measure the dynamic pressure directly. The static pressure is measured by the static port, and the dynamic pressure can be obtained as dynamic pressure = total pressure-static pressure.

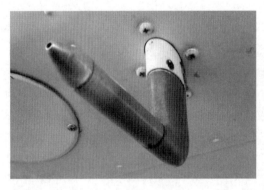

Figure 7.14 Airspeed tube and its location

3) Angle of attack sensors

The wings will lose lift if the angle of attack is too high, causing the plane to stall. Most modern civil aircraft have a stall warning system. When the actual angle of attack is close to the critical angle of attack, the stall warning system will send out various forms of warning signals. The angle of attack sensor provides data support for the stall warning system.

The angle of attack sensors are generally installed on the side of the aircraft near the nose position, which can be divided into the wind-vane sensors and null-seeking differential pressure sensors.

(1) The wind-vane angle of attack sensor

The wind-vane angle of attack sensor is composed of a symmetrical airfoil blade and an angle converter, which is shown in Figure 7.15. It has the advantages of simple structure and small volume. But wind vane is subject to small perturbations and location. It is very difficult to find a part with stable airflow on a high-speed aircraft.

Figure 7.15　Wind-vane angle of attack sensor

(2) Null-seeking differential pressure angle of attack sensor

The probe of the null-seeking differential pressure angle of attack sensor has two pairs of symmetrical air intake ducts facing the airflow. Figure 7.16 shows the null-seeking differential pressure angle of attack sensor. When the angle of attack is constant, the air pressure felt by each pair of ducts is equal. When the angle of attack changes, the air pressure felt by one pair of ducts increases, and the air pressure of the other pair decreases. The two air pressures generate an aerodynamic feedback torque in the same direction as the angle of attack changes, causing the probe to rotate until the symmetric plane of the air intake ducts is parallel to the direction of the airflow and the pressure of the two pairs of air intake ducts becomes equal again. At the same time, the electric brush of the probe has a corresponding angular displacement concerning the potentiometer, so that the output voltage signal is proportional to the change of the angle of attack. At present, the null-seeking differential pressure angle of attack sensor is widely used in all kinds of aircraft. It has the advantages of small errors and stable operation.

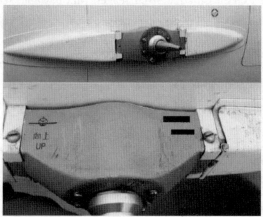

Figure 7.16　Null-seeking differential pressure angle of attack sensor

4) Sideslip angle sensor

The sideslip angle sensor, as shown in Figure 7.17, is generally installed in the nose part of the aircraft, and its principle is similar to the weathercock angle of attack sensor.

Figure 7.17 Sideslip angle sensors

7.2.2 Seeker

The seeker (known as homing eye and homing head) is a measuring and tracking component in missiles' homing guidance and control loop. It has different functions in different homing guidance modes, but the basic and main functions of it are as follows:

(1) Capture and track the target.

(2) Provide the measurements of target motion required to realize the guidance law.

E. g., the proportional navigation usually adopted by homing guidance and control loop requires the seeker to output the rotation rate of the line of sight (LOS), the closing velocity between the missile and target. The tracking method requires the seeker to output the angle between the LOS and missile velocity.

7.2.2.1 Classification of seekers

According to the relative relationship of the seeker's measure reference frame and missile's body reference frame, seekers can be classified as fixed seekers and movable seekers. Movable seeker also can be classified as movable non-tracking seeker and movable tracking seeker.

By receiving the energy radiated or reflected by the target, the seeker can get the relative position and kinetic characteristics between the missile and the target, and then form the guidance command. Depending on the position of the energy source, seekers can be classified as follows:

(1) Active homing head. It receives the energy reflected by the target, the source of energy is in itself.

(2) Semi-active homing head. It gets the energy reflected by the target, the source of energy is outside.

(3) Passive homing head. It receives the energy radiated by the target.

For the differences in physical properties of the energy received, seekers can be divided into radar seekers, imaging seekers, laser seekers, and multi-mode composite seekers.

1) Radar seeker

A radar seeker is shown in Figure 7.18. The principle of a radar seeker is to use the equipment on the missile to receive the radio waves radiated or reflected from the target, track the target, and form guidance instructions to guide the missile to fly to the target. The radar seeker has the functions of tracking the target position, calculating the control parameters, and outputting the control instructions, which can realize the precise attack of the missile to the target.

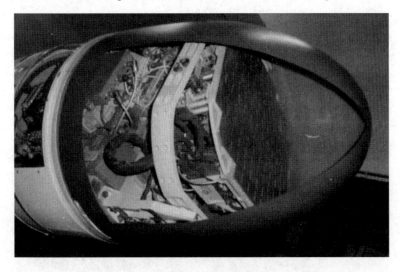

Figure 7.18   Radar seeker

According to the working wavelength of radar, commonly used radar seeker can be divided into centimeter wave radar seeker, millimeter wave radar seeker, and submillimeter wave radar seeker.

(1) Centimeter wave radar seeker

Centimeter wave radar is a microwave radar working in centimeter wave band detection. In general, a centimeter wave refers to the electromagnetic wave in the frequency domain of 3 ~ 30GHz (wavelength of 1 ~ 10cm). Centimeter wave radar seeker uses centimeter wave radar to detect and track targets. The technology is mature and the cost is low, but the anti-jamming capability is insufficient. Centimeter wave radar seeker is widely used in anti-ship missiles around the world, including AGM-84 "Harpoon" missile of the United States and "Exocet" missile of France.

(2) Millimeter wave radar seeker

Millimeter wave refers to the electromagnetic wave in the frequency domain of 30 ~ 300GHz (wavelength of 1 ~ 10mm). Millimeter wave radar seeker can receive the millimeter wave information reflected or radiated by the target, capture and track the target, and then generate guidance instructions to guide the missile to hit the target. Millimeter wave radar has the characteristics of small size, good resolution, strong penetrating ability, and short detection range. Millimeter wave radar is often used in air-defense missiles and anti-tank missiles, typical applications of which include the American AGM-114L missile, PAC-3 air-defense missile, etc.

(3) Submillimeter wave radar seeker

The operating frequency domain of submillimeter-wave radar is between 300 ~ 3000GHz (wavelength 0.1 ~ 1mm). Submillimeter wave radar seeker has higher resolution, rich target characteristics, and strong anti-jamming ability, which is beneficial to target detection and recognition.

According to the source of detection energy, radar seeker can also be divided into active radar seeker, semi-active radar seeker, passive radar seeker, and compound radar seeker.

2) Imaging seeker

The imaging seeker tracks the target image captured by the sensor, which is shown in Figure 7.19. When the distance between the missile and the target reaches the effective action distance of the seeker, the target is found and identified by the shooter or the identification algorithm through the image of the seeker. After stable tracking, the missile hits the target following the guidance law.

Figure 7.19 Imaging seeker

Television (TV) imaging seeker and infrared imaging seeker are two typical imaging seekers which have been widely used in the field of precision guided weapons.

(1) TV imaging seeker

The TV imaging seeker takes in the visible image information around the target through the TV camera and outputs the corresponding guidance law after processing for guidance. Its components include the TV camera, the sawtooth wave generator, the synchronization machine, the deflection scanning circuit, the gate circuit, the video processing circuit, the angle tracking mechanism, and the servo mechanism, etc. TV imaging seeker uses visible light for imaging, so it is also called visible light imaging seeker. TV imaging seeker has the advantages of strong anti-jamming ability, easy image processing, and high tracking accuracy, but it also has some disadvantages, such as being easily restricted by meteorological conditions. With the development of digital signal processing chip and real-time image processing technology, TV imaging seeker has made some progress in the direction of miniaturization and intelligence. TV imaging seeker is widely used in air-to-ground missiles and precision-guided bombs. American AGM-65A "Maverick", AGM-130 missiles, and Russian KH59 missiles are all using TV guidance mode.

(2) Infrared imaging seeker

An infrared imaging seeker uses the thermal radiation difference between the target and the background to form the image of the target and the surrounding scenery to realize the automatic guidance of the missile. Its components include an optical system, imaging system, infrared detection system, and servo system, which involves infrared image detection and automatic target recognition, and other technologies. According to the improvement of technology in related fields, the infrared imaging seeker will develop toward dual-color, high-resolution staring infrared, intelligence, light miniaturization, and generalization in the future. The infrared imaging seeker is often used in anti-tank missiles, air-to-air missiles, and anti-ship missiles. The more representative applications are the American FGM-148 "Javelin" anti-tank missile, the AIM-9X "Sidewinder" air-to-air missile, and the Norwegian "Penguin" MK2 series of anti-ship missiles.

3) Laser seeker

The working principle of the laser seeker is to receive the target diffuse reflection of the laser, track the target, and process the laser signal to get guidance instructions. As shown in Figure 7.20, it is composed of an optical system, photoelectric conversion device, electronic circuit, and servo system, etc. The laser seeker has high guidance precision and target resolution, but it is susceptible to cloud and smoke and can't fight in all-weather.

Figure 7.20 Laser seeker

According to the setting of laser irradiator, laser seeker can be divided into semi-active laser seeker and active laser seeker two kinds.

(1) Active laser seeker

The laser irradiator of the active laser seeker is installed on the seeker head, which can take the initiative to irradiate the target and receive the reflected laser signal, carry on the matching analysis to the laser imaging characteristics of the target, and output the guidance instruction to guide the missile to track and strike the target. The active laser seeker can be used in the field of cruise missiles and air-to-ground missiles. Typical applications include the American LOCAAS "low-cost autonomous attack system" cruise missile.

(2) Semi-active laser seeker

There is only a receiving system in the semi-active laser seeker. The laser signal is emitted to the target by the external laser transmitter. After diffuse reflection on the target, it is transmitted to the receiving system of the seeker through the atmosphere, and the guidance is carried out by processing the laser signal. Semi-active laser seeker is mainly used in air-to-ground missiles and laser-guided bombs, etc. The American AGM-114A "Hellfire" missile, the M712 "Copperhead" cannon launched guided projectile and the Chinese LT-2 guided bomb all adopt the semi-active laser-guided guidance mode.

4) Multi-mode composite seeker

Multi-mode composite guidance refers to the combination of a variety of guidance sensors in a certain way to form a more superior performance guidance system, to achieve the purpose of accurate homing in the complex environment. The multi-mode composite seeker can make up for the deficiency of single guidance, give play to the advantages of various sensors, and improve the detection probability of the target and the anti-jamming ability of the seeker. Based on making full use of the existing homing guidance technology, the multi-mode composite guidance acquires a variety of spectrum information of the target and improves the combat ability of the precision guided weapon through information fusion technology. Figure 7.21 shows the composition of the multi-mode composite seeker.

Figure 7.21　Multi-mode composite seeker

Common multi-mode compound seekers have active radar/infrared dual mode compound seeker and laser/infrared dual mode compound seeker. In recent years, with the development of information fusion technology, some progress has been made in the research of three-mode composite seekers.

(1) Active radar/infrared dual mode compound seeker

Active radar/infrared imaging seeker after startup, completed by radar search and position to the target, found to identify the target in infrared imaging seeker after the switch to infrared imaging guidance mode, in the process of target tracking, infrared imaging guidance failing to correctly identify the tracking target, switched to active radar seeker guidance work mode. Active radar/infrared imaging composite guidance can effectively enhance the anti-jamming ability of the seeker and improve the hit probability of the missile.

(2) Laser/infrared dual mode compound seeker

Laser/infrared compound guidance is a kind of optical multi-mode compound guidance detection means. Combining the advantages of the two kinds of guidance methods, laser/infrared compound guidance can improve the performance index of precision guided weapons under complex conditions. Infrared imaging seeker has advantages of high sensitivity, high reliability, good concealment, but infrared imaging seeker has a short operating range and is easy to be affected by climate conditions. The laser seeker has a long operating range, rich goal information, and good anti-jamming performance. The laser/infrared compound seeker can play the advantage of the two seekers, so the laser/infrared compound seeker has a broad application prospect in the field of guidance.

In recent years, many countries in the world have studied missiles witha multi-mode composite seeker, mainly focusing on cruise missiles, air-to-ground missiles, and air defense missiles, among which the typical applications are the American JAGM Joint Air-to-ground Missile and the Chinese FL-3000N Shipborne Air Defense Missile.

7.2.2.2 Homing head requirements

The seeker is an important device of homing guidance system, and its accurate measurement and tracking of the target is very beneficial to improve the guidance precision of the missile. So the main parameters of the seeker must meet some requirements.

1) The distance R of finding and tracking the target

For example, R of a ground-to-air missile adopting homing guidance mode in its whole flight is decided by the inequality as follows:

$$R \geqslant \sqrt{(d_{\max} + V_t t)^2 + H_m^2} \tag{7-29}$$

where, $d_{\max}$ is the largest launch distance, V_t is the velocity of the target, t is the missile flight time, H_m is the difference between the missile flight height and the target flight height.

If a missile adopts homing guidance mode in its terminal flight stage, now R is related to the distance from which the missile begins to terminal guidance and not decided by the missile's largest flight range.

2) The angle of field of view (FOV) Ω

The angle of FOV of the seeker is a three-dimensional angle and the seeker watches the target at this angle. For the optical seeker, it is decided by the parameters of the seeker's optical system, and for the radar seeker, it is decided by the antenna characteristics and wavelength.

If we want to make the seeker resolution to be high, the angle of FOV should be small. If we want the seeker to track the target of high speed, the angle of FOV should be large.

For the fixed seeker, the angel of FOV should be larger than a certain angle, which can make the target not exceed the FOV when there is a delay in the missile system, i.e.

$$\Omega \geqslant \dot{\varphi} \tau \tag{7-30}$$

where, $\dot{\varphi}$ is the rotation rate of LOS, τ is the delay time of the system.

For movable tracking seeker, since the axis Ox of the seeker measure reference frame coincides with the LOS, it changes along with the LOS, which may reduce the requirements for the seeker's angle of FOV i.e. the angle of FOV may be not very large.

3) The shortest distance of stopping missile guidance

In homing guidance system, as the distance between the missile and target becomes shorter, the rotation rate of LOS becomes larger and the signal received by the seeker becomes stronger. There exists the shortest distance. If the distance between the missile and the target is shorter than the shortest distance, the signal received by the seeker will be so strong that it exceeds the capability of the seeker and the seeker won't pick the target movement signal anymore. The area closer than the shortest distance usually is called the "blind zone". Before the missile flies into the blind zone, the automatic tracing loop of the seeker should be shut down.

4) The turning range of the seeker

The seeker usually is installed in a framework whose rotation relative to the missile body is limited by the space and mechanical structure, so the rotation angle is also limited and usually is in the range of $[-40°, 40°]$.

Other requirements include the capability of tracking the target rapidly and stably, the high measuring precision of guidance parameters, good electromagnetic compatibility and reliability, the capability of capturing the target rapidly and the high probability of capturing the target, the good suitability in the electromagnetic environment and variable flight environment.

7.2.3 Star sensor

The star sensor measures the angular position of the star relative to the spacecraft and compares it with the angular position parameters of the star in the ephemeris to determine the attitude of the spacecraft. The star sensor is the most accurate attitude sensor at present, which can provide inertial attitude information at the arcsecond level or even higher. The attitude information of the star sensor comes from two aspects: one is the direction of the star light direction vector in the inertial coordinate system, and the other is the direction of the star light direction vector in the sensor coordinate system. Due to the small field angle of the stars, after hundreds of years of astronomical observation, their position in the inertial space is accurately known, that is, the star catalog provides a high-precision inertial reference datum.

The star sensor has the characteristics of low power consumption, small size, high accuracy and strong autonomy. Moreover, it can be used to correct the inertial navigation system because it has no accumulated attitude error. It is currently widely used in satellites, ships and spacecraft. However, traditional star sensors usually have some problems for the microsatellites, such as large mass, high cost and complex internal algorithm.

According to its development stage, star sensors can be divided into three types: star scanner, gimbaled star tracker and fixed-head star sensor.

7.2.3.1 Star scanners

The star scanner, also known as the star map instrument, generally has a slit field of view, which is suitable for spinning satellites. It uses the rotation of the spacecraft to search for and capture the target star.

The star scanner includes optical system, imaging device (such as slit), detector and electronic circuit, etc. The structure diagram is shown in Figure 7.22. The star scanner uses the spin of the spacecraft to scan celestial bodies. A star is detected when it passes through the optical system and passes through the slit in the focal plane. If the signal exceeds the threshold, the electronic device generates a pulse to indicate the appearance of the star. By detecting the time and signal amplitude of the star passing through the slit, and then combining the ephemeris and the spin speed of the spacecraft, the spacecraft's attitude information can be calculated.

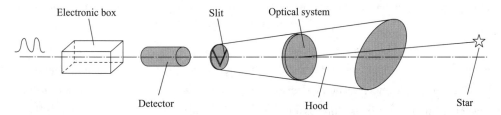

Figure 7.22　Structure of the star scanner

The star scanner has no moving parts, so it hasa simple structure and high reliability. However, due to the low signal-to-noise ratio of the system, it is seriously limited in engineering practice and has been basically eliminated.

7.2.3.2　Gimbaled star trackers

The gimbaled star tracker, which installs the sensitive head on a rotatable gimbal, searches and captures the target star by rotating the gimbal.

The gimbaled star tracker includes optical system, imaging device, detector, electronic circuit, gimbal with two degrees of freedom, angle encoder and servo mechanism, etc. The structure diagram is shown in Figure 7.23. The star is imaged on the detector through the optical system and imaging device, then the position and size of the star image in the field of view are detected by the processing circuit. According to the detection results, the mechanical frame is rotated driven by the servo mechanism in order to keep the star image in the center of the instantaneous field of view. Finally, the attitude of the spacecraft is determined on the basis of the information of the star as well as the size and direction of the frame angle.

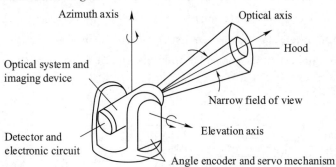

Figure 7.23　Structure of the gimbaled star tracker

The gimbaled star tracker has a small instantaneous field of view, but a large tracking field of view. The gimbal motion is used to expand the tracking field of view, but with the introduction of mechanical movable parts, the structure is complex and the reliability is low.

7.2.3.3　Fixed-head star sensors

The sensitive head of the fixed-head star sensor is fixed relative to the spacecraft, and it has the ability of searching and tracking in a certain field of view. The fixed-head star sensor uses the photoelectric conversion device to sense the star, and the processing circuit scans and searches the field of view to obtain and identify the star, and then determine the attitude of the spacecraft. This type of star sensor has a cone-shaped field of view, which is easy to determine the position of the star image, and there are no mechanical moving parts, so it has high reliability and wide application prospects.

At present, the CCD star sensor, which uses charge-coupled devices as photosensitive devices, has been widely used in engineering because of its good image quality, high resolution, and relatively mature technology development. The most important components of the CCD star sensor are the image sensor, imaging circuit and image processing circuit, which are the most important parts of the star sensor that affect the measurement accuracy. In addition, the hardware consists of a hood, an optical lens, a power supply and a data interface, as shown in Figure 7.24.

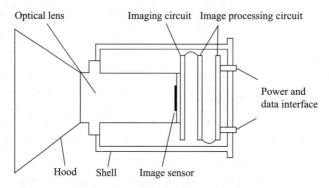

Figure 7.24 Structure of the CCD star sensor

The working process of the CCD star sensor is described in Figure 7.25. The star sensor images the star through the lens and converts the energy of the star light into the analog electrical signal. Then the digital image is obtained by analog-to-digital conversion. In the shooting process, a lot of noise will be produced, so the star map is denoised by preprocessing first. Based on the star map with a low noise level, star points are extracted to obtain the position information of star points. Then the star information is compared with the star database, and the star information is obtained by star map recognition. Finally, the information of the current spacecraft is calculated by attitude calculation.

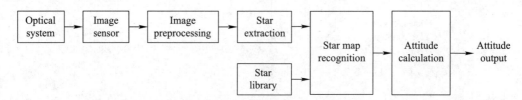

Figure 7.25 The attitude measurement principle of the CCD star sensor

With the development of large-scale integrated circuit technology and CMOS technology, active pixel sensor (APS) technology is applied to star sensors. Figure 7.26 and Figure 7.27 show two kinds of star sensors. The new CMOSAPS star sensor is the development direction of the star sensor due to its high integration, no charge conversion, and large dynamic range.

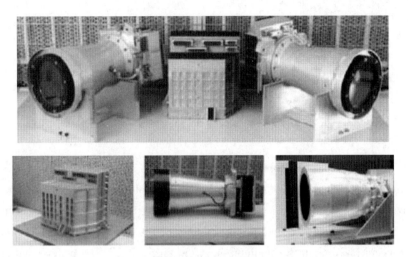

Figure 7.26　The HAST star sensor

Figure 7.27　The MAI-SS star sensor

7.2.4　Actuation

The control output command signal is amplified, transformed, and then transmitted to the actuation devices which finally change the control surface or thrust vector to implement the control function. So actuation devices are important components of the control system whose characters will usually directly determine the flight performance and the design of the control system.

The actuation process is illustrated in Figure 7.28. And its main parts consist of one or more control surfaces and a rudder loop. The control surfaces are usually elevators, rudders and ailerons. The rudder loop is composed of amplifier, convertor, drive device, operating mechanism and transducer for a feedback signal. Especially, the combination of amplifier, convertor and derive device is called a rudder.

According to the working principles, the actuating parts are divided into three types, which are proportional actuators, relay-operated actuators and pulse-width modulation actuators. According to the used power sources, there are about four kinds of actuators as follows, atmospheric pressure actuator, hydraulic pressure actuator, magnetic actuator and motor-driven actuator.

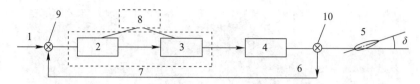

Figure 7.28 Schematic block diagram of actuation process

1—Control command; 2—Amplifing-convertion device; 3—Drive device; 4—Operating mechanism;
5—Control element; 6—Feedback loop; 7—Rudder actuator; 8—Energy; 9,10—Synthesizer.

7.2.4.1 Atmospheric pressure actuator

There are some typical characteristics of the atmospheric pressure actuator. That is, the construction is relatively simple, and it has high reliability. But it reacts slowly, and with large delay.

Here, two types are divided according to different air sources. One is the cold air type and the air is from the compressed air in a container shown as Figure 7.29. The compressed air is from the container to drive the actuator to work. Then the control surface is forced to turn. In a stored gas system, working pressures are typically about $5MN/m^2$ ($1MZ/m^2 = 9.87$atm) and this is much higher than the effective pressure obtainable in any electric motor, which makes for a small final control surface.

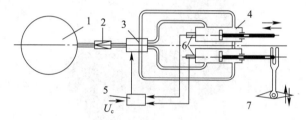

Figure 7.29 Atmospheric pressure actuator using compressed air

1—Compressed-air container; 2—Pressure reducing valve; 3—Pilot valve; 4—Actuator; 5—Sensor element;
6—Potentiometer; 7—Control element.

Another type is that the driving gas is from an engine or a gas generator. The advantage of burning cordite or some other mono-fuel such as isopropyl nitrate and using the hot gas to drive the actuator is the reduction of the size and hence the weight of the fuel container. The cordite can burn at the supply pressure and need not be stored at ten times the supply pressure. Since cordite burns at a greater rate at a higher pressure, a relief valve must be used to allow generated gas to escape to the atmosphere when demand is low. Figure 7.30 shows a typical arrangement of a hot gas servo using an equal area double-acting piston.

7.2.4.2 Hydraulic pressure actuator

The basic work process of the hydraulic pressure actuator is similar to the atmospheric pressure actuator but the energy source is from the hydraulic oil.

Hydraulic natural frequencies are high on account of the much greater bulk modulus of hydraulic fluid compared with that of nitrogen used in the atmospheric pressure actuator. Very high-performance servos can be designed using hydraulic power, for example, bandwidths of 50Hz or more can be obtained. And hydraulic pressure actuator can have a very large power so it is very suitable to drive parts with large inertia like moving wings. Besides, the hydraulic pressure actuator reacts fast

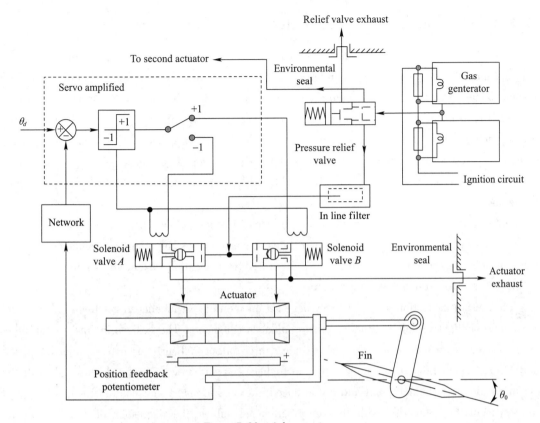

Figure 7.30 A hot gas servo

and with a small delay. But it is usually more expensive, relatively more complicated in construction and hence with bigger size and weight. Hydraulic servos are not made commercially below about 8kW and tactical missile servos rarely require such large powers. So the actuating part with hydraulic pressure type is mainly used on the medium and long-range strategic missiles.

7.2.4.3 Magnetic actuator

The actuating part with magnetic type has the characteristics of simple construction, light weight, and high reliability. But the power is relatively small because the energy source is from magnetic fields. In general, it is used in small-scale flight vehicles.

7.2.4.4 Motor-driven actuator

The actuating part with motor-driven type has a power source of the motor with D. C. or A. C. Therefore, it can output large power. It is simple, convenient to manufacture, and with high reliability. But it reacts slowly. In the past, owing to the disadvantages of heavy weight and slow reaction, the uses are restricted. With the development of science and technology, it has been improved greatly.

7.3 Guidance system

Generally, there are four types of guidance systems. They are autonomous guidance system, remote guidance system, homing guidance system, and combination guidance system.

7.3.1 Autonomous guidance system

An autonomous guidance system means that a missile is guided onto the target autonomically by preprogramming before launching or reference basis from outer fixed reference points just like celestial bodies, satellites, and geographic region. It means that the missile does not receive the command from outside during the flight procedure. That is to say, it depends on itself. This is the reason that we call it an autonomous guidance system. Usually, missiles with autonomous guidance systems are used to attack the target at a fixed location or the target whose flight path has been known. The main advantage of using this guidance system is that it has the ability of anti-interference.

There are several kinds of autonomous guidance systems, which include inertial navigation system, celestial navigation system, Doppler radar navigation system, planning guidance system, terrain-matching system, Global Positioning System (GPS) navigation system and so on.

7.3.1.1 Inertial navigation system

The inertial navigation system (INS) depends on inertial instruments to get flight velocity and coordinates of flight vehicle and hence create guidance commands. The basic principle is that, three accelerometers are used to measure the missile's three orthogonal accelerometers projected in the inertial space, and through two integrators the flight speed and range can be determined; with the known coordinate of launch point and initial speed, the missile speed and coordinates can be got at every moment; by comparing with the pre-programmed values, the displacement can be got and transformed to guidance commands, which can lead the missile flight to the target following the pre-planning trajectory. The schematic block diagram of the inertial guidance system is shown in Figure 7.31.

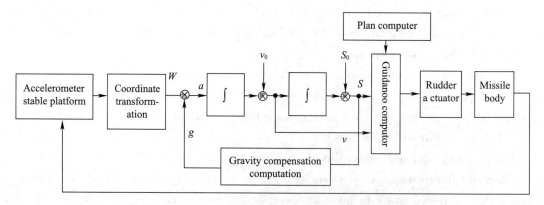

Figure 7.31 Schematic block diagram of inertial guidance system

During the flight, to a degree, the missile's attitude is always changing. It isn't proper to mount the accelerometers directly on the missile body because if that, the sensitive axis of accelerometers will be changing, which is not allowed for inertial guidance except the missile attitude is known at every moment. So the accelerometers are usually installed on a gyroscope-stabilized platform which can supply a fixed and stable reference frame relative to inertial space. A gyroscope-stabilized platform equipped with three accelerometers is shown in Figure 7.32.

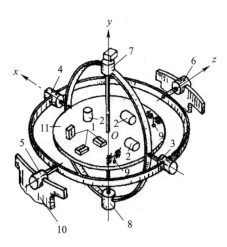

Figure 7.32 Gyroscope-stabilized platform
1—Accelerometers; 2—Gyros; 3,5,7—Angle transducers; 4,6,8—Torque motors;
9—Gravity pendulum; 10—Mounting base; 11—Platform.

The gyroscope-stabilized platform mainly consists of three rate integrating gyroscopes, three torque motors, three-angle transducers, a platform and a set of gimbals. The three rate integrating gyroscopes are orthogonally equipped on the platform and are used to sense the change rate of the three attitude angles of the missile. Then the output signals of gyroscopes are used by the torque motors to turn the gimbals so to keep the platform fixed and stable relative to the inertial space. To make the rotation of torque motors more accurate, three-angle transducers are necessary. If the three axes of the gyroscope-stabilized platform are consistent with those of the launch reference frame, the accelerometers mounted on the gyroscope-stabilized platform will gain the missile's accelerations along the axis of the launch reference frame during the whole flight, which are important for the inertial navigation system.

There are two main types of inertial navigation systems in modern use, they are, inertial navigation system stable relative to inertial space and inertial navigation system stable relative to the local geographic level. In the later type, the platform is isolated from the disturbance of the missile body as well as is controlled by the computer to be consistent with the local geographic level. The former type is commonly used in strategical missiles and carrier rockets, and the latter type is usually used in cruise missiles. With the rapid development of computer technology in recent years, a new inertial navigation system named strapdown inertial navigation system appears which doesn't need the gyroscope-stabilized platform anymore. In this system, the gyroscopes and accelerometers are directly fixed on the missile body, and the computer on board is used to handle the effect on the measurement of accelerometers, in other words, the computer plays the role of a digital stabilized platform. Due to the attitude change and vibration of the missile body, strapdown inertial navigation system is hard to get as high navigation accuracy as the conventional inertial navigation system with a gyroscope-stabilized platform gets. But the strapdown inertial navigation system is much cheaper and has higher reliability, so it is used commonly in some new cruise missiles.

The inertial navigation system can be used to guide the missile all on its own. So it has a strong ability of anti-interference. While, because of the gyroscope drift errors and calculation errors, the accumulated errors will be large for missiles with long flight time. So inertial navigation system is not usually used alone but combined with other kinds of guidance systems.

7.3.1.2 Celestial navigation system

By measuring the stars, the celestial navigation system is used to determine the position of the missile. Generally, a star or two stars are used. Figure 7.33 shows a celestial navigation system by means of two stars. The sextant is used to measure the angles between the stars and the local level which are called elevation angles. So the sextant must be installed on a gyroscope-stabilized platform, which will supply a constant local horizontal plane. Using the measured two elevation angles, certain azimuth information of the missile and the known data of the relative positions between the earth and the stars, the position of the missile is computed by the computer.

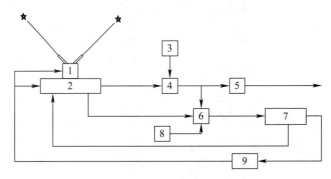

Figure 7.33 Celestial navigation system using two stars
1—Sextant; 2—Stabilized platform; 3—Pre-data; 4—Computer; 5—Terminal control device;
6—Autopilot; 7—Missile body; 8—Altimeter; 9—Motion relation.

The celestial navigation system works autonomically, so it has a strong ability of anti-interference and its navigation accuracy only lies on the errors of instruments. But for the celestial navigation system, if the sextant can't see the stars, the system will fail to be used. So the celestial navigation system is usually combined to use with other guidance systems, e. g. inertial navigation system, in which case, the celestial navigation system plays the role of correcting the error of inertial navigation system. Due to the complex composition and large weight, the celestial navigation system is usually used only on missiles with long range and some spacecrafts.

7.3.1.3 Doppler navigation system

The principle of the Doppler navigation system is on the basis of Doppler effects. The distance variation between the vibration source and the observation point has a relation with the change of the vibration frequency measured at the observation point. And the quantity of the frequency variation has a proportional relation with the rate of distance variation.

Figure 7.34 illustrates the principle of the Doppler navigation system. Assuming the missile cruises at a fixed altitude with the speed of V, through the antenna, the transmitter on the missile transmits the radio signal to the ground which has a frequency of f and a direction of φ_b relative to the missile longitudinal axis. The reflected wave with the frequency of f' is received by the receiver on board and the Doppler frequency f_d, which is the change of frequency between the transmitted and received radio, is obtained by the mixer. As the distance change rate is $V\cos\varphi_b$, so we have

$$f - f' = f_d = \frac{2V}{\lambda_d}\cos\varphi_b \qquad (7-31)$$

where λ_d is the wavelength.

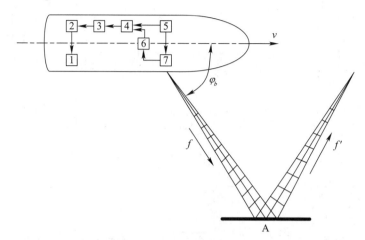

Figure 7.34 Schematic block diagram of the Doppler navigation system
1—Integrator; 2—Frequency meter; 3—Amplifier; 4—Mixer; 5—Transmitter; 6—Receiver; 7—Antenna.

From Eq. (7-31), we can see that the missile's cruising speed is in direct proportion to the Doppler frequency. The frequency meter is used to produce a voltage signal which is in direct proportion to the Doppler frequency, and by integrating this signal from the launching moment, the flight range is obtained.

If only one radio signal is transmitted, the flight speed and range can be obtained. If two radio signals are used, both the longitudinal speed and horizontal speed can be determined. And if four, the missile rotation rate can be obtained, too.

The Doppler navigation system can be used at any climate and terrain conditions. Due to errors of measuring instruments, the accumulated guidance error will become larger with the increase of flight range. Besides, the Doppler navigation system has low anti-interference ability because of its transmitting radio. So it is usually combined to use with other guidance systems and plays a role in correcting guidance errors.

7.3.1.4 Planning guidance system

The flight planning is pre-programmed at the planning guidance system. There is an example as following.

As seen in Figure 7.35, the planning is pre-programmed for controlling the pitching movement. Here, the pitching attitude angle can be as the planning.

During the cruise flight, the flying altitude should keep constant. If there is a deviation of flight altitude, the deviation will be reduced according to the planning guidance system shown as Figure 7.36.

7.3.1.5 Terrain-matching guidance system

Terrain-matching guidance system utilizes the known area to be as the pre-fixed terrain under the flight path, and the trajectory is corrected according to the comparisons between the measured

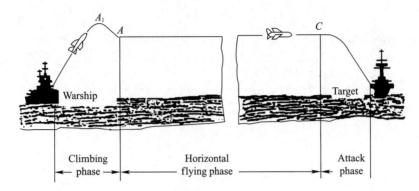

Figure 7.35 Flight trajectory

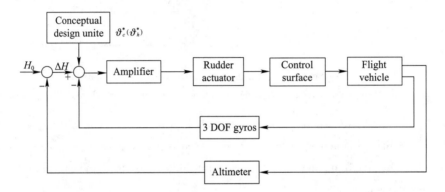

Figure 7.36 Planning guidance system loop

and pre-fixed terrain. Figure 7.37 shows the principles of the terrain matching guidance system. Terrain matching guidance system is useless in areas without terrain difference, e. g. the sea level and plain field.

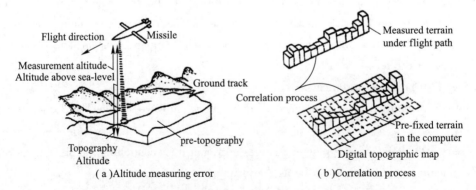

Figure 7.37 Terrain-matching guidance system

7.3.1.6 Satellite navigation system

By means of the satellite system, the position and speed of any object on the ground surface and low earth orbit can be determined. The satellite system is a new and powerful navigation system.

The typical satellite navigation system is the U. S. Global Positioning System, which has been widely used in military, civil and commercial fields all over the world. Besides, there are other satellite navigation systems running or planned such as the Russian GLONASS satellite navigation system, the European Galileo navigation satellite system and the Chinese BeiDou (Compass) navigation satellite system.

The Global Positioning System (GPS) is a space-based global navigation satellite system that provides reliable location and time information in all weather and at all times and anywhere on or near theearth when and where there is an unobstructed line of sight to four or more GPS satellites. It is maintained by the United States government and is freely accessible by anyone with a GPS receiver.

GPS was created and realized by the U. S. Department of Defense and was originally run with 24 satellites. It was established in 1973 to overcome the limitations of previous navigation systems. It consists of three parts: the space segment, the control segment, and the user segment. The U. S. Air Force develops, maintains, and operates the space and control segments. GPS satellites broadcast signals from space, which is used by GPS receivers to calculate its three-dimensional location (latitude, longitude, and altitude) plus the current time. The space segment is composed of 24 to 32 satellites in medium earth orbit and also includes the boosters required to launch them into orbit. The control segment is composed of a master control station, an alternate master control station, and a host of dedicated and shared ground antennas and monitor stations. The user segment is composed of hundreds of thousands of U. S. and allied military users of the secure GPS Precise Positioning Service, and tens of millions of civil, commercial, and scientific users of the Standard Positioning Service.

Figure 7. 38 shows the basic principles of the receiver measurement system using GPS L-band signals. Figure 7. 39 illustrates the principles of the track and measurement system using GPS S-band signals.

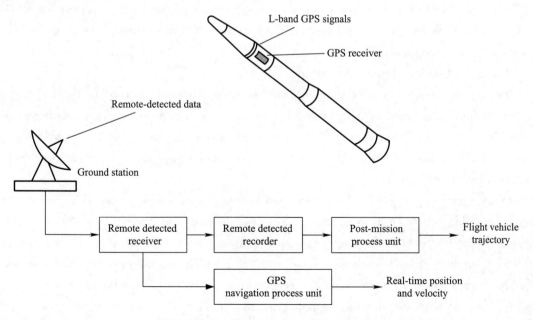

Figure 7. 38　Receiver measurement system

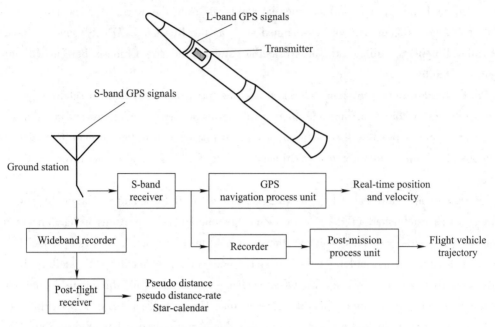

Figure 7.39 Track and measurement system

GPS has been used in many kinds of missiles. A lot of designers focus on the development and application of the GPS.

7.3.2 Remote guidance system

A remote guidance system means that the command to control the missile is from the remote guidance station. The guidance equipment is partly installed on the missile, and the main parts are installed at the guidance station, which can be on the ground, in the airplane, or on the warship. The remote guidance system can be separated into the command guidance system, wave beam guidance system and radio navigation system.

7.3.2.1 Command guidance system

To form the guidance command, the guidance station must continually compute the deviation of the missile flight path based on the guidance law and the motion parameters of the missile and the target. Then the guidance command will be transmitted to the missile. So the command guidance system is mainly composed of three parts which are observation equipment, command forming device and command transmission part.

Observation equipment is used to observe and measure the motion parameters of the missile and target. According to differences in target characters, missile types and guidance law, the observation equipment may be visual equipment, optical equipment, radar equipment or TV equipment. Depending on different guidance laws, in some cases (e.g. three-point guidance law) the observation equipment just needs to measure the relative position between the missile and the target, but in other cases, it may need to measure the accurate position and motion parameters of the missile and target. Command forming device forms the command by computer based on the observed motion parameters of the missile and target and the guidance law adopted. The command transmission part transmits the command signals to the missile through such ways as power cable or radio and so on.

Figure 7.40 gives the schematic block diagram of a typical command guidance system.

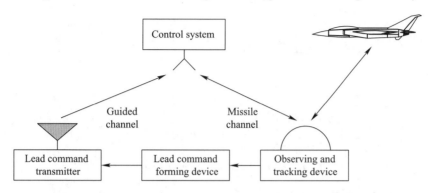

Figure 7.40 Schematic block diagram of command guidance system

The command guidance system is divided into manual command guidance system, semi-autonomic command guidance system and autonomic command guidance system. They are illustrated as follows.

Figure 7.41 shows the basic principles of the manual command guidance system. On the basis of the human brain, eye-sight and hands, the command is computed to control the missile onto the LOS to target.

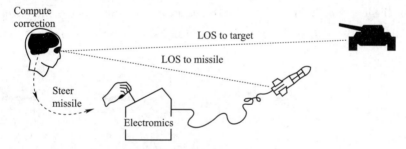

Figure 7.41 Manual command guidance

Figure 7.42 illustrates that the guidance command is formed by the computer and command signal generator according to the man's actions. The shooter only needs to aim the sight at the target.

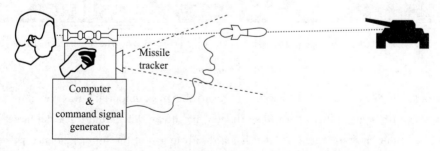

Figure 7.42 Semi-autonomic command guidance

As seen in Figure 7.43, the guidance command is automatically formed by the computer, missile and target tracker without the participation of man during the process of the command formation. Missile and target are observed and tracked automatically. The position data of the missile and target are passed to the computer, then calculated and coded to form commands which are sent automatically to the missile over the command link.

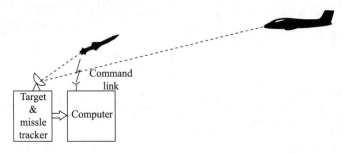

Figure 7.43　Autonomic command guidance

7.3.2.2　Beam guidance system

It canbe divided into radio beam guidance system and laser beam guidance system. The missile is expected to fly in the center of the beam. If there is displacement between the missile and the center line of the beam, the sensor on the missile will measure the displacement and then form control command to "pull" the missile back onto the center line. The beam tracks and irradiates the target, so the missile will be lead onto the target finally.

Figure 7.44 shows two kinds of beam guidance systems. One is to observe and track the missile and target with single beam, the other is with two beams. In the latter system, the target is found by the warning radar first, and then the guidance command is computed by means of another beam system.

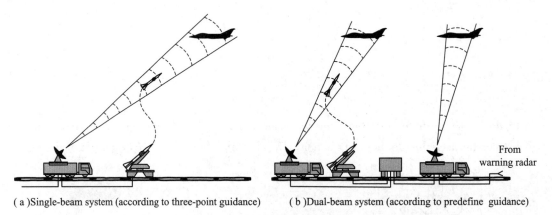

(a) Single-beam system (according to three-point guidance)　　(b) Dual-beam system (according to predefine guidance)

Figure 7.44　Beam guidance system

Using radar to detect and track targets is one of the most important ways in the radio beam guidance system. It has the ability of anti-disturbance, and it can detect and track the target with high speed and can handle multiple targets. But it has a disadvantage of the lower guidance precision with the increase of the flight range.

The laser beam guidance system is an important way of beam guidance system. It has the advantages of higher guidance precision and being difficult to be disturbed. But it is limited by the power factor, so the laser beam guidance system is only utilized for the short-range missile.

7.3.3 Homing guidance system

A homing guidance system receives specific energy from the target by means of the device installed on the missile, with which the relative position of the missile and the target is determined. At the same time, the control signals are produced to guide the missile onto the targets automatically.

Based on the different signal sources, homing guidance system can be divided into three basic categories which are active homing guidance system, semi-active homing guidance system and passive homing guidance system as illustrated in Figure 7.45. The transmitting and reflection waves in homing guidance system can be radar wave, infrared wave, sound wave, laser, visible light and so on.

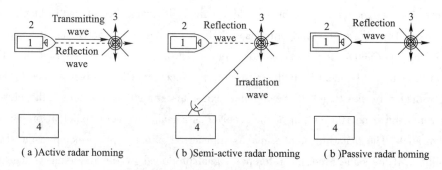

(a) Active radar homing (b) Semi-active radar homing (b) Passive radar homing

Figure 7.45 Autonomic homing guidance system

1—Guidance unit in missile; 2—Missile; 3—Target; 4—Control station.

Figure 7.46 shows the principles of the active homing guidance system. The missile is equipped with the power source on board beams energy to target and receives the reflected wave to guide to missile onto the target.

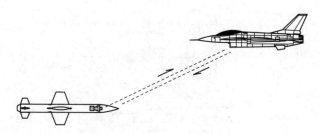

Figure 7.46 Active homing guidance

In Figure 7.47, the principles of the semi-active homing guidance system are given. There is only the receiver on the missile. The missile receives the reflected signal from the target while the signal is beamed from a device, which is not on the missile. Usually, the beam source may be on the ground, airplanes or other guidance stations.

A passive homing guidance system means that the receiveron the missile can attain the energy radiated from the target and there is no extra beam source outside.

According to different physical characteristics of the targets, such as radar, infrared, laser, TV and so on, the relative different homing guidance systems are utilized.

In the radar homing guidance system, the radar seeker is used to measure the position and

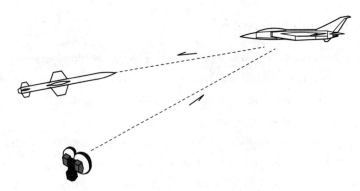

Figure 7.47 Semi-active homing guidance

motion states of the target relative to the missile. Figure 7.48 gives the schematic block diagram of the radar seeker. The transmitter and antenna switch is needed in an active radar seeker but not necessary in the semi-active and passive radar seeker. The antenna switch is used in active radar seeker to let the antenna be used by both transmitter and receiver. The receiver is used to receive the radio wave reflected by the target. If the antenna axis aims straight at the target, the output of the receiver is zero. And if not, the output of the receiver is a maladjustment signal in proportion to the displacement angle. The maladjustment signal is transferred into the phase-comparison circuit after being amplified. At the same time, the reference signal produced by the reference voltage generator and with the same frequency as the antenna is also transferred into the phase-comparison circuit after being amplified. After comparison between the maladjustment signal and the reference signal, the deviation signal is obtained which will be used by both autopilot and the drive device of the antenna. The autopilot controls the missile to fly to the target and the drive device of the antenna is used to keep the antenna aiming at the target.

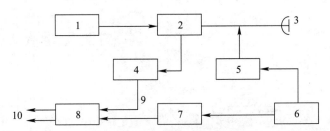

Figure 7.48 Schematic block diagram of radar seeker

1—Transmitter; 2—Antenna switch; 3—Antenna; 4—Receiver; 5—Drive device; 6—Reference voltage generator;
7—Amplifier; 8—Phase-comparison circuit; 9—Error signal; 10—Control signal.

In the semi-active radar homing guidance system, the transmitter is usually mounted on the ground, an airplane or a ship, which will supply large transmitting power. So compared with the active radar homing guidance system, the maximum working range of the semi-active radar homing guidance system is much larger. Obviously, the disadvantage of homing guidance system is being easy to be interfered.

7.3.4 Combination guidance system

Generally speaking, each kind of guidance system has its own advantages and disadvantages. If combining two or more kinds of guidance systems, we can make full use of these advantages and design such guidance systems with high performance, high reliability, high adaptability and wide use conditions. In some cases, e. g. for medium-range and long-range missiles, the combination guidance system is essential. For example, after the missile is launched the autonomous guidance system is used in the initial navigation to lead the missile to a certain required area; the remote command guidance system is used in the midcourse guidance to lead the missile near to the target accurately; in terminal guidance, the homing guidance system is used to attack the target.

Figure 7.49 illustrates the combination of the inertial navigation system and terrain-matching guidance system, which is widely used on cruise missiles. At the same time, there are many other different combination guidance systems, e. g. inertial guidance combined with Doppler navigation system, celestial navigation system combined with Doppler navigation system and so on.

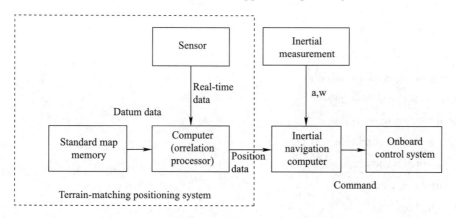

Figure 7.49 A combination guidance system

In the working process of the combination guidance system, one important thing is that the switch of guidance systems must be smooth. Based on the different combination methods during the whole flight phase, the combination guidance system can be classified into three types: series combination guidance system, parallel combination guidance system and series-parallel combination guidance system. Series combination guidance means that a single different guidance system is used in a different flight phase. Parallel combination guidance means that more than one different guidance systems are used together during the same flight phase. And the series-parallel combination guidance means that series combination guidance and parallel combination guidance are both included during the whole flight phase.

7.4 Guidance laws

7.4.1 Autonomous guidance law

Autonomous guidance means that the missile seeker receives energy radiated or reflected by the target, and then the guidance command leading the missile to the target is formed automatically. In the autonomous guidance system, the relative location between the missile and target is shown in Figure 7.50.

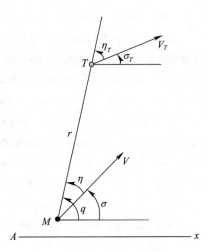

Figure 7.50 Relative location between the missile and target

Define:

M is the position of the missile,

T is the position of the target,

Ax is inertial reference line,

r is range between missile and target,

q is line of sight angle,

V is missile velocity,

V_T is target velocity,

σ is missile flight path angle, that is, angle between the missile velocity vector and inertial reference line,

σ_T is target flight path angle,

η is pre-setting angle of missile velocity, that is, the angle between the missile velocity vector and LOS,

η_T is pre-setting angle of target velocity.

According to the relation between the missile and target described in Figure 7.50, the relative movement equations between them are determined as follows:

$$\begin{cases} \dfrac{dr}{dt} = V_T\cos\eta_T - V\cos\eta \\[4pt] r\dfrac{dq}{dt} = V\sin\eta - V_T\sin\eta_T \\[4pt] q = \sigma + \eta \\ q = \sigma_T + \eta_T \\ \varepsilon_1 = 0 \end{cases} \qquad (7-32)$$

The first and second equations of Eq. (7 – 32) describe the change rate of range between the missile and target and the change rate of LOS angle respectively. The third and fourth equations of Eq. (7 – 32) describe the geometry relation between the missile and target. The last equation is the guidance relation equation decided by the guidance method.

There are several typical autonomous guidance methods as follows:

(1) Tracking Law (Pursuit). The direction of the missile velocity vector points to the target during the whole attacking process. The guidance relation equation is expressed as

$$\eta = 0 \tag{7-33}$$

(2) Parallel Approaching Law (Constant-Bearing Law). The change rate of LOS equals zero, i. e. LOS keeps parallel in the space. The guidance relation equation is

$$\frac{dq}{dt} = 0 \tag{7-34}$$

(3) Proportional Navigation Low. The angular rate of the missile velocity vector is proportional to the angular rate of the LOS. The guidance relation equation is

$$\frac{d\sigma}{dt} = k \frac{dq}{dt} \quad (1 < k < \infty) \tag{7-35}$$

where k is the proportional coefficient. Furthermore, based on the equation $q = \sigma + \eta$, we can get

$$\frac{dq}{dt} = \frac{d\sigma}{dt} + \frac{d\eta}{dt} \tag{7-36}$$

Combining Eq. (7-35) and Eq. (7-36), we can get the other two expressions of proportional navigation law, which are

$$\frac{d\eta}{dt} = (1-k) \frac{dq}{dt} \tag{7-37}$$

$$\frac{d\eta}{dt} = \frac{1-k}{k} \frac{d\sigma}{dt} \tag{7-38}$$

As seen from Eq. (7-37), if $k = 1$, we get $d\eta/dt = 0$, which means that η is a constant, which is constant lead angle guidance law. The tracking law ($\eta = 0$) is a particular case of constant lead angle guidance law. If $k \to \infty$, we get $dq/dt \to 0$, which means that q is a constant, and this is parallel approaching law. So we can find that tracking law and parallel approaching law are both special cases of proportional navigation law. The value of k expresses that proportional navigation law is one kind of law, which is between tracking law and parallel approaching law. And in proportional navigation law k is selected in the range of $1 < k < \infty$, and the range is usually desired to be [2,6]. Trajectory characteristics of the three guidance laws are given in Figure 7.51.

Tracking law has the advantage of being simple to be realized, but it requires missiles to be of largely available overload and large maneuverability. The trajectory of tracking law is more curving. Parallel approaching law has the advantage of straighter trajectory and lower required overload, but the realization is so difficult that now parallel approaching law is barely adopted. Meanwhile, the characteristics of proportional navigation law are between them, i. e. simpler realization than parallel approaching law and much straighter trajectory than tracking law, so proportional navigation law is widely used. We must note that the choice of proportional coefficient k depends on many factors such as the missile's available overload, convergence condition of the angular rate of LOS and so on.

Besides, for the proportional navigation law, its variants have been treated extensively and used

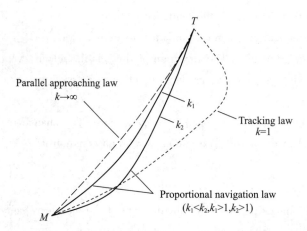

Figure 7.51 Trajectory characteristics of the three guidance laws

widely, which include biased proportional navigation (BPN), true proportional navigation (TPN), generalized proportional navigation (GPN), augmented proportional navigation (APN) and optimal proportional navigation (OPN).

7.4.2 Remote guidance law

The remote guidance missile is guided by a guidance station (control station), which is stationary or moving, so the movement of the missile is relative not only with the movement of the target but also with the movement of the guidance station. Here, three points are discussed, which are the locations of the missile, target and control station, defined as M, T and C respectively. Figure 7.52 describes the relative location between the missile, target and control station. R_M and R_T are the distance between the control station C and the missile M and the target T respectively. q_M and q_T are the angles of the sight of line CM and CT with the reference line Ax respectively. σ, σ_T and σ_C represent the angles between the speed vector V, V_T and V_C with Ax respectively.

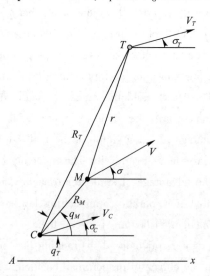

Figure 7.52 Relative location of missile, target and control station

According to Figure 7.52, the relative movement equations among a missile, a target and a guidance station are presented as follows:

$$\begin{cases} \dfrac{dR_M}{dt} = V\cos(q_M - \sigma) - V_C\cos(q_M - \sigma_C) \\[6pt] R_M \dfrac{dq_M}{dt} = -V\sin(q_M - \sigma) + V_C\sin(q_M - \sigma_C) \\[6pt] \dfrac{dR_T}{dt} = V_T\cos(q_T - \sigma_T) - V_C\cos(q_T - \sigma_C) \\[6pt] R_T \dfrac{dq_T}{dt} = -V_T\sin(q_T - \sigma_T) + V_C\sin(q_T - \sigma_C) \\[6pt] \varepsilon_1 = 0 \end{cases} \qquad (7-39)$$

Similarly, the last equation of Eq. (7-39) is the guidance relation equation determined by guidance law.

Typical remote guidance laws are given as follows:

(1) Three-point Law. In the whole process of missile intercepting target, guidance station, missile and target are on the same line. The guidance relation equation is

$$q_M = q_T \qquad (7-40)$$

(2) Pre-setting Law. The line of missile and guidance station is always ahead of the line of target and guidance station, and the angle between them is changing according to some laws. The guidance relation equation is

$$(q_M - q_T) = C_Q(R_T - R_M) \qquad (7-41)$$

where C_Q is a function, which tries to make the trajectory straighter.

The advantage of the three-point law is simple to be realized, but the trajectory is rather curving. Adopting pre-setting law may make the missile trajectory much straighter, but the realization of pre-setting law is more complex than that of three-point law.

7.4.3 Optimal guidance law

The guidance laws introduced above are all classical guidance laws. Generally speaking, the classical guidance laws have the advantages of simple structure, depending on little information and being easy to be realized, so classical guidance laws or their variants are used in most tactical missiles. But to the target with high maneuverability, the classical guidance laws are usually not competent. With the rapid advancements of computer science and technology, some modern guidance laws based on modern control theory have speedy development such as optimal guidance law, differential game guidance law, adaptive guidance law, differential geometry guidance law, feedback linearized guidance law, neural network guidance law, H_∞ guidance law and so on.

Compared with classical guidance laws, modern guidance laws have many advantages, e.g., small miss distance, attacking the target with required attitude, strong robustness, conquering target with high maneuverability, reasonable distribution of trajectory required overload, large combat area and so on. So, it is a very good choice to adopt modern guidance law in tactical missiles to attack targets with high speed, super maneuverability, and strong interference source on the future battlefield. While in practical application, the modern guidance laws usually have some difficulties because of their complicated structures and more measurement parameters needed. Fortunately, this situation is being improved due to the rapid development of microcomputers.

The advantage of the optimal guidance law is that it can reach the optimal solution based on given performance indexes considering various constraints. The performance indexes different forms based on different requirements. The performance indexes of the tactical missile mainly include minimum terminal miss distance, minimum control energy and minimum impact time, etc. The homing guidance system usually chooses a quadratic performance index, whose general form is

$$J = \int_0^T G(\boldsymbol{c}, \boldsymbol{u}, \boldsymbol{r}, t) \, \mathrm{d}t \tag{7-42}$$

where, $\boldsymbol{c}$ is the output of the system, $\boldsymbol{r}$ is the input of the system, $\boldsymbol{u}$ is the controlled variable, $G(\boldsymbol{c}, \boldsymbol{u}, \boldsymbol{r}, t)$ is called loss function, which represents the change of the actual performance versus the ideal performance over time.

The optimal control problem is to determine the controlled variable u to minimize the performance index J when the system is constrained.

However, it is difficult to solve this optimal control problem as the guidance law of the missile is a nonlinear problem with variable parameters and random disturbances. Therefore, the process of intercepting the target is usually linearized to obtain the approximate optimal solution of the system, which is easy to realize in engineering and is close to the optimal guidance law in performance.

The relative movement between the missile and target is shown in Figure 7.53. The quadratic linear optimal guidance problem is described below.

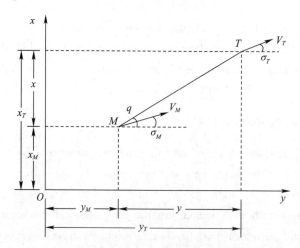

Figure 7.53 Relative movement between the missile and target

According to the relative motion relationship between the missile and target

$$\begin{cases} \dot{x} = \dot{x}_T - \dot{x}_M = V_T \sin\sigma_T - V_M \sin\sigma_M \\ \dot{y} = \dot{y}_T - \dot{y}_M = V_T \cos\sigma_T - V_M \cos\sigma_M \end{cases} \tag{7-43}$$

where, (x_T, y_T), (x_M, y_M) are the position of target and missile separately.

Because σ_T and σ_M are small, we suppose that $\sin\sigma_T \approx \sigma_T$, $\sin\sigma_M \approx \sigma_M$, $\cos\sigma_T \approx 1$, $\cos\sigma_T \approx 1$, therefore

$$\begin{cases} \dot{x} = V_T\sigma_T - V_M\sigma_M \\ \dot{y} = V_T - V_M \end{cases} \quad (7-44)$$

Let

$$\begin{cases} x_1 = x = x_T - x_M \\ x_2 = \dot{x} \end{cases} \quad (7-45)$$

The linearized missile's motion state equation can be written as

$$\dot{x} = Ax + Bu \quad (7-46)$$

where

$$x = (x_1, x_2)^T, A = \begin{bmatrix} 0 & 1 \\ 0 & 0 \end{bmatrix}, B = (0,1)^T$$

For any guidance system, the most important thing is that the miss distance of impact time t_f is minimal. Since the selected index is quadratic, it should be expressed as the square of the miss distance, i.e. $[x_T(t_k) - x_M(t_k)]^2 + [y_T(t_k) - y_M(t_k)]^2$.

To simplify the analysis, we usually choose the value of x when $y = 0$ as the miss distance, which means the smaller the value of x at time t_f, the better. Because the available overload of the missile is limited, the controlled variable u should also be limited. Therefore, the quadratic performance index function is selected as

$$J = \frac{1}{2}cx_M^2(t_k) + \frac{1}{2}\int_0^{t_k} Ru^2 dt \quad (7-47)$$

where c and R are positive real numbers.

Based on the optimal control theory, the optimal guidance law can be obtained as

$$u = -R^{-1}B^T P x \quad (7-48)$$

where P is determined by Riccati equation

$$A^T P + PA - PBR^{-1}B^T P = P \quad (7-49)$$

After P is obtained, the optimal guidance law can be obtained as

$$u = -\frac{(t_k-t)x_1 + (t_k-t)^2 x_2}{\dfrac{R}{c} + \dfrac{(t_k-t)^3}{3}} \quad (7-50)$$

To make the miss distance equal to zero, select $c \to \infty$, then

$$u = -3\left[\frac{x_1}{(t_k-t)^2} + \frac{x_2}{t_k-t}\right] \quad (7-51)$$

Based on the relative motion relationship between the missile and target, we can obtain

$$\tan q = \frac{x}{y} = \frac{x_1}{V_C(t_k - t)} \qquad (7-52)$$

if q is small, $\tan q \approx q$, then

$$q = \frac{x_1}{V_C(t_k - t)} \qquad (7-53)$$

$$\dot{q} = \frac{1}{V_C}\left[\frac{x_1}{(t_k - t)^2} + \frac{x_2}{t_k - t}\right] \qquad (7-54)$$

Substituting Eq. (7-54) into Eq. (7-51), we can get

$$u = -3V_C \dot{q} \qquad (7-55)$$

Based on the relationship between the controlled variable and the overload of the missile, we can obtain

$$u = -a = -V\dot{\sigma} \qquad (7-56)$$

So

$$\dot{\sigma} = \frac{3V_C}{V}\dot{q} \qquad (7-57)$$

It can be seen from Eq. (7-57) that when the inertia of the missile body is not considered, the optimal guidance law of the homing guidance is proportional navigation law, and its proportional coefficient is $3V_C/V$, which also proves that proportional navigation law is a good guidance method.

7.4.4 Differential game guidance law

Differential game theory is a theory to study the optimal strategy of both sides or multiple sides. It is an effective combination of optimal control theory and game theory and has obvious advantages in dealing with offense-defense confrontation problems. Differential game guidance law is different from optimal guidance law. The essential difference between them lies in the different assumptions about target maneuver trajectory and maneuver capability. The optimal guidance law assumes that the target maneuver strategy is completely known, while the differential game guidance law does not make specific assumptions about the target maneuver strategy. When intercepting the target, the missile can choose different pursuit strategies, and the target can also choose the corresponding countermeasures. This strategy confrontation process can be described by mathematical model:

Let the relative motion equation of missile and target be

$$\dot{x} = f(x, u, v, t) \qquad (7-58)$$

where, x is the system state vector, u is the missile control vector, v is the target control vector.

The initial condition is

$$x(t_0) = x_0 \qquad (7-59)$$

The performance index is

$$J(u, v) = \Phi[x(t_f), t_f] + \int_{t_0}^{t_f} L(x, u, v, \tau) \, dt \qquad (7-60)$$

For the missile, the optimal strategy $\bar{u}(t)$ should be sought to make $J(u,v)$ the minimum value, while for the target, the optimal strategy $\bar{v}(t)$ should be sought to make $J(u,v)$ the maximum value.

Differential games based on a simple model can be solved analytically. By using the Hamilton equation, the analytic solution of the problem can be obtained by solving a two-point boundary value problem. The analytical methods mainly include the Hamilton-Jacobi equation, the Bellman-Isaacs principle, etc. However, in most cases, the problem of confrontation is complicated and changeable, which makes obtaining the analytical solution a difficult task. In this case, numerical calculation methods including indirect methods and direct methods can be used to solve the problem. Indirect methods are the most commonly used methods, which convert the problem into a two-point boundary value problem through the first-order necessary conditions and then solve the problem by numerical methods. Direct methods search the expression of optimal strategies $\bar{u}(t)$ and $\bar{v}(t)$, and find the optimal control strategy point after several iterations.

7.4.5 Cooperative guidance law

Cooperative guidance law generally refers to the guidance method which can make multiple missiles fly to the target or predetermined position simultaneously, time consistency is the core of cooperative guidance. Now the cooperative guidance is mainly divided into open-loop cooperative guidance and closed-loop cooperative guidance.

7.4.5.1 Open-loop cooperative guidance law

The open-loop cooperative guidance law is the time-constrained guidance law. Before the missile is launched, the expected impact time of each missile is designated. The missile hits the target according to the expected impact time, and the cooperative attack of multiple missiles on the target is completed. There was no communication between the missiles during the entire flight.

The typical guidance law form of open-loop cooperative guidance is

$$a = kV\dot{q} + k_\varepsilon \varepsilon_T \qquad (7-61)$$

where, a is normal acceleration, $\varepsilon_T = t_{go} - \hat{t}_{go}$ is the error feedback term of the time-to-go. $\hat{t}_{go} = t_d - t$ is the desired remaining flight time, where t_d is the expected impact time and t is the current time. t_{go} is the estimated time-to-go. k_ε is the coefficient, which is one of the important factors affecting guidance performance, and can be designed according to different scenarios and tasks. It should be noted that a reasonable value of t_d is the key to cooperative guidance. If t_d is too big, it means the missile takes a detour and wastes energy. If t_d is too small, some slow missiles may not be able to arrive in time. The estimated time-to-go t_{go} directly affects the guidance accuracy, which can be calculated by analytical methods or numerical integration.

It can be seen from the expression that the guidance law is transformed into proportional navigation law when $\varepsilon_T \rightarrow 0$, which means the missile achieves two constraints of miss distance and impact time.

7.4.5.2 Closed-loop cooperative guidance law

In the closed-loop cooperative guidance law, there is no need to set the expected impact time before the missile is launched, and the cooperative attack of the missile is realized through the information exchange among each missile so that missiles can hit the target in cooperation. Closed-loop cooperative guidance can be divided into centralized cooperative guidance and distributed cooperative guidance according to the different communication structures between missiles.

Centralized cooperative guidance means that there are one or several central nodes in the whole cluster which can interact with all missiles during the guidance process. The leader-follower structure is a typical centralized structure, as shown in Figure 7.54.

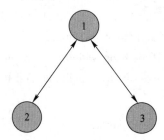

Figure 7.54 Leader-follower structure
1—Leader; 2,3—Follower.

The Leader can utilize the classical proportional navigation law. The followers can utilize the guidance law with the similar formation with open-loop cooperative guidance.

$$a_i = k_i V_i \dot{q}_i + k_{\varepsilon i} \varepsilon_{Ti} \qquad (7-62)$$

where, $\varepsilon_{Ti} = t_{go,i} - \hat{t}_{go,i}$ is the time error of i th follower, $t_{go,i}$ is the estimated time-to-go of ith follower, $\hat{t}_{go,i}$ is the expected time-to-go. Unlike the open-loop cooperative guidance, the expected time-to-go is not specified, but determined according to the leader, which can be represented as

$$\hat{t}_{go,i} = \hat{t}_{go}^L \qquad (7-63)$$

where $\hat{t}_{go}^L$ is the time-to-go of the leader.

Distributedcooperative guidance has the advantages of low communication cost and strong robustness. The missile in the cluster only needs to interact with several adjacent missiles for information exchange, as shown in Figure 7.55. The directed graph $G(V,E,A)$ is used to describe the communication relationship between missiles, where V is the set of nodes, E is the set of edges. $A = [a_{ij}]$ represents the weighted adjacency matrix. $a_{ij} = 1$ means that the i th missile can receive the information of j th missile. $a_{ij} = 0$ $(i \neq j)$ means that the i th missile cannot receive the information of j th missile. In order to realize time-coordinated guidance, the expected time-to-go of each missile is taken as the coordination variable

$$\xi_i = \hat{t}_{go,i} \qquad (7-64)$$

The distributed consistency coordination algorithm is given as

$$\dot{\xi}_i = - \sum_{j=1}^{n} a_{ij} (\hat{t}_{go,i} - \hat{t}_{go,j}) \qquad (7-65)$$

The control command coordination part of each missile involves the state information of all the missiles that can communicate with it. Through the interconnection of communication structure, the

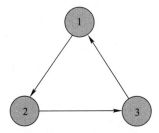

Figure 7.55   Distributed structure

state information is shared indirectly. Through the consistency algorithm, the flight time of all missiles is asymptotic, and the time cooperative attack of multiple missiles is realized.

7.5   Control system

As discussed before, for guided flight vehicles, the control system is an inner closed loop contained in the guidance loop as shown in Figure 7.1, which is composed of the autopilot and missile body. The autopilot is the controller while the missile body is just the controlled object.

Based on the relative motion between the missile and target and by means of the guidance law, after the work of the lead system, the control command is given. Then control system is used to produce control signals according to control commands to drive actuating mechanism and alter control surface, so to change or stabilize flight states.

7.5.1   Control modes

Pitching, yawing and rolling are three kinds of attitudes of the missile, which usually are called three channels to control. To realize the control for attitudes of the missile, attitude control modes could be classified into three kinds, namely, single-channel control, double-channel control and three-channel control.

7.5.1.1   Single-channel control

Some small-scale missiles whose diameters are small, rolling around its longitudinal axis can be controlled by only one channel, and this control way is called single-channel control.

In the single-channel control system, the missile rotates around the longitudinal axis with a constant angular velocity. The signal of the missile attitude angle is produced by the position gyroscope and transferred to the control equipment, which then sends the control signals synchronous with missile rotation. Generally, the control signal changes four times in one period of missile rotation, and it makes servo work to drive the control surface to swing from one limit position to another limit position like a relay. The swing of the control surface just depends on the polarity of the control signal and has nothing to do with its magnitude. Though adjusting the frequency of control force produced changes its direction and the alternating position, the missile may get the average control force needed, by which the missile motion in pitching and yawing can be both controlled. In the single-channel control system, the rolling of the missile around the longitudinal axis is necessary. If not, the control mode only can control the missile to move in one plane rather than in three-dimension space.

For there is only one servo, not too many pieces of equipment are needed on the missile, so its structure is simple and has high reliability. However, owing to the space motion is controlled by only one pair of control surfaces, many special problems must be considered for the guidance system. They mainly are as follows:

(1) The velocity reference frame and body-fixed reference frame are usually used to study the motion of the missile without rolling, while for rolling missiles, we must deal with the rolling reference frame. As we must study the relationship between missile and target in the fixed reference frame, some transformations between reference frames are needed which are usually more complex owing to the rolling.

(2) Movement coupling will be caused by the missile's rolling, which must be considered when we design the guidance system.

(3) The choice of rolling angular velocity and some other parameters should be especially considered when we design the guidance and control system.

(4) The maximum periodic equivalent control force of the rolling missile is just 64 percent of the instantaneous force produced by the control surface, which means control efficiency is low. In addition, the control surface must be of little inertia and be changed quickly, which are more strict requirements for the control surface.

7.5.1.2 Double-channel control

Usually, double-channel control means that pitching channel control and yawing channel control are used, and the rolling channel doesn't need to be controlled but is stabilized. This control way is called double-channel control, namely, rectangular coordinates control.

The schematic block diagram of double-channel control is shown in Figure 7.56, the principle of operation is as follows. Detecting devices measure motion parameters of missile and target in measuring coordinate system, and according to the guidance law, control commands in both pitching channel and yawing channel are formed, which includes calculations of guidance law, dynamic error and gravity compensation, etc. Control commands in two channels are used by the control system to alter the two control surfaces respectively.

The characteristics of double-channel control are as follows.

When we build the math model of the missile, we may regard that the equations are nonlinear and the disturbed trajectory is very near to the undisturbed trajectory. So the disturbed motion equations can be linearized on the undisturbed trajectory. Ignoring some small variables we can divide the disturbed motion equations into two parts, which are pitching disturbed motion equations and yawing disturbed motion equations. If the configuration of the missile is axisymmetric, the three-dimension motion of the missile can be described in pitching, yawing and rolling channel respectively.

If the missile configuration is axisymmetric, the same design parameters may be adopted in the pitching channel and yawing channel, which will cut down the design workload and make the physics concept very clear for system analysis and design.

Compared to single-channel control, the design of double-channel control is more complicated.

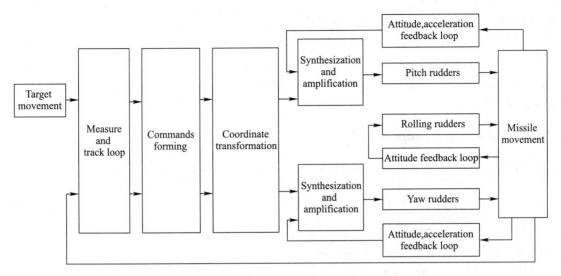

Figure 7.56　Schematic block diagram of double-channel control

7.5.1.3　Three-channel control

The control mode that pitching, yawing and rolling channel are all controlled is called three-channel control. The schematic block diagram of the three-channel control is shown in Figure 7.57. The principle of operation is that motion parameters of missile and target are measured by detecting devices, based on which control commands in three channels are formed. These control commands are used by three sets of actuating mechanisms to control the three channels' movements respectively.

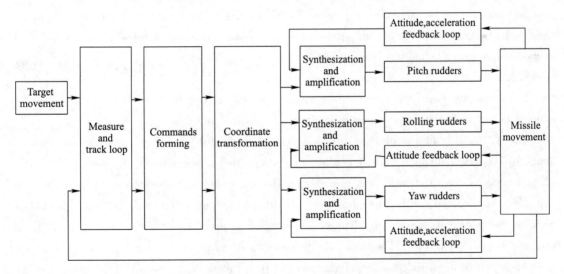

Figure 7.57　Schematic block diagram of three-channel control

When designing the guidance system of three-channel control, we should consider some special questions as follows:

(1) Considering the forming of the control command, the objectives of adopting three-channel control are usually different, so the command forming must satisfy different requirements. Control command is not only relative with guidance method, limitation to guidance system from other subsystems of weapon system, but also with the computation speed of command computation

equipment.

(2) The coupling of pitching, yawing and rolling channels is serious, so the design method of the multi-variable system should be used. When designing control command, by choosing the structure and parameters of control loop, we should try our best to reduce the coupling between three channels.

7.5.2 Skid-to-turn (STT) and bank-to-turn (BTT)

According to the different turning maneuver modes, the control methods can be divided into STT and BTT.

7.5.2.1 STT

STT is mainly used for missiles with axisymmetric aerodynamic configuration. It is the main control mode of air-to-air missiles in service.

For the axisymmetric missile, it is equipped with two pairs of wings uniformly distributed in the circumferential direction. STT relies on manipulating the deflection (angle δ_z) of the elevator to adjust the angle of attack α, thereby changing the direction and magnitude of normal force, and controlling the pitching movement of the missile. The change of the angle of sideslip β is realized by manipulating the deflection angle δ_y of the rudder, so that the direction and magnitude of the lateral force can be changed to control the yaw motion of the missile. The deflection angle δ_x of aileron is manipulated to keep the rolling channel stable.

In STT control mode, the coupling among the three channels of missiles is small, and the structure and design of autopilot are relatively simple. Since the pitch and yaw channels are simultaneously controlled, the dynamic response of STT control is rapid. At present, the STT control method is quite mature and has been widely used in active missiles.

7.5.2.2 BTT

BTT is mainly used for missiles with surface-symmetric aerodynamic configuration. It's generally adopted for modern medium and long-range air-to-air missiles.

The BTT missile first manipulates the deflection angle δ_z of the elevator to generate the angle of attack α. Then the deflection angle δ_x of aileron is manipulated to generate the roll angle γ. The main lift generated by the wings also rolls accordingly. Through orthogonal decomposition, the lift in the side plane is obtained, so as to realize bank to turn. While the rolling and pitching channel is working, the motion in the yawing channel is controlled to make the angle of sideslip and lateral overload as small as possible. The ideal condition has no sideslip.

According to the aerodynamic configuration and power plant of the missile, BTT missiles can be divided into three types: BTT-45, BTT-90, and BTT-180.

1) BTT-45

The BTT-45 missile is an axisymmetric missile. For the axisymmetric missile, the rudder surface and wing surface of the " × – × " type are more efficient and maneuverable than the " + – + " type. Through the control of the rolling channel, the missile is always in " × – × " type state in the whole flight process, and the target is always in the central plane between the two control surfaces, that is, the maximum lift surface. The maximum control roll angle of BTT-45 is 45°.

2) BTT-90

BTT-90 is applied to surface-symmetric missiles. The rudder surface and wing surface of the missile are arranged in " – – ×" type, and there are only two effective lift surfaces, so the lateral maneuvering ability is limited. In order to aim the maximum lift surface at the target, the maximum control roll angle of BTT-90 is 90°.

3) BTT-180

BTT-180 is applied to surface-symmetric missiles. Due to the requirements of long-range attack and defense missions, the ramjet is applied to BTT-180 missiles. As a result, BTT-180 missiles can only provide positive angle of attack and positive lift. The maximum control roll angle of BTT-180 is 180°.

The performance of BTT missiles are significantly improved in many aspects compared to that of STT missiles, such as mobility, stability, lift drag ratio and compatibility with advanced ramjet engines. However, the BTT missile is a complex system with aerodynamic coupling and control coupling, which brings great difficulties to the design of control system.

7.5.3 Autopilot

7.5.3.1 Introduction

In the whole flight process of a missile, as the flight conditions e. g. the flight velocity and flight height are constantly changing, the parameters expressing a missile's dynamic characteristics such as the time constant and damping coefficient of the missile transfer function are changing, too. However, for most kinds of missiles, the aerodynamic configuration is fixed. Thus the requirement of stability, dynamic response of missile system can't be ensured to be satisfied in any flight condition, under which condition an autopilot is needed.

As we all know, the guidance system of a missile is a system of multi-loop. The main loop is composed of observation and tracking equipment, command forming equipment, actuators and missile body, which is generally called guidance loop. Another loop is the attitude stabilization loop, also called stabilization control system, which contains autopilot and missile body. In the stabilization control system the missile body is the controlled object, and when the control surface is rotated or the thrust vector is changed, the missile body maneuvers correspondingly. If there is such inertia measurement equipment as the accelerometer and/or gyroscope in the missile, their outputs form additional feedback to the missile control system to correct the control command and so to modify the missile movement. Generally speaking, the actuator, sensitive components and control electric circuit are combined to be called an autopilot.

The basic work of an autopilot is to make the attitude movement stable and control the missile to fly correctly and quickly according to the guidance command. Concretely speaking, they are as follows:

(1) An autopilot eliminates the attitude change of a missile caused by any disturbance so to make the trajectory not be affected by disturbance and to keep the flight attitude unchangeable.

(2) An autopilot controls the actuator to be deflected or changes the direction of the thrust vector according to the control command to change missile attitude and make the missile fly following the trajectory set in advance.

The above two types of work situations of the autopilot are called stability work situation and control work situation. An attitude stabilization loop containing an autopilot and the missile body is given in Figure 7.58.

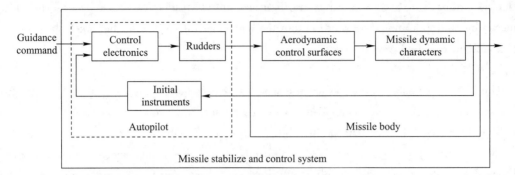

Figure 7.58 Attitude stabilization loop with autopilot

Generally speaking, if an autopilot is used to control the missile's movement in pitching and yawing channels, it is called pitching and yawing autopilot; if an autopilot controls missile's roll movement around the longitudinal axes, it is called a roll autopilot. To a symmetrical missile with " + " or " × " type wing, the pitching autopilot and yawing autopilot are equivalent if the gravity effect in the pitch channel is neglected, so both of them may be called lateral autopilot. Especially, for some rolling and small missiles, there is usually not a roll autopilot but just a lateral autopilot used to control missile motion. And so this lateral autopilot is called a single-channel autopilot.

Autopilots are classified into skid-to-turn (STT) autopilot and bank-to-turn (BTT) autopilot according to their control mode. Based on the relationship among channels of pitching, yawing and rolling, autopilots can be classified into the autopilot in which three channels are independent and the autopilot in which there are coupling among the three channels. Additionally, there are some kinds of autopilots for special use e.g. the autopilot for vertical launch system, the autopilot for altitude control and the autopilot for azimuth control using inertia technology. The classification of autopilot is shown in Figure 7.59.

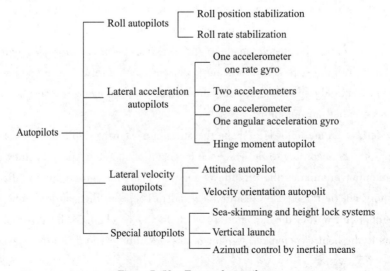

Figure 7.59 Types of autopilots

We must note that not every missile needs an autopilot. If a missile is of high static stability, a little move of pressure center and/or gravity center cannot cause a large change of static stability. Many anti-tank missiles with nearly constant flight velocity and flight height when they attack the target of low velocity belong to the type. Under this condition the problem of keeping the aerodynamic gain unchangeable is not serious, so now an autopilot is not needed.

7.5.3.2 Functions of autopilot

The autopilot and missile body form the stabilization loop of a missile, whose main function is making a missile attitude stable and control the missile to fly according to control command. So the main functions of an autopilot include the following aspects.

(1) Stabilize angle position and angle rate of the longitudinal axis of the missile in space.

The guidance system of the non-rolling missile generally requires keeping the roll angel and roll angular rate equal to or nearly equal to zero. As we all know, the roll movement of the missile is not of static stability, so equipment that can stabilize the missile roll movement must be used. The autopilot can be used to achieve this goal.

(2) Improve the dynamic and static characteristics of a missile.

As the flight height and flight velocity of a missile are changing, the aerodynamic parameters are changing, too. So the dynamic and static characteristics of a missile will also change. The changes of parameters make the design of the whole control system more complicated. Adopting an autopilot can make sure that the changes of dynamic and static characteristics are in allowable ranges under any flight condition.

(3) Increase the damping coefficient of the missile body and improve the transient process of the guidance system.

The relative damping coefficient of the missile body is decided by aerodynamic damping coefficient, static stability coefficient and missile movement parameters. To the missile with high static stability and high flight altitude, its damping coefficient of the missile body generally is about 0.1, which means the missile body is badly underdamped. Now when there is any disturbance and the missile executes the guidance command, the dynamic characteristic is not acceptable, that is, there is a serious surge in transient process and the overshoot and adjusting time are not short enough. So an autopilot is needed to improve the damping coefficient.

(4) Stabilize the statically unstable missile.

The control object of an autopilot i.e. missile body may be statically stable, neutrally stable or statically unstable. To the missile which is neutrally stable or statically unstable, it may depend on the autopilot to assure stability in the flight process.

(5) Increase the natural frequency.

A high natural frequency is essential for the stability of the guidance loop. As in the guidance loop, the transfer function expressing the relative movement between missile and target consists essentially of two integrations and a D.C. gain, so we have 180° phase lag at all frequencies even though there are no else dynamic lags in the loop. To improve the situation of the phase lag of the loop, we can introduce a phase advance network in the stabilization loop, but it is usually not enough unless an autopilot is adopted.

(6) Execute the guidance command and control missile to fly according to normal trajectory.

The stabilization loop containing an autopilot receives guidance command, and then the command is amplified to deflect the actuator or change the direction of the thrust vector, which makes the missile body produce the normal overload needed.

Some autopilots with different functions are shown as follows.

1) Autopilot of keeping flight stability

A flight stability system with autopilot is described in Figure 7.60. By means of the autopilot, the closed loop is formed to drive the control surface so as to eliminate the disturbance and keep the flight stable.

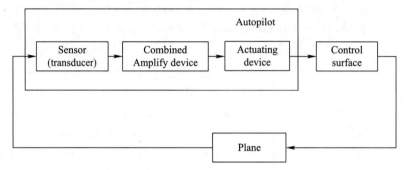

Figure 7.60　Flight stability with an autopilot

2) Autopilot of controlling flight

Figure 7.61 illustrates the closed loop of the flight control by an autopilot. If the command is given, through the autopilot, the control requirements will be achieved well.

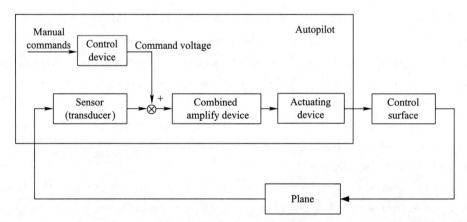

Figure 7.61　Flight control by an autopilot

3) Autopilot of controlling and stabilizing flight altitude

Figure 7.62 describes the flight altitude controlled and stabilized by an autopilot. According to this diagram, the disturbances inducing variations of flight altitude can be eliminated automatically. In another aspect, to change flight height, that is, to change from one altitude to another, the expected flight altitude can be reached automatically and gradually.

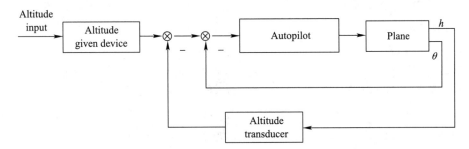

Figure 7.62 Flight altitude stability and control by an autopilot

Questions

1. Please describe the relations and differences between lead system and control system.
2. Enumerate the types of guidance systems.
3. What is the Doppler Effect?
4. Explain the main characteristics of a gyroscope.
5. Try to give some applications of different types of sensitive components.
6. Explain main functions of seekers and some requirements for seekers.
7. Briefly describe the attitude determination principle of star sensors.
8. Try to establish the relationship between the active/semi-active/passive homing head and guidance laws.
9. Try to draw a guidance system principle diagram of a ground-to-air missile, and give the working principles.
10. Try to establish the equations of a missile in the vertical plane with proportional navigation law.
11. Try to establish the equations of amissile with three-point navigation law.
12. Give the differences of the typical guidance laws.
13. Try to give several application scenarios of the optimal guidance law and compare their performance index functions.
14. Elaborate the advantages and disadvantages of centralized and distributed coordinated guidance.
15. Please describe the relations and differences between BTT and STT.
16. What are the main functions of autopilots?

Words and phrases

absolute altitude 绝对高度
A. C. (Alternating Current) 交流电
accelerometer 加速度计
actuating mechanism 执行机构
altimeter 高度计
autopilot 自动驾驶仪
autonomous guidance system 自主制导系统
barometric altimeter 气压高度计
carrier mobility 载流子迁移率
closed loop system 闭环系统
CAS (Control Augmentation System) 控制增稳系统

coincide 一致,相符合	orthogonal 直角的,直交的
cordite 无烟火药	perpendicular 垂直的,正交的
Coriolis force 科氏力	pivot 枢轴,支点,中心点
D. C. 直流电	potentiometer 电位计,分压计
D. C. gain 稳态增益	precess 使产生进动
denominator 分母	precession 陀螺进动
differential game 微分对策	rigidity 陀螺定轴性
differential geometry 微分几何	projection 投影
discriminability 分辨能力	pulse-width modulation 脉宽调制式
Doppler navigation 多普勒导航	rate gyroscope 速率陀螺仪
electric charge 电荷	rate integrating gyroscope 速率陀螺仪
elevation angle (星光)高度角	reference frame 参考坐标系,坐标系
flexure pendulous accelerometer 挠性摆式加速度计	resultant force 合力
	rotor 转子
fixed axis character 定轴性	SAS(Stability Augmentation System) 增稳系统
gimbal 万向节	
gyro, gyroscope 陀螺仪	sealed 密封的
hermetically 密封地	sextant 六分仪星座
homing guidance 自寻的制导	spring-mass accelerometer 弹簧架式加速度表
hydraulic 液压的	stall 失速
hysteresis 滞后作用,磁滞现象	strapdown inertial navigation system 捷联惯性导航系统
inertial guidance 惯性导航	
integrator 积分器	stiffness 坚硬,硬度
intercept 拦截	symmetry 对称,匀称
isopropyl (化)异丙基	tangent 切线
liquid-floated pendulous accelerometer 液浮摆式加速度计	tangential force 切线力
	terrain-matching system 地形匹配制导
longitudinal 纵向的,经度的	toppling 环架自锁
moment of momentum 动量矩,角动量	torsion bar 扭杆
neural network 神经网络	transducer 传感器,变频器,变换器
nitrate (化)硝酸盐,硝酸钾	vertical 垂直的,竖的
normal force 法向力	viscous 黏性的,黏滞的,胶黏的
normal trajectory 标准弹道,理想弹道	

References

[1] Nelson R C. 飞行稳定性和自动控制[M]. 顾俊晓,译. 北京:国防工业出版社,2008.

[2] 江加和. 导弹制导原理[M]. 北京:北京航空航天大学出版社,2012.

[3] 吴森堂. 飞行控制系统[M]. 北京:北京航空航天大学出版社,2013.

[4] 汤晓云,樊小景,李朝伟. 相控阵雷达导引头综述[J]. 航空兵器,2013(3):25-30.
[5] 秦玉亮,王建涛,王宏强,等. 弹载合成孔径雷达技术研究综述[J]. 信号处理,2009,25(04):630-635.
[6] 高烽. 合成孔径雷达导引头技术[J]. 制导与引信,2004(1):3-6.
[7] 王海峰,吴宏宇. 被动雷达导引头发展历程及技术综述[J]. 飞航导弹,2013(1):78-80.
[8] 李一辰. 主动/半主动复合雷达导引头抗干扰仿真系统研究[D]. 南京:南京理工大学,2019.
[9] 邹汝平,陈士超,陈韵,等. 一种图像寻的空地导弹的制导控制方法[J]. 战术导弹技术,2021(1):93-98,114.
[10] 邓宏伟. 电视导引头图像处理系统研究[D]. 南京:南京理工大学,2006.
[11] 陈成,赵良玉,马晓平. 激光导引头关键技术发展现状综述[J]. 激光与红外,2019,9(2):31-136.
[12] 刘智颖,邢天祥. 激光半主动导引头光学系统设计[J]. 激光与红外,2016,46(5):527-531.
[13] 吴丰阳,沈志,胡奇. 复杂场景下多模复合制导关键技术研究[J]. 航空兵器,2018(1):3-7.
[14] 孙少军. 主动雷达/红外成像复合制导抗干扰技术[J]. 舰船电子工程,2016,36(2):61-63,82.
[15] 李国通,尚琳,常家超. 导航卫星系统自主运行技术[M]. 北京:科学出版社,2020.
[16] 提舒雯. 基于星敏感器的星图预处理与星点提取技术研究[D]. 哈尔滨:哈尔滨工程大学,2020.
[17] 沙鑫宽. 面向微小卫星的星敏感器工程化设计与实现[D]. 杭州:浙江大学,2020.
[18] 房建成,宁晓琳,田玉龙. 航天器自主天文导航原理与方法[M]. 北京:国防工业出版社,2006.
[19] 何家维. 高精度全天时星敏感器关键技术研究[D]. 长春:中国科学院研究生院(长春光学精密机械与物理研究所),2013.
[20] 祁载康. 战术导弹制导控制系统设计[M]. 北京:中国宇航出版社,2018.
[21] 施广慧,赵瑞星,田加林,等. 多导弹协同制导方法分类综述[J]. 飞航导弹,2017(1):85-90.
[22] 赵建博,杨树兴. 多导弹协同制导研究综述[J]. 航空学报,2017,38(1):22-34.
[23] 赵昱. 新型导弹的BTT/STT切换控制研究[D]. 哈尔滨:哈尔滨工程大学,2019.
[24] 于秀萍. 基于动态逆系统和神经网络理论的BTT导弹控制系统研究[D]. 哈尔滨:哈尔滨工程大学,2004.
[25] 李朝川. BTT导弹控制系统研究[D]. 哈尔滨:哈尔滨工业大学,2008.

Chapter 8 Ground equipment and launch

To realize the function of the flight vehicles, the important is to launch them successfully and enter the predetermined area. Different flight vehicles have different platforms and environment according to their categories and missions. Generally, the platforms launching flight vehicles locate on the ground, in the air, under the water, and others. At the same time, there are great differences among the platform and launch environment for them. The three types of the flight vehicles such as carrier rocket & space vehicle launch, missile launch, airplane launch on the ground are mainly discussed in the following.

8.1 Carrier rocket & space vehicle launch

8.1.1 Space launch base

Space launch base is used for launching carrier rockets and shuttles into orbits around Earth or interplanetary trajectories. It takes on the task of organization and command of launching mission, service, and support for satellites testing, rockets and satellites reshipment and transport, system docking, observation and control of rocket, propulsion injection, and provides weather condition service, communication service, and other tech-support system.

Generally, the space launch base can be divided into three main parts, i. e. , launching site, flying site, and recovery site. The launching site is the place for launching carrier rockets, the flying site can be utilized for multistage rockets to fall off, track, measure, and control, and the retrieving site is used for the carrier rockets and shuttles testing to reentry to return the atmosphere.

The launching site is a main part of the space launch base. The launching area is the special area for launching and testing, and the launching tower is the most important part of the launching site, as seen in Figure 8.1.

Figure 8.2 gives the carrier rocket launching site which consists of the launching tower, launching control center, power supply station, examination center, liquid hydrogen and nitrogen storeroom, and other auxiliary equipment.

The procedure for launching is briefly introduced as follows. At first, the rocket is tested to make the preparation. If it is qualified, the rocket will be transported to the launching site where it is installed on the launching platform. Then the satellite being launched is connected with the rocket, the whole system is examined and synthetically tested, and the propellant is added. In the end, the rocket with satellite will be pushed into space.

There are some famous space launch bases all over the world. In Russia, Baikonur and Plesetsk are the famous space bases. Baikonur is the world's first space launch base for orbital and manned

Chapter 8　Ground equipment and launch

Figure 8.1　The launching tower of Centre Spatial Guyanais

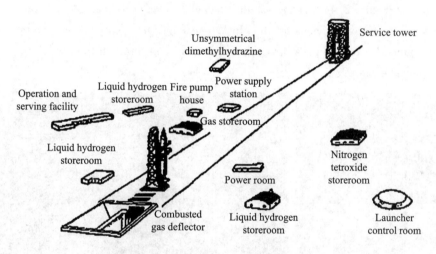

Figure 8.2　Carrier rocket launching site

launch, started as a Soviet military rocket range in 1955. It achieved the first orbital flight (Sputnik 1) in October 1957. Kennedy Space Center and Canaveral Corner are the two places for launching the shuttles and satellites in the United States. Guyana, belonging to France, is a famous place for launching satellites for ESA, Europe, and other countries.

There are four China Satellite Launch centers; They are Jiuquan Satellite Launch Center, Xichang Satellite Launch Center, Taiyuan Satellite Launch Center and Wenchang Satellite Launch Center.

As shown in Figure 8.3, Jiuquan is the earliest satellite launch center in China, and the main base for testing and launching Changzheng (CZ) series launch vehicles, various test satellites in low and medium orbits, application satellites, manned spacecraft, and rocket missiles. DFH-1, the first man-made satellite of China, was launched successfully here in 1970. And the first continental rocket of China was launched successfully in 1971. In 1975, the first returned satellite of China was launched. The unmanned spacecraft Shenzhou-I was launched in 1999. In addition, Jiuquan has also undertaken the launch of the Shenzhou series manned spacecraft, the Chinese space station

Tiangong, and numerous commercial space missions.

Figure 8.3　Jiuquan satellite launching center

As shown in Figure 8.4, Xichang satellite launching center is mainly to launch the low dip angle orbit satellite and synchronous satellite. In 1984, China's first geosynchronous orbit satellite was successfully launched. And China's first communication and broadcasting satellite, DFH-2, was successfully launched in 1986.

Figure 8.4　Xichang satellite launching center

As shown in Figure 8.5, Taiyuan satellite launching center is mainly to launch the scientific research and meteorological satellites. One rocket with two satellites was launched in 1999. In addition, it has also launched China's first sun-synchronous orbit meteorological satellite "Fengyun 1", the first China-Pakistan Resources 1 satellite, the first marine resources survey satellite, and so on. That is to say, it has created a number of "China's first".

As shown in Figure 8.6, Wenchang is the first open coastal space launch base in China and one of the few low-latitude launch sites in the world. As a low-latitude coastal launch site, Wenchang can reduce rocket fuel consumption efficiently (up to 10% more capacity for the same type of rocket) by taking advantage of the greater linear velocity close to the equator and the centrifugal phenomenon brought about by inertia, and solve the problem of transporting giant rockets by sea and to improve

Figure 8.5 Taiyuan satellite launching center

the safety of wreckage fallout. The launch site is capable of launching CZ-5 series rockets and CZ-7 launch vehicles, mainly for spacecraft such as geosynchronous orbit satellites, large mass polar orbit satellites, large tonnage space stations, and deep space explorer. In 2020, Tianwen 1, China's first Mars probe and Chang'e 5 Lunar sample and return probe were launched here.

Figure 8.6 Wenchang satellite launching center

8.1.1.1 RLV mission procedure

Sending payloads into space at lower cost and higher efficiency has always been the pursuit of the space launch field, and the reusable launch vehicle (RLV) is an effective tool to achieve this goal. Since the middle of the 20th century, various concepts and actual products of the fully or partially RLV have been proposed around the world.

According to their takeoff and landing methods, RLV can be divided into vertical takeoff horizontal landing reusable launch vehicle (VTHL-RLV), vertical takeoff vertical landing reusable launch vehicle (VTVL-RLV), and horizontal takeoff and horizontal landing reusable launch vehicle (HTHL-RLV). HTHL-RLV, represented by the Skylon, is based on a unique hybrid engine, the Synergistic Aspirated Rocket Engine (SABRE). The complex structural design of SABRE is very difficult to achieve with the current level of technology, so only the mission procedure of VTHL-RLV and VTVL-RLV is described here.

8.1.1.2 VTHL-RLV mission procedure

In particular, VTHL-RLV, represented by the space shuttle and X37B test vehicle, adopts the design scheme of rocket bundle launch and horizontal return of the first stage. This program can effectively enhance the space response capability, while achieving a long stay or maneuver in orbit, and is an effective way to enhance space military power and space launch capability.

However, the reentry process of the VTHL-RLV has serious thermal protection problems, long ground maintenance cycles, high maintenance costs, and low real cost savings.

Figure 8.7 illustrates the launching procedure of a shuttle. At first, the shuttle is installed with a solid propeller and orbit vehicle. Then the examination is made for the whole system. At E point, the shuttle is launched, and the propeller is separated from the shuttle at F point. At H point, the orbit-vehicle is sent to orbit. And at K point, the velocity of the orbit vehicle is decreased after retrofire. From L point, the shuttle escapes from the operation orbit and reentering the atmosphere. Till point O, the shuttle is landed, and the tasks are finished.

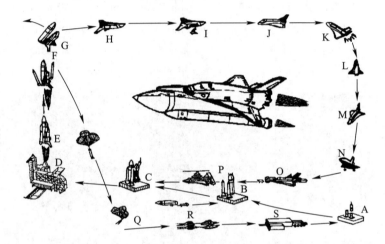

Figure 8.7 Mission procedure of a shuttle

A—Prepare for final assembly; B—Fix solid booster and external tank; C—Fix orbiter; D—Test before launching; E—Shuttle launching; F—Solid booster separation; G—External tank separation; H—Orbiter gets into running orbit; I—Operation in orbit; J—Turn; K—Retrorocket ignite to decelerate orbiter; L—Deorbit; M—Atmosphere reentry; N—Glides into the land area and prepare for landing; O—Landing, and checkout; P—Disassemble cargo cabin and overhaul; Q—Solid booster landing over sea surface; R—Recovery solid booster; S—Check and clean solid booster, fill propellant.

8.1.1.3 VTVL-RLV mission procedure

Comparing with VTHL-RLV, VTVL-RLV, represented by Space X Falcon 9, is based on the traditional rocket configuration with minor modifications (with an RCS and grid fin) and adopts the second stage orbit entry and the first stage vertical landing scheme. This solution has a low cost and high success rate, which greatly facilitates the development of private space companies and commercial space exploration.

Meanwhile, the high technological inheritance of VTVL-RLV relative to mature conventional launch vehicles, and it has a weaker impact on launch mission execution and smaller landing site requirement, which make it highly competitive in the commercial space launch market and promising for a wide range of applications in future interplanetary exploration such as the Moon, comets, and Mars.

Although VTVL-RLV has been reused many times, there are still key issues to be solved, such as precise soft-landing guidance and attitude control, high precision return navigation, whole life cycle fault detection and health management, and so on.

Figure 8.8 illustrates the mission procedure of Falcon 9. At first, the Falcon 9 takes off from the launch site powered by a liquid rocket propulsion system with thrust adjustment capability and actuators such as RCS and grid fins. After the ascent flight, the first stage and second stage complete separation, and the second stage continues to ignite to deliver the payload into the intended orbit. The first stage goes through the flight segments of flip maneuver and attitude adjustment, reentry deceleration and ignition, aerodynamic deceleration, and so on. Finally, the first stage achieves vertical landing on the landing site.

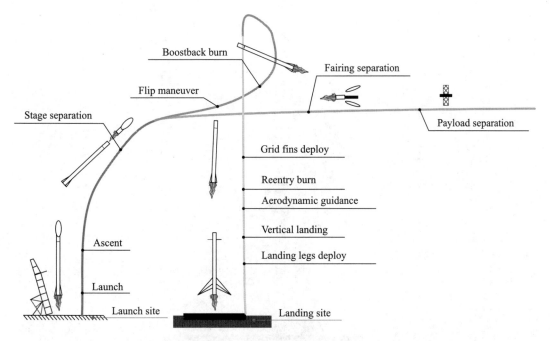

Figure 8.8　Mission procedure of Falcon 9

8.1.2　Space vehicle return & reentry

Deviating from space orbit and entering the atmosphere, a spacecraft safely lands on the ground. The above-mentioned procedure is called the spacecraft's return. Reentry means a spacecraft flies from outer space into the dense terrestrial atmosphere. Reentry spacecraft, such as photo-reconnaissance satellite, biological satellite, manned spacecraft, space shuttle, and lunar or planetary probes, needs a part of the entire vehicle return to Earth after flying in orbit for some time.

Spacecraft is slowed down gradually from a high speed in space orbit to a safe landing speed when closing to the ground during the whole return procedure as a deceleration process. Despite the thickness of the terrestrial atmosphere's dense part reaching only a few dozen kilometers, the atmosphere can generate a considerable resistance force whose magnitude is big enough to decelerate the spacecraft. Therefore, an economical and feasible method of using air resistance force is proposed. Decelerated remarkably under the influence of huge air resistance force, the spacecraft's speed reduces from the cosmic speed when the spacecraft just enters the atmosphere to subsonic as much as two hundred meters per second with a height less than 15km. Finally, further deceleration measures such as using parachutes are brought into effect to slow down the spacecraft to a safe landing speed.

Generally speaking, the return process that uses the atmosphere resistance force includes 4 stages with deorbit, free-fall (transition phase), reentry to the atmosphere (reentry phase) and landing. Figure 8.9 shows the "Soyuz" spacecraft's return process.

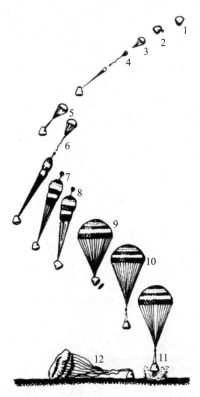

Figure 8.9 The "Soyuz" spacecraft return process

1—Capsule; 2—The cover of eject parachute; 3—Lead parachute with the large area; 4—Lead parachute with the small area; 5—Stable parachute; 6—Stable parachute separates and draws the main parachute out; 7—Main parachute unfurls; 8—Parachute unfolds completely; 9—Bottom heatproof shield separates; 10—Turn to symmetrical hanging status; 11—Buffer rocket fires; 12—Main parachute separates.

8.2 Missile launch

8.2.1 Missile launch modes

There are many different modes to launch missiles. The launch category can be described according to the combination of launching point, launching power, or launching attitude, seen in Figure 8.10.

1) Launching point

Chapter 8 Ground equipment and launch

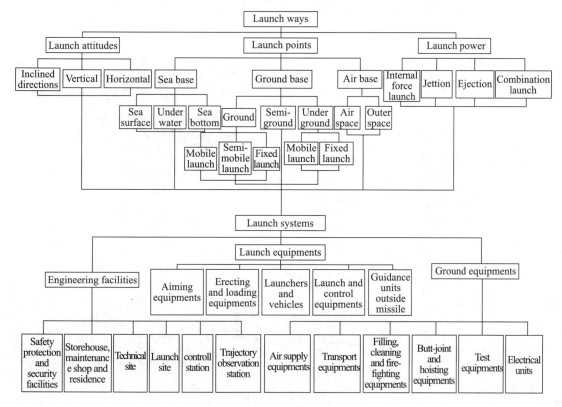

Figure 8.10 Missile launching styles

For the launching points, ground base, sea base, and air base are divided for different purposes of missiles. If the launching point is from the ground base, the missiles can be launched on the ground, underground, and semi-underground. At the same time, the launching points can be fixed, or movable. If the launching point is from the sea base, the missiles can be launched from the warships on the sea, the submarines under the sea, and on the bottom of the sea. The launching point is movable, or fixed. If the launching point is from the air base, the missiles can be launched from the airplanes in all directions. Same as the sea base, the launching point from air base is movable, or fixed.

2) Launching power

There are four modes from the point of the launching power. It means that the different powers are used at the launching time. The first mode is that the missiles are launched by self-support, without the help of the others. The second mode is that the missiles are launched through throwing using the weight. The third mode is that the missiles are launched through ejecting, by using ejected equipment. The fourth mode is that the missiles are launched by the combination of two or more modes, for example, the combination of self-support and ejection.

3) Launching attitude

As seen in Figure 8.11, three ways are separated according to the launching attitude. The launching direction is horizontal, such as air-to-air missiles, air-to-ground missiles, and so on. The launching direction is with the vertical, like the ballistic missiles, carrier rockets. A lot of missiles are launched with inclined directions, that is, a launching angle is given for the flight vehicles. The

351

ground-to-air missile is one of the kinds with inclined direction.

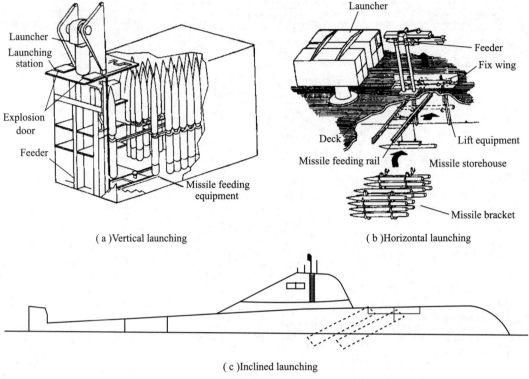

Figure 8.11　Launching modes according to launching attitudes

8.2.2　Missile launch equipment

The functions of the missile launching equipment are preparation, support, sight and orientation, and launching control. Besides, ground equipment is utilized to control the missile tracking the target by the control box, test the missile system, and maintain the whole missile system, as well as the auxiliary devices. There are many different kinds of launching equipment. Some examples are as follows.

8.2.2.1　Tactical missiles

Generally speaking, the launching equipment is relatively simple for tactical missiles. For example, lots of anti-tank missiles have the launching equipment with the tube-launching or frame-launching.

About the ground-to-air missiles, they are launched by frame-launching in ordinary. The launching equipment is shown as Figure 8.12, which is one of the parts of the missile system composed of a missile, control station including radar, command formation, coordinates transformation devices, and launching equipment. It gives the launching direction with a certain angle. And the launching angle can be changed with the inclination of the frame-launching.

As given in Figure 8.13, the launching system of seashore-to-ship missiles is described. As seen, the seashore-to-ship missiles are prepared for launching with frame-launching. The launching system is composed of not only the frame-launching and the missiles, but also the command center

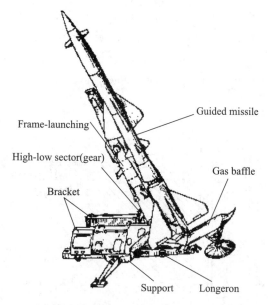

Figure 8.12　Frame-launching equipment of ground-to-air missile

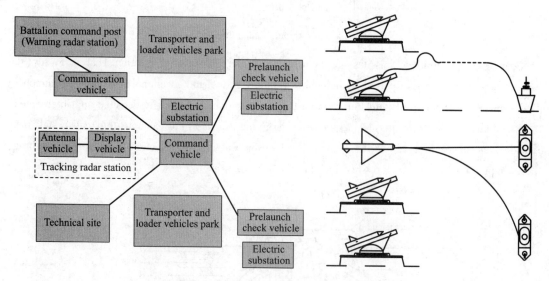

Figure 8.13　Launching system of seashore-to-ship missiles

including the testing vehicles, a communications vehicle, control station, technical positions, and so on.

8.2.2.2　Strategic missiles

Compared to Tactical missiles, the launching equipment of the strategic missiles is relatively complicated. According to the launching ways, they are launched with vertical directions as usual. The launching powers are divided into two forms. One is the hot launching, which means that the strategic missiles are launched using the engines, relying on themselves. Owing to the gas coming from the engines, the exhausting pipes must be used. The other is the cold launching, which illustrates that the strategic missile is ejected using the ejection equipment, and has the advantages of without exhausted pipes, as well as launching fast repeatedly.

Usually, the strategic missiles are launched with a ground base. In most cases, the launching point is fixed. For example, the strategic missiles located underground well as seen in Figure 8.14, which has good environment, with insurance, and the ability of anti-nuclear. At the same time, it reacts fast, and with high hitting probability. But it has some disadvantages, that is, it is easy to be detected, dangerous to be hit, and easy to be intercepted.

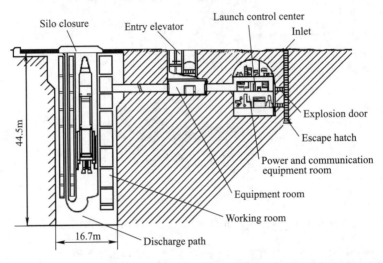

Figure 8.14 Underground launching equipment

Sometimes, the strategic missiles are launched with the maneuvered position. That is to say, the strategic missiles can be transported by launching vehicles, trains, and so on. So that they can launch in different places. Without fixed launching position, the missiles cannot be easily detected and intercepted.

Especially, the strategic missiles are launched from the sea base. For example, they can be taken by submarines as shown in Figure 8.15, launched under the water, using ejection equipment. Something, e. g. water pressure, must be noted when launching under the sea.

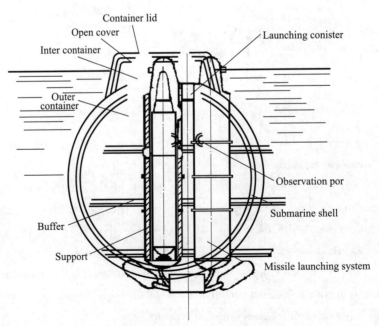

Figure 8.15 Submarine launching equipment

As stated before, there are many types of launching ways. According to the different styles of missiles, different purposes of the missiles, the missile performances, and some characteristics of the missiles, the selection of launching ways is determined.

8.3　Airport and ground equipment

8.3.1　Airport

The airport is the place where the airplanes takeoff, land, stop, and move with some flight activities. It is divided into military airports and civil airports. Most of the airports are permanent, but there are also some temporary airports.

Generally speaking, an airport is divided into two parts, one is in space, the other is on the ground, where it consists of flight ground, and service area.

The equipment of the flight ground is as follows. One is the runway, which is about 1000 ~ 5000m long, 45 ~ 100m wide, to make sure that the airplane can slide as the taxi does. At the end of the runway, the airplanes can take off safely. The others are the sliding way, safety way, forced area, and airplane parking area.

The functions of the technical service area are the control tower and auxiliary area. The former is used to guide the airplane, and to give the command so that the airplane can slide, take off, and land on the runway. And flight activities of the airplanes to some extent can be managed. The latter is an area to be as the auxiliary place so that the airplanes can be repaired, oiled, and so on.

Figure 8.16 illustrates the sketch of an airport in which there are two runways, a control tower, a technical service area, and a flight ground. For a better understanding of the airport system, a schematic diagram of the operation of a typical airport is provided in Figure 8.17.

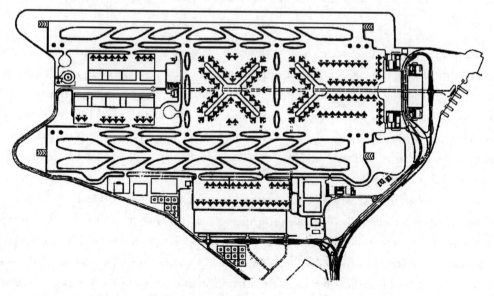

Figure 8.16　Overview of an airport

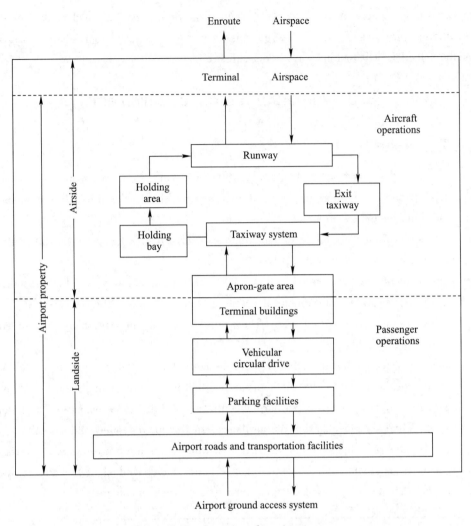

Figure 8.17　Operation of an airport system

　　Additionally, there is also such a way to describe the components of an airport on the ground, that is, an airfield and terminals. A typical airfield is composed of a runway as well as two or one parallel taxiing lane. The angle of the connecting lanes between the runway and the taxiing lanes permits the quick exit from the runway for planes that have just landed. Two of three exiting options are provided per landing direction depending on the plane's size. A small airplane will take less distance to break than a large one and has thus the opportunity to short exit the runway, freeing valuable takeoff or landing slots.

　　Generally, most terminals are designed as the three categories seen in Figure. 8. 18. The linear orientation of terminals allows several planes to transit passengers at the same time and represents one of the most common terminal designs. The islet terminal is an answer to this type of problem while permitting the stowage of several planes on a smaller surface. Some airports select the shuttles terminal, which enables to reduce the size of the terminal and maximize the number of planes that can be serviced. But the design may involve longer boarding times.

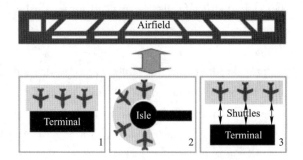

Figure 8.18 Airport Components
1—Linear orientation terminal; 2—Islet terminal; 3—Shuttles terminal.

8.3.2 Ground equipment

There is various ground equipment to guarantee the normal operation of the airport and airplanes. For example, aviation oil must be stored at the airport so that the aircraft can be filled with if it needs. And some tests must be examined for landing gears, flaps, and other necessary parts. At the same time, electricity, cold air, and oxygen should be supplied. Some equipment for first-aided must be prepared. Here, only some important ground equipment is introduced, such as air traffic control, navigation aids, guidance signs, lightings, and weather observations.

8.3.2.1 Air traffic control

At some busy airports, or airports with other special requirements, the air traffic control (ATC) system will be needed to direct airplanes via radio or other communications links. Air traffic control is usually composed of at least two main areas: ground and tower, though a single controller may work both stations. The busiest airports also have clearance delivery and other specialized ATC stations.

Ground control is responsible for directing all ground traffic including planes, baggage trains, snowplows, fuel trucks, and a wide array of other vehicles in designated movement areas, except the traffic on runways. Ground control will indicate these vehicles on which taxiways to use, which runway they will use, where they will park, and when it is safe to cross runways. When a plane is ready to takeoff it will stop short of the runway, at which point it will be turned over to tower control. After a plane is landed, it will depart the runway and be returned to ground control.

Tower control is in charge of the airplanes on the runway and the airspace surrounding the airport. Tower controllers may use radar to locate the position in the three-dimensional space of airplanes, also they may rely on pilot position reports and visual observation. They coordinate the sequencing of airplanes in the traffic pattern and direct airplanes on how to safely join and leave the circuit. The airplanes which are only passing through the airspace must also contact tower control to assure that they remain clear of other traffic.

8.3.2.2 Navigational aids

There are many aids available to pilots, though not all airports are equipped with them. A Visual Approach Slope Indicator (VASI) helps pilots fly the approach for landing. Some airports are equipped with a VHF (very high frequency) omnidirectional range (VOR) to help pilots find the direction to the airport. VORs are often accompanied by distance measuring equipment (DME) to determine the distance to the VOR. VORs are also located off airports, where they serve to provide airways for aircraft to navigate upon.

8.3.2.3 Guidance signs

Airport guidance signs provide direction and information to taxiing airplanes. Some small airports have few or no signs, but some airport diagrams and charts. There are two categories of guidance signs usually, one is operational guidance signs including location signs, direction or runway exit signs, and others. The other is mandatory instruction signs which show entrances to runways or critical areas, for instance, runway signs, frequency change signs, and holding position signs.

8.3.2.4 Lightings

Lightings are needed to help guide planes using the runways and taxiways at night or in rain or fog. Different colors of lightings have different meanings, and it is standard to use these lightings. On runways, green lights indicate the beginning of the runway for landing, while red lights indicate the end of the runway. Runway edge lighting consists of white lights spaced out on both sides of the runway, indicating the edge. The purposes of the lighting include marking hazards, giving pilots a visual aid, and meant to be visible to pilots and not a disturbance to people on the ground. Figure 8.19 gives an overview of the airport lighting systems.

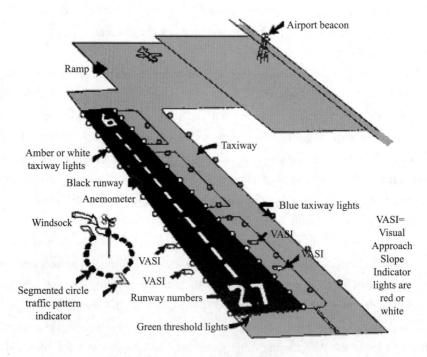

Figure 8.19 Overview of airport lighting systems

8.3.3 Weather observation

Weather condition is one of the most important influence factors to the safety of airplanes, so weather observation is indispensable. It is crucial to make the takeoffs and landings of airplanes safely. No matter it is large or small, every airport has weather observation apparatus all over the world. The forms of weather observation can be divided into automated airport weather stations, human observers, and a combination of both.

Questions

1. What the functions are of different parts of space base?
2. Explain the different launching modes of missiles.
3. Give the classification of RLV and its advantages and disadvantages.
4. Describe the launching powers of missiles.
5. Describe the launching procedure of VTVL-RLV.
6. List at least three ground equipment of an airport and illustrate their main functions.

Words and phrases

auxiliary equipment　辅助设备
Baikonur　拜科努尔
biological satellite　生物卫星
buffer　缓冲器
Canaveral angle　堪纳维尔角
capsule　航天舱
carrier rocket　运载火箭
Centre Spatial Guyanais　法属圭亚那太空中心
clearance delivery　许可颁发
cosmic speed　宇宙速度
deorbit　离轨
docking　入坞
frame-launching　架式发射
grid fins　栅格舵
Guyana　圭亚那
heatproof shield　防热罩
hitting probability　命中率
Hydrogen　氢
interplanetary trajectory　星际轨道
Kennedy Space Center　美国肯尼迪航天中心
launching site　发射场
low dip angle　低倾角
lunar　月球的
mandatory instruction　强制性指令
meteorological　气象的
multistage rockets　多级火箭
Nitrogen　氮

omnidirectional range　全向信标
overhaul　大修
pipes　小管
Plesetsk　普列谢茨克发射基地
reaction control system（RCS）　反作用控制系统
reconnaissance satellite　侦察卫星
recovery site　回收区
reshipment　货物重装
retrorocket　制动火箭, 减速火箭
reusable launch vehicle（RLV）　可重复使用运载器
Skylon　英国云霄塔空天飞机
slot　空位
snowplows　扫雪机
Soyuz　联盟号飞船
space orbit　航天轨道
stowage　堆装物
synchronous satellite　地球同步卫星
taxiway　滑行道
terminals　枢纽
transit passenger　过境旅客
tube-launching　管发射
underground well　地下井
universal velocity　宇宙速度
VASI　目视进场下滑指示器

References

[1] 郝继光,谭大成,姜毅. 航天器发射技术[M]. 北京:北京理工大学出版社,2020.

[2] 宋征宇. 运载火箭地面测试与发射控制技术[M]. 北京:国防工业出版社,2016.

[3] Baker D. Nasa Space Shuttle Manual:An Insight into the Design,Construction and Operation of the Nasa Space Shuttle [M]. UK:Haynes Publishing,2011.

[4] Grantz A. X-37b Orbital Test Vehicle and Derivatives [C]. AIAA SPACE 2011 Conference & Exposition, USA,2011.

[5] Space X. Falcon 9 User's Guide:Rev 2 [M]. Space Exploration Technologies Corp,2015.

[6] Davis L A. First Stage Recovery [J]. Engineering,2016,2(2):152-153.

[7] 周宏宇. 组合动力可重复使用运载器三维轨迹优化与在线制导方法研究[D]. 哈尔滨:哈尔滨工业大学,2019.

[8] 田栢苓,李智禹,吴思元,等. 可重复使用运载器再入轨迹与制导控制方法综述[J]. 航空学报,2020,41(11):6-30.

[9] 崔乃刚,吴荣,韦常柱,等. 垂直起降可重复使用运载器发展现状与关键技术分析[J]. 宇航总体技术,2018,2(2):27-42.

Chapter 9　Payloads

For aircraft transportation or space exploration, the payload is the important parameter carrying capacity of aircraft or spacecraft, including passengers, cargos, scientific instrument or experiment equipment. External fuel, when optionally carried, is also considered into a part of the payloads.

Different from aircraft and spacecraft, missile has its special payload, i. e. , the warhead, which is used to destroy and damage the targets. In other words, that is the main difference between missiles and other flight vehicles.

9.1　Rocket payloads

According to various usages, rockets can be divided into military rockets, launch vehicles, sounding rockets and so on. Military rocket's payload is warhead, just like a missile.

In spaceflight, a launch vehicle or carrier rocket is a rocket used to carry a payload from the Earth's surface into outer space. A launch system includes the launch vehicle, the launch pad and other infrastructure. Usually, the payload is an artificial satellite, spacecraft, part of a space station, and deep space probe or other spacecrafts. Figure 9.1 shows a typical launch vehicle including image satellite as a payload. Some spaceflights are sub-orbital while others enable spacecraft to escape Earth orbit entirely. Sounding rocket is a rocket that carries out exploration and scientific test in the near-earth space with its payload. The structural components and parameters of each layer of the atmosphere can be detected at altitude. It can fly higher than a sounding balloon and lower than a low-orbit earth satellite.

In general, payload is located on the top of the launch vehicle, covered by payload fairing. Strictly speaking, payloads do not belong to any part of the rocket's structure.

Payload fairing is one of the main components of a launch vehicle. As seen in Figure 9.2, the fairing protects the payload during the ascent against the impact of the atmosphere (aerodynamic pressure and aerodynamic heating). More recently, an additional function is to maintain the cleanroom environment for precision instruments. Outside the atmosphere the fairing is jettisoned, exposing the payload. At this moment mechanical shocks and a spike in acceleration might be observed. The standard payload fairing is typically a cone-cylinder combination, due to aerodynamic considerations, however specialized fairings are in use as well.

Now different launch vehicles' capabilities of carrying payloads are given in the following. A small lift launch vehicle is capable of lifting up to 2000kg of payload into low earth orbit (LEO). A medium lift launch vehicle is capable of lifting between 2000 to 10000kg of payload into LEO. A medium-heavy lift launch vehicle is capable of lifting between 10000 to 20000kg of payload into

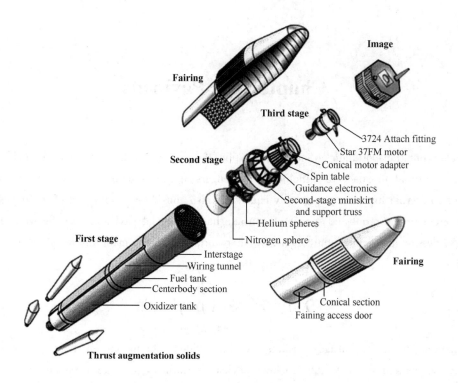

Figure 9.1 A typical launch vehicle

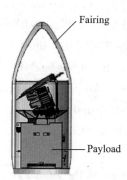

Figure 9.2 Payload fairing

LEO. A heavy lift launch vehicle is capable of lifting between 20000 to 50000kg of payload into LEO. A super-heavy lift vehicle is capable of lifting more than 50000kg of payload into LEO.

A sounding rocket is basically divided into two parts, a solid fuel rocket motor and the payload. Figure 9.3 shows Dongfang-1 and SS-520, typical sounding rockets. Payloads of sounding rocket are all kinds of detection equipment, test equipment and practical life-forms installed in the head of rocket for appointed task of exploration and experiment. As primary object for detecting and testing, payload occupies a dominant status in sounding rocket system. Sounding rocket payload is determined by the degree of exploration's content and widely varying level of scientific instruments, usually a few kilograms, tens of kilograms and several hundred kilograms. The role of a sounding rocket is to transport payload to a certain height to achieve space exploration, experimentation and testing.

(a) Dongfang-1　　　　　　　　　　　　　(b) SS-520

Figure 9.3　Sounding rockets

Payloads are generally produced and provided by their users. Sometimes a user relegates engineering design of payload to the department of sounding rocket development because of the restriction of condition. For example, a biological life support system in a biology experimental rocket, which belongs to a part of payload essentially, can also be commissioned to produce by the department of rocket development.

Now take the rocketsonde as an example to give a brief description of requirements for the rocket payload design. As a kind of sounding rocket, the rocketsonde is a system for atmospheric observations including a rocket which launches instruments making weather observations up to an altitude of 75000m. A typical rocket is 3.7m tall and is powered by a 10cm diameter solid fuel rocket engine. The meteorological instruments, which are used to record data on temperature, moisture, wind speed and direction, wind shear, atmospheric pressure, and air density, are the main payloads. Position data (altitude and latitude/longitude) recorder may also be a payload.

9.2　Missile payloads

9.2.1　Warhead composition

Usually, a warhead is composed of four parts, that is, powder, shell, fuse, and explosive transmission series.

The powder is called payload. It is the charge to destroy the targets. After the charge is detonated, the excitation is used by means of external energy, which can be separated into four kinds, the chemical energy, the mechanical energy, the electrical energy and the explosive energy. Then the targets are destroyed by the excitation.

The shell is a container to load the charge. It is the basic part connected with other parts.

The fuse is a detonator to set off an explosive charge according to different targets adequately. At the same time, it has the safety and arming device to ensure the flying safely. Besides, it can be used to destroy itself.

The explosive transmission series are the amplifier of the energy, whose function is to realize the energy transmission so that the starting energy can be converted into the explosive waves and flames.

That is to say, the four parts are formed an integrity from the initiator, booster, to main charge, till the explosion.

9.2.2 Typical warheads

Generally speaking, warheads can be divided into three categories. They are conventional warheads, nuclear warheads and special warheads. The conventional warheads include the blast warheads, fragmentation warheads, shaped-charge warheads and other warheads. Nuclear warheads utilize the nuclear energy to destroy the targets. Special warheads contain laser warheads, X-rays warheads, incendiary warheads, toxic warheads and so on.

1) Blast warheads

Blast warheads are used to destroy the targets from the ground or water surface, underground and underwater, such as military establishments, warships, submarines and so on, by means of the high explosive waves with a large quantity of gas with high temperature and high pressure. The fragmentations are the 2nd effects of the warheads.

The charge is designed to achieve its effects and damage level according to different targets in different spaces. It has the characteristics of big volume, and large proportion charge relative to the whole weight of the warhead. The explosive reacting procedure is given in Figure 9.4.

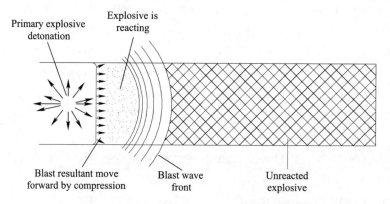

Figure 9.4   Explosive reacting procedure

Figure 9.5 gives 9M133ФM-3 with blast warhead. It is equipped with a thermobaric warhead, TNT equivalent of 7.7kg. It is composed of a large number of high explosives in metal shell, which destroy the target by the blasting effect of the main charge. The expansion of high temperature and high-pressure gas causes the shell of warhead to rupture and generates a shock wave in the air around the warhead. The shock wave reaches its maximum pressure in a few microseconds and then drops to atmospheric pressure in a few hundred seconds. Then the pressure dropped below atmospheric pressure and finally returned to normal. The change of pressure from positive to negative has a destructive "push-pull" effect on the target.

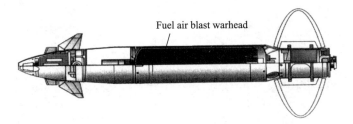

Figure 9.5　9M133ФМ-3

2) Fragmentation warheads

A fragmentation warhead is used to attack the target in the air and on the ground, like aircraft, automobile, radar equipment, and other weapons. It utilizes the explosive charge surrounded by pre-formed metal so that it can be effective to both air and surface targets. If the target in the air is attacked, the warhead will be detonated near the target, rather than hit the target.

Fragmentation warheads include fragmenting warheads, semi-prefabricated fragmentation warheads and others. For fragmenting warheads, the destroyed effectiveness depends on the density of the fragments, effective lethal radius, and the space distribution of the fragments. Figure 9.6 shows Standard-4 missile with Mk125 high energy fragmentation warhead. The warhead has a mass of 135kg, which can set the explosion height during the vertical dive attack. A large number of high-speed armor fragments produced by the warhead can kill multiple targets and destroy important targets in the enemy area.

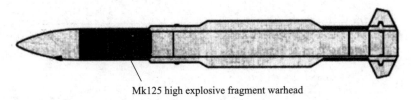

Figure 9.6　Standard-4 missile

3) Shaped-charge warheads

The shaped-charge warhead is used to attack the target on the ground, like tank, armored vehicle, and airport and so on. Usually, it has the hollow charge, i. e. , shaped-charge, accumulating the energy effects by the metallic jet so that the shaped-charge warhead can penetrate the armored vehicle, rather than the kinetic energy. Figure 9.7 illustrates the forming procedure of the metallic jet of the shaped-charge warheads with the time-varying. Figure 9.8 shows the perforation capacity of different charge structures.

4) Nuclear warheads

By using the releasing nuclear energy while it occurs to nuclear reaction, the nuclear warhead is used to attack the strategic target. According to the different ways of releasing nuclear energy, the nuclear warheads are divided into atomic warheads, hydrogen warheads and neutron warheads.

Atomic warheads use the nuclear energy from atomic nuclear fission, especially a heavy nucleus such as an isotope of uranium, splits into fragments, and usually two fragments of comparable mass, with the evolution of from 100 million to several hundred million electron volts of energy. Figure 9.9

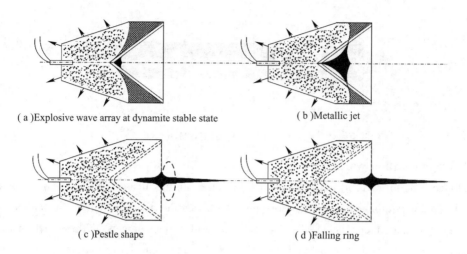

(a) Explosive wave array at dynamite stable state
(b) Metallic jet
(c) Pestle shape
(d) Falling ring

Figure 9.7　The forming procedure of the metallic jet

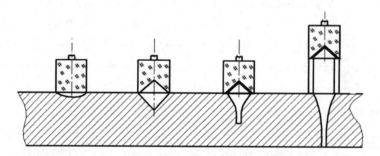

Figure 9.8　Perforation capacity of different charge structures

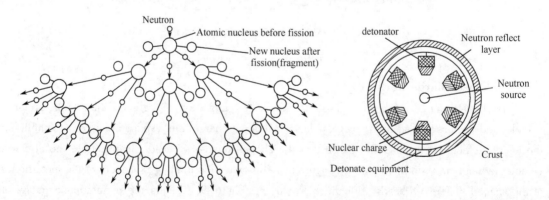

Figure 9.9　Atomic warhead

shows atomic warhead structure.

　　Hydrogen warheads utilize the nuclear energy from the nuclear fusion in which nuclei combine to form more massive nuclei with the simultaneous releasing of energy. Same as the hydrogen warheads, neutron warheads utilize the nuclear energy from the nuclear fusion too. A neutron warhead would produce great numbers of neutrons rather than little blast and thus it would destroy life rather than spare property.

Figure 9.10 shows separate nuclear warheads of LGM-118, named Peacekeeper. The main instruments and engines of the autonomous control system are installed in the instrument cabin, which can ensure the sequential separation of 10 warheads. Quasi heavy decoy is used in the penetration system of missile, which can simulate the performance of most atmospheric and extra atmospheric orbital warheads. Other nuclear warheads installed will produce nuclear explosions in time to destruct the target.

5) Special warheads

In order to achieve some special purposes, missiles will also use other special warheads. As seen in Figure 9.11, AGM-114R9X is a special missile. In order to reduce the harm to innocent people, only the blades are used as the effective killing parts. In addition, the graphite warhead can carry carbon fiber wires, which can be attached to the electrical equipment when attacking the target, resulting in short circuit or even burning. EMP warhead uses low-frequency or high-frequency EMP, microwave devices and other equipment to destroy each other's electronic and information platforms, causing a fatal blow to reconnaissance, communication, command and information systems in a certain area.

Figure 9.10 Nuclear warheads

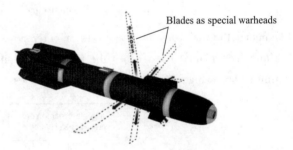

Figure 9.11 AGM-114R9X

9.2.3 Fuses

The function of a fuse is to release the insurance, and detonate the warhead. The requirements followed should be considered.

(1) Reliability.

(2) Instantly detonate, without delay.

(3) Insurance, i.e., removing off by means of two levels, 1st and 2nd level.

Figure 9.12 gives the fundamental composition of typical fuse, connection of its parts and together with Figure 9.13 describing its relationship with target and warhead.

Action process of fuse means that the whole process from launching to detonating the warhead. The fuse must be kept in the safety condition before launching. After the warhead is

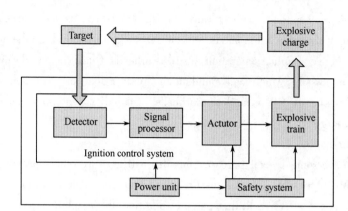

Figure 9.12 Fundamental composition of fuse

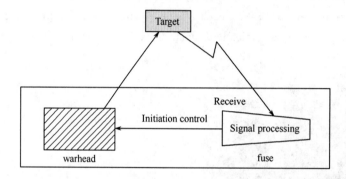

Figure 9.13 Relationships between target, fuse and warhead

launched, the fuse uses its own power or outer power to make it in a state that once there are any priming commands or it reaches the priming time, it will detonate the warhead. This state is called awaiting state. Figure 9.14 gives the action process of fuse.

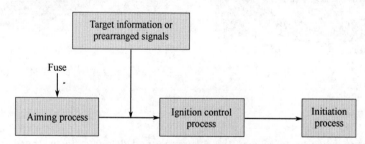

Figure 9.14 Action process of fuse

There are many types of fuses. The fuses are different because there are many different kinds of missiles, and various purposes for attacking the targets. Usually, two kinds of the fuses are separated as noncontact fuses and contact fuses according to the manners of reacting on targets.

1) Non-contact fuses

According to the source of physics field which transfers target information, non-contact fuses can be divided into three types, which are active non-contact fuses, semi-active non-contact fuses and passive non-contact fuses, as Figure 9.15 ~ Figure 9.20 show.

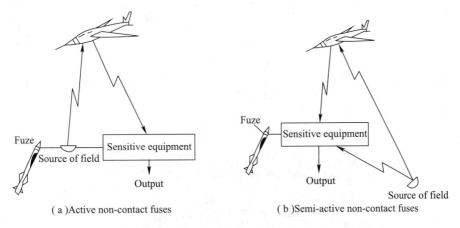

Figure 9.15 Active non-contact fuses and Semi-active non-contact fuses

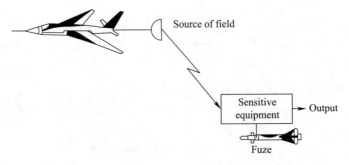

Figure 9.16 Passive non – contact fuses

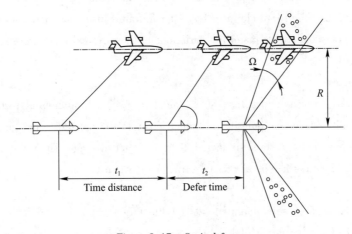

Figure 9.17 Optical fuses

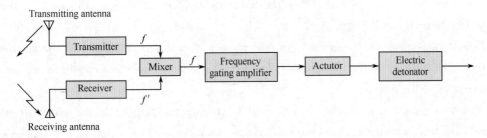

Figure 9.18 Radio fuses

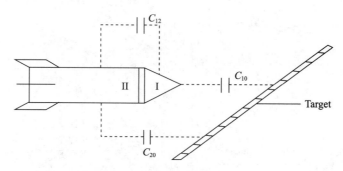

Figure 9.19 Capacitance fuses

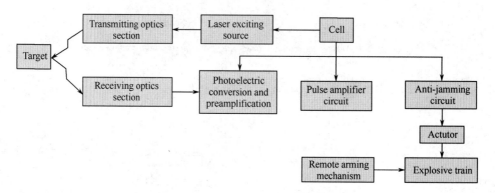

Figure 9.20 Laser fuses

According to the operating principles of non-contact fuses, the typical fuses can be divided into radio fuses, optical fuses which include laser fuses and infrared fuses, magnetic fuses, acoustic fuses, capacitance fuses and so on. By means of the variations of the physical characteristics, different kinds of fuses are designed.

2) Contact fuses

Of contact fuses, a piezoelectric fuse is a typical case. The fundamental principle is that the generation of electricity or electric polarity in dielectric crystals subjected to mechanical stress, or the generation of stress in such crystals subjected to an applied voltage. Then the explosion will be detonated. Figure 9.21 shows a typical piezoelectric fuse structure.

3) Fuse selection

Generally speaking, fuse selection depends on the target category and its environment. If the targets are in the air, for example, airplanes or missiles, non-contact fuses are adopted. If the missile is guided near the target which is within the lethal radius, the warhead is detonated by the non-contact fuse to destroy the target.

If the targets locate on the ground (or sea) surface, according to the different characteristics of the targets, the different warheads are used, therefore, the different fuses are chosen by means of the characteristics of the targets.

As stated before, blast warheads are used to destroy the targets on the ground (or sea) surface. Generally, the contact fuses are utilized to detonate the blast warheads. A shaped-charge warhead is used to attack the armored targets. And the piezoelectricity fuse is utilized to detonate

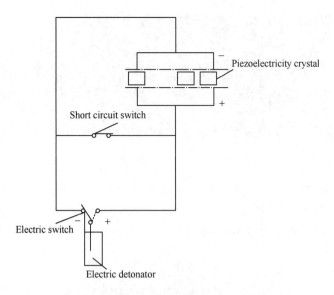

Figure 9.21 Piezoelectric fuses

it. The missiles with nuclear warheads usually have the special fuses to detonate the warheads.

In short, to meet the requirements of the various battles, the fuses are considered and chosen according to the different manners of reacting on the targets. Besides the detonation, safety and arming device must be discussed in order to ensure the safety and reliability.

The laser fuse configuration of AIM-9L missile is seen in Figure 9.22. Pulse Doppler or linear frequency modulation can be used to obtain the target range and velocity information at the same time. It has the advantages of high sensitivity, rich information and effective filtering of stray background light, and has been widely used and rapidly developed in military, surveying and mapping, communication and other fields. Figure 9.23 shows radio proximity fuse.

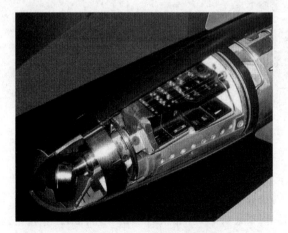

Figure 9.22 Laser fuse of AIM-9L

As discussed before, a warhead is very important for a missile, which can realize to destroy the targets. With the development of missile weapon technology, some new types of warheads appear. For instance, submunitions, kinetic warheads, and so on.

Figure 9.23 Radio proximity fuse

The warhead technology is very important and sophisticated. It should be suited to other systems on missiles. With the development of high and new technology, more categories of warheads will be invented. The warheads will be multipurpose, multi-mission, smaller, more efficient and more intelligent in the future.

9.3 Air vehicle payloads

Air vehicle weights are composed of a large number of various accessories, purchased parts (equipment or adjuncts), standard parts and all kinds of working fluid. In general, aircraft payloads include available fuel, cooling liquid, anti-icing fluid, safety equipment, passengers, daily necessities and cargo, etc. It should be noted that the type and weight of air vehicle payloads are selected according to technical requirements of the aircraft. For instance, a fighter's payloads include guns, flares, missiles and their launching devices as seen in Figure 9.24. And the payload of airborne warning and control system is the radar.

Figure 9.24 J-20 and the missiles PL-10E and PL-15

Unmanned aerial vehicle (UAV) can be used as weapon platform to carry missiles and bombs. In addition, it is a photoelectric payload platform. Its payloads can be divided into three types of photoelectric loads, infrared devices (thermal imager, scanning instrument and imaging spectrometer), visible light devices (CCD camera and low light level TV), and laser devices. Figure 9.25 shows photoelectric devices on unmanned aerial vehicle.

There is a natural trade-off between the payload and the range of an aircraft. Figure 9.26 shows the relationship between them. A payload range diagram (also known as the "elbow chart") illustrates the trade-off.

The top horizontal line represents the maximum payload. It is limited structurally by maximum

Figure 9.25 Photoelectric devices

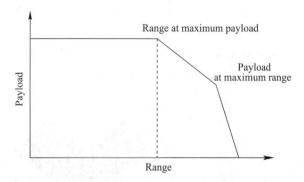

Figure 9.26 Relationship between payload and range

zero fuel weight (MZFW) of the aircraft. Maximum payload is the difference between maximum takeoff weight and maximum fuel weight (OEW). Moving left-to-right along the line shows the constant maximum payload as the range increases. More fuel needs to be added for more range.

Weight in the fuel tanks in the wings does not contribute as significantly to the bending moment in the wing as weight does in the fuselage. So even when the airplane is loaded with its maximum payload which the wings can support; it can still carry a significant amount of fuel.

The vertical line represents the range at which the combined weight of the aircraft, maximum payload and needed fuel reaches the maximum takeoff weight (MTOW) of the aircraft. If the range is increased beyond that point, payload has to be sacrificed for fuel.

The maximum takeoff weight is limited by a combination of the maximum net power of the engines and the lift/drag ratio of the wings. The diagonal line after the range-at-maximum-payload point shows how reducing the payload allows increasing the fuel (and range) when taking off with the maximum takeoff weight.

The second kink in the curve represents the point at which the maximum fuel capacity is reached. Flying further than that point means that the payload has to be reduced further, for an even lesser increase in range. The absolute range is thus the range at which an aircraft can fly with maximum possible fuel without carrying any payload.

9.4 Space vehicle payloads

System consisting of instruments, equipment or supplies that are used for performing a specific space mission is usually called space vehicle payload. The system is made up of space application device or special materials, other than space vehicle's guarantee system for maintaining flight.

Spacecrafts for different purposes are equipped with different payloads. The payload system with particular purpose is often named by its specific function.

1) Satellite payloads

Satellites are generally made up of two parts, payloads and platforms. Different satellites carry different payloads depending on their own usages. Payloads of the communication satellite are composed of the communication repeater and antenna, known as the communication system. Figure 9.27 shows Shijian-20 satellite. It is a new technology verification satellite in geosynchronous orbit. It will verify the key technologies of the new generation of large satellite platform of DFH-5, implement a number of new technology verification work, and carry out geosynchronous orbit communication and broadcasting business. Payloads of a navigation satellite include high-stability clock, data receiver, navigation data storage, broadcast navigation signal transmitter and directional antenna, named as navigation system.

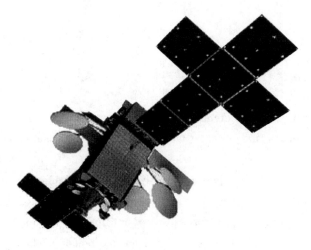

Figure 9.27　Shijian-20

2) **Probe payloads**

Probes have been launched to explore the celestial body in or outside the solar system, including the sun, planets and asteroids. Some space vehicles have been to the moon and the Mars. A probe space vehicle can be divided into orbiter and lander. The payload of the orbiter is mainly photoelectric reconnaissance equipment, similar to artificial earth satellite. As for the lander, its payload is mainly the detection and sampling equipment after landing. For example, Chang'e-5 is equipped with a variety of payloads. It includes panoramic camera and its turntable, landing camera, lunar spectrum analyzer, lunar soil structure detector, payload data processor, etc. Figure 9.28 shows a γ-ray spectrometer and Figure 9.29 shows stereo camera and interference imaging spectrometer.

3) **Transfer vehicle payloads**

Shuttles are used for transportation between the Earth and space station. For instance, the Dragon ship delivered cargo to the station, including experimental equipment, spare parts, clothes, food and refrigerators needed by space station astronauts, and returned with other items, including

Figure 9.28 γ-ray spectrometer

Figure 9.29 Stereo camera and interference imaging spectrometer

hundreds of blood and urine samples of space station astronauts for studying the nutritional status of astronauts and how to eat, to alleviate the adverse effects of long-term space environment on astronauts' health. Figure 9.30 shows the typical structure of automated transfer vehicle.

4) Manned space vehicle payloads

For manned space vehicle, the payload refers to the space test system or the materials loaded, which also includes astronaut system. The payload of manned space vehicles can be divided into two categories. One is the scientific payload for astronomy, life science, physics, planetary science and other basic theoretical research, such as the space laboratory in a shuttle. The other is application payload, mainly for the practice of science and technology, such as Earth observation, communication and navigation, materials and processes, etc. As for environmental control and support system (ECLSS), it has to create a comfortable living environment for passengers and carrying out scientific research effectively. Oxygen plant and trace pollutant control system ensure sufficient oxygen and clean atmosphere, and ensure the environmental pressure. The ventilation device is used to mix the air properly to ensure the atmospheric environment of the whole cabin. Ventilation and water-cooling system ensure the cabin temperature and humidity are appropriate. Wastewater and urine are treated

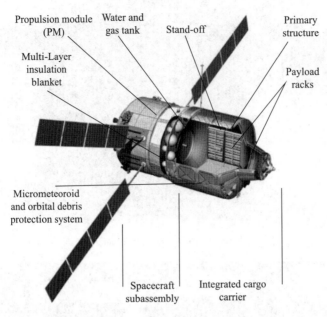

Figure 9.30 Automated transfer vehicle

by water treatment unit to realize water recycling. Smoke detector ensures the emergency treatment. These devices make manned spacecraft more complex than unmanned spacecraft.

Space vehicle payload is a very important sub-system in spacecraft system. Payload directly affects the capacity of accessing to information and transmission of aerospace equipment, consequentially affects the completion of the task. At the same time, payload directly influences overall performance of aerospace mission, so payload design is an important part of the overall design and evaluation of spacecraft. United States, Russia and the European Space Agency have invested a lot of money for research and development of the payload design and applied the payload design to overall design of satellite systems widely.

Questions

1. What is the difference between payloads of rocket, missile, air vehicle and space vehicle?
2. Give the usage of the warheads.
3. How to control the fuses not to detonate too early?
4. Explain the relationship between the payload and the range of an aircraft.
5. Why is manned spacecraft more complex than unmanned spacecraft?
6. What's the relationship of warheads, missiles and fighters?
7. What's your opinion about that payload also has its load?
8. What's the difference between satellite and space station?
9. Give some other special warheads.
10. Give your opinion about the space stations.

Words and phrases

arming device　解除保险装置
artificial satellite　人造卫星
blast warhead　爆破战斗部
buffeting　抖振
capacitance fuse　电容引信
charge　装药
communication satellite　通信卫星
contact fuse　触发式引信
conventional warheads　常规战斗部
crystal　晶体
detonate　引爆
detonator　起爆器
Earth resources satellite　地球资源卫星
explosive transmission series　传爆系列
fairing　整流罩
fragmentation warhead　破片战斗部
isotope　同位素
jettison　投弃货物
gust　阵风
launch vehicle　运载火箭

manned spacecraft　载人飞船
metallic jet　金属射流
meteorological satellite　气象卫星
navigation satellite　导航卫星
nuclear fission　核裂变
nuclear fusion　核聚变
piezoelectric fuse　压电引信
polarity　极性
powder　火药
radio fuse　无线电引信
reconnaissance satellite　侦察卫星
remote sensing system　遥感系统
rocketsonde　气象火箭
safety　保险,保险装置
semi-prefabricated fragmentation warhead　半预制破片杀伤战斗部
shaped-charge warhead　聚能破甲战斗部
sounding rocket　探空火箭
toxic warhead　毒气弹
Uranium　铀

References

[1] 姜雪红,张雨薇. 俄罗斯短号-ЭM 多用途导弹系统[J]. 飞航导弹,2012(2):49-50,55.
[2] 魏圣军,吴法文,张琳,等. 美国标准导弹发展演化分析[J]. 飞航导弹,2020 (10):16-21,53.
[3] 吴小宁,张秀刚. SS-18 导弹研制技术综述[J]. 飞航导弹,2019(5):13-19.
[4] 刘锡民,张建华,杨德钊,等. 相干激光引信综述[J]. 红外与激光工程,2018,47 (3):31-36.
[5] 徐孝彬. 脉冲激光引信近程周向探测技术研究[D]. 南京:南京理工大学,2017.
[6] 张凯华,蒋祎,廖俊. 临近空间慢速飞行器载荷概述[J]. 航天返回与遥感,2017,38(6):1-10.
[7] 张若岚,陈洁. 从 UAV 到 UAS——无人机系统及光电载荷发展的思考[J]. 红外技术,2014,36 (8):601-608,664.

Chapter 10 Flight vehicle system research and development

A flight vehicle system is a very large system composed of several subsystems. Therefore, the flight vehicle system research and development (R&D) is the complex process. During the flight vehicle system research and development, the concepts of the system and the views of system engineering are the fundamental of system design and system analysis of flight vehicle system. Then the research and development are integrated for the overall system.

10.1 Flight vehicles design concepts

No matter what the aircraft systems, satellite systems, missile systems or other flight vehicle systems are, the design concepts and philosophy are consistent. The missile system R&D is discussed as follows.

10.1.1 The R&D process

How does one begin to research and development a flight vehicle? Since R&D is really just a creative problem solving, the classic problem-solving method often called the scientific method is a great place to start. The principal steps in the scientific method are commonly understood as Figure 10.1 given.

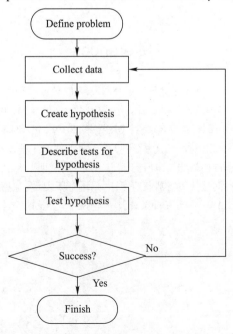

Figure 10.1 Principal activities in the R&D cycle

As seen in Figure 10.1, for the scientific research of a complex system, just from the definition of the problem, the construction of a hypothesis, till the verification of the hypothesis, the process is a cycle including the creative synthesis in any directions, rational analysis, and decision making. Finally, one answer for the proposed question will be given.

There are many parts of research and development for a missile system. Generally speaking, the missile system contains the missile itself, launch and control equipment. Whereas, the missile itself is composed of body and wing, power equipment, guidance and control, warhead and fuse parts, and other subsystems. Therefore, missile system R&D means that it is a complicated system R&D. In other words, the missile system R&D should be made according to the points of view of System Engineering.

10.1.2 System analysis

As stated before, the flight vehicle system consists of some subsystems which have the interactions among them. When the flight vehicle system is researched and developed, system analysis or system engineering must be discussed at first by means of the system analysis methods.

10.1.2.1 System analysis method

It was proposed by RAND which stands for Research and Development Corporation.

Briefly speaking, combining mathematical methods with engineering methods, the system analysis method is constructed. From the point of view of system, the project is researched and analyzed by using the basic methods and theories, then a plan is selected. Steps of the system analysis method are to be discussed in the following.

(1) Goal analysis.
(2) Modeling establishment.
(3) System optimization.
(4) System evaluation.

10.1.2.2 System simulation

System simulation is very important for the flight vehicle system R&D. At first, the mathematical model of the flight vehicle system must be established according to the basic theories and principles of mathematics, physics and other fields. And the program for the mathematical model is compiled with the computer languages. Then the digital simulation is made by using digital computers. Furthermore, half-object-simulation or object-simulation of the flight vehicle system will be made.

10.2 Basic stages for R&D

For a flight vehicle system, the procedure of R&D is given as Figure 10.2, which includes the following research and development stages.

As seen in Figure 10.2, at first, the ideas are put forward. Then the tactical and technical requirements the flight vehicle system must be proved. That is to say, the conceptual design is constructed. Furthermore, the preliminary design is made for some key techniques. Both the conceptual design and preliminary design are called the research in advance. After the key techniques have been solved, the technical design, i.e., detail design will be gone forward with advanced technology. At the

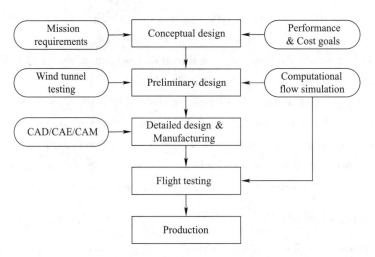

Figure 10.2 The flight vehicle system R&D process

same time, manufacturing and flying tests are developed. Through the variations of the finality of design, the production of the flight vehicle will be manufactured by the corporations and factories. At the end, the flight vehicle system will be put into utilization by the corporations.

10.2.1 Idea proposition

A new design concept for a flight vehicle system is proposed at first. Usually, there will be the requirements, or a new concept is proposed by the development sections. Combined with these two cases, the new concept of a flight vehicle system is put forward.

10.2.2 Proof of tactical and technical requirements

After the new concept is being proposed, the requirements for utilization, the conditions of technological and economical sides must be considered.

New concepts for a flight vehicle system should be specified. Furthermore, the reasonable tactical indexes and technical requirements are determined.

Usually, the idea proposition and the proof of the tactical and technical requirements are named the conceptual design stage.

10.2.3 Preliminary design stage

During this stage, more details about the flight vehicle system must be researched, for example, the system composition and analysis, plan evaluation, performance analysis, parameter calculation, and so on. According to these parts, the overall plan of the flight vehicle system and compositions for different subsystems will be constructed.

At the same time, the key techniques and questions for this flight system must be analyzed and researched. The special subject will be furthermore studied.

Sometimes, the work at the preliminary design stage is combined with that at the conceptual design stage.

10.2.4 Technical design stage

The detail works for a flight vehicle system will be discussed during this stage. In fact, the technical design for this system must be studied. That is to say, a lot of work will be done, especially for test research, such as the wind tunnel test, simulation test of guidance and control system, the engine hot test, strength and vibration test of the flight vehicle, impact test of components, ground test, and so on.

The calculation and analysis about aerodynamics, flight mechanics, strength, reliability, and so on must be done and the complete drawing is plotted.

10.2.5 Trial-manufacture and test stage

The trial-manufacture for the product and elements, parts is completed in this stage. As well the tests must be made to prove the design results. In this stage, the primary works is to design the sample of each subsystem and total flight vehicle system, and complete ground tests and flight tests adopted kinds of samples.

The flight tests for different purposes must be made. For tests, there will be two results, success and failure. If failure, the design would be checked and redesigned. Till the test is successful, this stage will be completed.

10.2.6 Finality of design stage

After the flight tests, the revised design must be made. The flight vehicle system is to be finished by finality of design after the overall tactical indexes and technical requirements are qualified. Then the tests for finality of design will be examined according to the utilization requirements. At the same time, the design documents will be finished. At the end, the meeting for finality of design is held.

10.2.7 Manufacture and utilization stage

At last, the flight vehicle system will be produced by the product departments. Furthermore, the products will be delivered for the utilization departments.

Till now, the circulation of the design, research and manufacture of the flight vehicle system will be finished.

10.3 Basic design requirements

10.3.1 Design requirements for flight vehicle systems

During the design process, some basic requirements must be met. For instance, tactical requirements, technical requirements, utilizing and maintaining requirements, economical requirements, system and subsystem requirements should be considered carefully. Finally, the design balance will be reached. Figure 10.3 illustrates the design balance during the flight vehicle research and development.

As seen in Figure 10.3, the considerations in flight vehicle design balance include launch

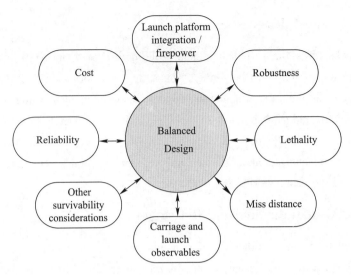

Figure 10.3 Design balance sketch

platform integration, robustness, warhead lethality, miss distance, carriage and launch observables, other survivability considerations, reliability and cost. Launch platform integration including firepower and launch control system will give some constraints on the flight vehicle design, e. g. , the tube-launch has strict requirements on the missile diameter, wing and tail. The sufficient robustness is necessary to confront the measurement error, system delay, uncertainty and adverse flight environments. Warhead lethality should destroy the attacked targets. Miss distance, which relates to circular error probability (CEP), impacts the required flight performance. Carriage and launch observables affect on the survivability of the launch platform. Other survivability considerations are standoff distance, radar cross section (RCS), insensitive munitions and so on. The system reliability depends upon the individual subsystem reliability and approximately equals with the product of the individual subsystem reliability.

10.3.2 Quality management for flight vehicle system R&D

Quality management is very important for the research and development of the flight vehicle system. That is to say, the quality management must be considered at first during the R&D procedure. Quality management is determined as the management activities for realizing the quality goal of the flight vehicle system R&D, including quality plan, quality control, quality assurance and quality improvement. From the elements to parts, from subsystems to whole system, the quality management should be made strictly, so that the system has the good quality, the performances of the system will be met the design requirements.

Quality control is one part of the quality management activities. Some technical measurements and management measures should be adopted in order to meet the quality requirements.

If there were the quality accidents, the measures of the space systems-Closed Loop Problem Solving Management in China are used the aerospace quality management double return to zero. It is the closed loop for the quality management during the R&D process. And it is called double five return to zero, i. e. , the technology return to zero and management return to zero.

The technology return to zero mainly includes the five points as follows, just from the side of the technology. ①finding the accurate location of the fault took place. ②being clear for the principles of the fault appeared. ③recurring the fault under the same conditions. ④being effective for the measures to avoid the faults. ⑤inferring other things from the one fact.

And the management return to zero consists of the other five points as follows from the side of the management. ①being clear for the procedure of the fault took place. ②knowing clearly the duty for the fault appeared. ③being practicable for the measures to be adopted. ④be dealt with severely. ⑤completing rules and regulations.

Combing with the both sides of the technology and management, the measures of the double five return to zero are employed, to be sure the good quality for the space vehicles. It is proved that it is a very good quality control method.

10.4 Design methods

As discussed before, based on the theory of the system engineering, by means of the system analysis method and the integrated design method, the flight vehicle research and development can be made. And the flight vehicle system is constructed. Figure 10.4 describes the sketch of integrated design of a flight vehicle.

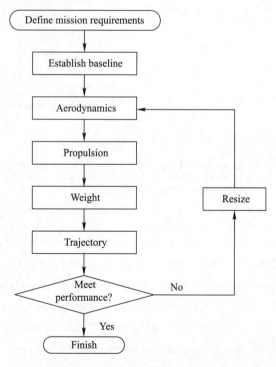

Figure 10.4 The integrated design of a flight vehicle

By using the diagram in Figure 10.4, a circulation is constructed. To begin with, the mission requirements of a flight vehicle are defined. Then the baseline of the flight vehicle is established. After the configuration and parameters are described, the aerodynamic characteristics of the flight vehicle can be calculated and analyzed. Combined with the model of the power system and other subsystems, the construction and mass characteristics of the flight vehicle will be determined. Furthermore, the mathematical model of the flight vehicle system is established. By means of the mathematical methods and computer programming language, the digital simulation of the flight vehicle movement will be researched. At the same time, the flight performances will be attained. By verifying the computed results and design constraints, if they meet all the requirements, one circle is completed. Otherwise, the variations of the configuration and parameters, as well construction arrangements will be redesigned. Then the new circulation will be started. Till the results meet the requirements, the circulation will be finished at the end.

Similar to Figure 10.4, Figure 10.5 shows the integrated process of the design validation and technology development of a flight vehicle system. In the same way, Figure 10.6 describes the iterating procedure of the flight vehicle design.

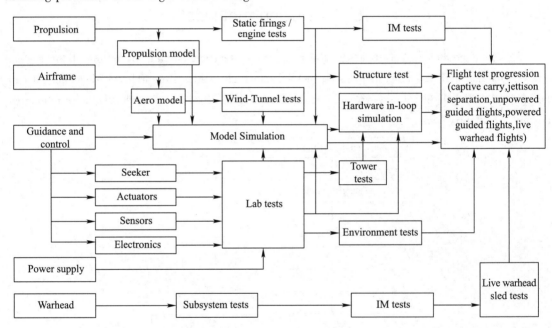

Figure 10.5　The integrated process for design validation and technology development

The typical design method, as shown in Figure 10.4, is a serial design mode, which makes the designed flight vehicle system be hardly to obtain the optimum result and decrease the synthetical performance due to the disadvantages of longer design cycle, higher economical cost and neglecting the coupling relationship of multidisciplinary. In order to handle the above problems of the typical design method, a new design method of flight vehicle, which called Multidisciplinary Design Optimization (MDO). Based on concurrent engineering, MDO develops the design potential by considering the coupling relationship of multidisciplinary, evaluates the scheme by analyzing the flight vehicle system synthetically, improves the reliability by the comprehensive disciplinary tradeoff, and enhance the performance of flight by integrated design.

The aspects referred in the MDO of flight vehicles include geometry model of the flight vehicle, analysis model of each discipline, analysis model of system and integration optimization model of

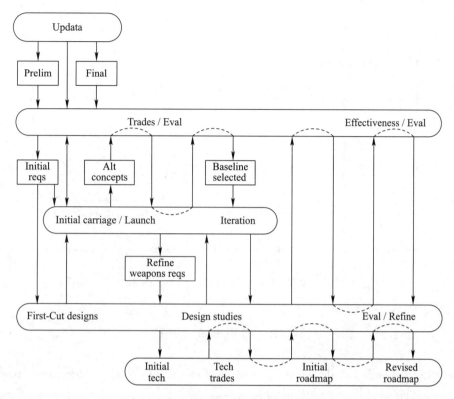

Figure 10.6 The iterating procedure

multidisciplinary.

The geometry model is the basis of the MDO of flight vehicles, and automatically updates once the geometry parameters change. There are two methods to build the geometry model by the software. One is to independently develop the CAD software, the other is to secondly develop the based on the existing CAD software.

The analysis model of each discipline is the precondition of the MDO of flight vehicles. The common analysis models contain analysis model of aerodynamics discipline, structure discipline, propulsion discipline, weight discipline, guidance and control discipline and trajectory discipline, and so on.

The analysis model of the flight vehicle system is built according to considering the coupling relationship of multidisciplinary. The coupling relationship can be expressed by the design structure matrix (DSM). Figure 10.7 gives the typical DSM of the flight vehicle. Each discipline is on the diagonal location of the DSM. The output data of each discipline is expressed by the transverse line of the DSM, the input data of each discipline is expressed by the longitudinal line of the DSM, and the coupling relationship is given by the point of intersection.

The integration optimization model of multidisciplinary integrates the analysis model of each discipline and adopts appropriate optimization method of multidisciplinary. The integration optimization design of multidisciplinary is completed based on the special MDO frame, which consists of MDO frame developed independently and the commercial MDO frame.

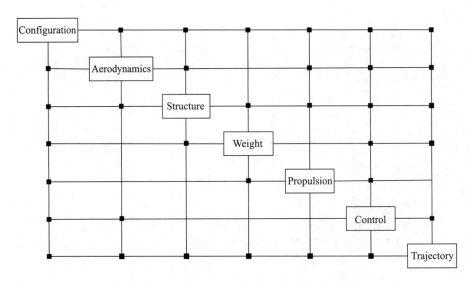

Figure 10.7　The typical DSM of a flight vehicle

10.5　Typical tests

During the research and development process of the flight vehicle system, some typical tests must be made, like wind tunnel tests, propulsion tests, guidance and control system simulation tests, warhead tests, intensity tests, flying tests, environmental tests, and others.

10.5.1　Wind tunnel test

Wind tunnel test is a very important method for flight vehicle research and development. For example, the aerodynamic coefficients and parameters are calculated in engineering at first. The results by computation must be combined with the results from the wind tunnel test, so that the used aerodynamic coefficients will be close to the real instances. By means of the wind tunnel test, the flow field and pressure distribution can be analyzed. According to the force and moment acted on the flight vehicle, the aerodynamic characteristics can be given.

10.5.1.1　Subsonic wind tunnel

The schematic diagram of subsonic wind tunnels is shown as Figure 10.8. A subsonic wind tunnel works like a venturi. A venturi tube is the best example of Bernoulli's theorem, which relates speed and pressure in a tube when no energy is added to the fluid. If the air reaches the restriction in the tube, the velocity increases. The increase in velocity causes a reduction in the static pressure, measured perpendicular to the direction of flow. Since the velocity is low, the air density and temperature remain essentially constant, which is not so in the case of transonic venturis.

The change in cross-sectional area of the tunnel from the fan section to the test section is called the contraction ratio. If you reduce the cross section of your cardboard wind tunnel to one-fifth, a contraction ration of 5, the airspeed in the test section will be five times the airspeed produced by the fan. So, a fan with a 61cm diameter moving air at a speed of 24km/h would produce 112km/h in a test section about 27cm in diameter. On the other hand, if one wants a test section that is 1.52m in

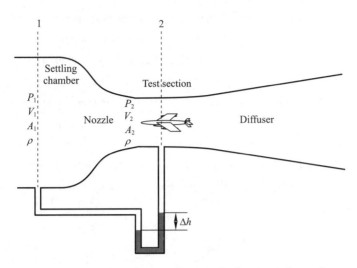

Figure 10.8 The schematic diagram of subsonic wind tunnel

diameter, it will be needed a fan of 3.3m in diameter. A more practical solution is to increase the fan's airspeed and reduce the contraction ratio to achieve a larger test section at the same speed.

10.5.1.2 Supersonic wind tunnel

Supersonic wind tunnels operate differently than subsonic and transonic wind tunnels. Firstly, because fans are inefficient at supersonic speeds, they must run subsonic and the air must make a transition from subsonic to supersonic speeds. Secondly, supersonic wind tunnels require an enormous amount of power. Supersonic wind tunnels require so much power that if run during periods of peak electricity demands they can cause a regional brown-out. Very few facilities have continuous supersonic wind tunnels for this reason.

The key to make a supersonic wind tunnel is to employ a supersonic venturi. Figure 10.9 shows a schematic of a closed-circuit supersonic wind tunnel. The fan moves the air in a subsonic channel. During startup the subsonic section has been pressurized while the test section remains at a static pressure of 1 atmosphere. The air accelerates in the first venturi until the speed at the throat becomes Mach 1. As the channel opens up, since the air is flowing into a region of lower pressure it accelerates, producing the supersonic flow in the test section. After the test section the airflow goes through a second venturi. Here the speed decreases until it becomes Mach 1 at the throat. Since the air is going into a region of higher pressure, as the channel opens up the flow slows down, becoming subsonic again.

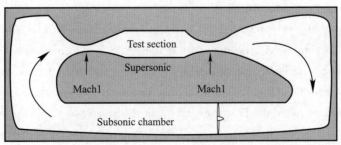

Figure 10.9 The sketch of supersonic wind tunnel

The great amount of power required for supersonic wind tunnels means there are very few continuous wind tunnels and they are not very large. A 1m × 1m test section is considered very large and requires half a million horsepower (375MW) to operate at Mach 3. But there are other methods to test supersonic aircraft.

One method is the "blowdown" supersonic wind tunnel depicted in Figure 10. 10. A huge tank is filled with high-pressure air and then exhausted through a venturi. This kind of wind tunnel works quite well but will only allow a few minutes testing. However, a carefully planned test can gather a tremendous amount of data in a very short time. With this technique the energy required is generated and stored ahead of time. This type of wind tunnel requires very little power but requires quite a long time between tests.

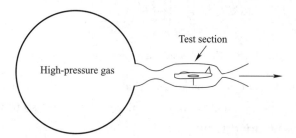

Figure 10. 10 Blowdown supersonic wind tunnel

Another option, which is more common, is the vacuum supersonic wind tunnel shown schematically in Figure 10. 11 Rather than pump a chamber to a high pressure, which is dangerous, the chamber is evacuated and the airflow is in the other direction through the test section. Thus, the upstream reservoir of air is just the atmosphere and the air is being drawn through the throat and test section into a vacuum.

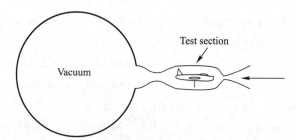

Figure 10. 11 Vacuum supersonic wind tunnel

10. 5. 1. 3 Hypersonic testing

With the incredible power required for supersonic wind tunnels, how can anyone expect to create hypersonic flow conditions, typically above a Mach 5, in a test environment? The only effective method to do this with a stationary model is with the blowdown method, lots of preheating of the air, and a very small test section. The key word in that last sentence was stationary. Some hypersonic facilities use a combustion gun, where gases combust in the breach to propel the model. The problem with this technique is that the desired measurements must be made on a nonstationary model, one that is moving very fast.

But there is another trick up an engineer's sleeve. Hypersonic flight implies that the Mach number is typically greater than Mach 5. Up to this point we implicitly assumed that to achieve hypersonic speeds we must increase the speed in the test section or of the model. What if we were to decrease the speed of sound instead? Sound speed differs for different gases. The speed of sound decreases as the weight of the gas molecules increases. So, instead of using air for our working gas, we could look for a heavier gas, like carbon dioxide, although this will only decrease the sound speed by 14 percent. The advantage of using an alternate gas is that the true speeds can be kept reasonable, while the Mach number is high.

10.5.2 Simulation tests

With the development of the computer technology, simulation or system simulation has become one of the very important measures to analog the actual process by means of the model established. According to the models with different kinds, the system simulation can be divided into digital simulation, physical simulation, hardware-in-loop simulation.

The digital simulation is usually made during the research and development. In conceptual design stage, a simplified model is used to analyze the flight performances. With the deeper research, the 6 DOF digital simulations will be made. Furthermore, hardware-in-loop simulation will be researched. Figure 10.12 shows the simulation test of the guidance and control system.

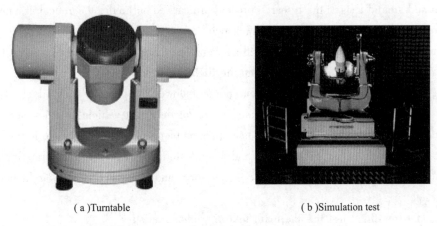

(a) Turntable (b) Simulation test

Figure 10.12 Simulation test on a turntable

10.5.2.1 Digital simulation

Digital simulation is the simulation based on mathematical model of the flight vehicle system, and it gets the system data of the flight vehicle by using the computer program which converted from the mathematical model of the flight vehicle.

The digital simulation system consists of simulation computer, simulation model and processor and display equipment. According to regulations, the system has the functions such as printing data, plotting curve, statistical results.

10.5.2.2 Hardware-in-loop simulation

Hardware-in-loop simulation is based on both mathematical model and physical model. The hardware-in-loop simulation system consists of material object, simulation computer and the other equipment. Figure 10. 13 illustrates the hardware-in-loop simulation test of homing missiles.

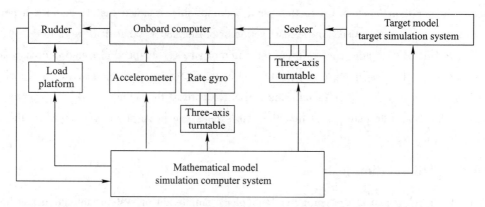

Figure 10. 13　Hardware-in-loop simulation test of homing missiles

10.5.3　Flight test

It refers to the flight test of the flight vehicles under real flight environment conditions. Flight test is the approach to test the flight vehicle system at the different R&D stages, such as, development, design finalization, production evaluation, equipment uses, etc. The purpose of flight test is to verify the results of theory and ground tests and to identify the design indexes and service performances. The design and ground tests in the research process are carried out under theoretical assumptions and non-comprehensive simulation conditions, which need to be verified by flight tests.

According to different categories of the flight vehicles, there are some different flight tests. So, it is divided into aircraft flight test, rocket & missile flight test, and so on.

Here, the model free-flight test is given in the following. It refers to a kind of experimental method in a real atmosphere by use of a scaled model of the flight vehicle, to study aerodynamics, aerothermodynamics, flight mechanics, and some other problems of the flight vehicle in the sky. It can be used for zero-lift drag test, longitudinal and lateral static derivative test, aircraft layout study, spinning study, powered aircraft model test, penetration and reentry tests, the influence of wind experiment, etc.

The model free-flight test has the main following characteristics.

(1) The model is not constrained. Without the constraints and interference of walls and sting supports of the wind tunnel, the model is not restricted by external space.

(2) The test height range is large. Flight experiments can be carried out in various altitude ranges. Also, the effect of altitude changes can be studied by free-flight tests.

(3) Wide range of test Mach number. In a single test, it is possible to obtain test data with a wide range of Mach numbers.

(4) The effects of different atmospheric conditions can be simulated. The influence of wind, rain and other atmospheric conditions on the flight vehicles can be carried out.

(5) The experimental Reynolds number is high. The real atmosphere is larger than the wind tunnel model, if an appropriate flight altitude is chosen, the experimental Reynolds number can be higher, closer to the real flight vehicle test Re.

(6) Strong comprehensive simulation ability. Model free-flight test can synthetically simulate the flight of real flight vehicle, and it is much simpler and less expensive than a real flight test. It is convenient to carry out experiments on the flight vehicle characteristics and performances. As well it could provide reliable test results for flight vehicle development, save development costs and shorten the development cycle.

Model free-flight test has become an important research method in the process of flight vehicle development because of its real simulation conditions and reliable test results. That is to say, compared with the wind tunnel test and theoretical calculation, the structure of the scaled aircraft model flying in the sky is more complex and there is more internal test equipment. Generally, it is required to carry out the wind tunnel test as well as the model free-flight test, in order to study the test parameters, effectively reduce the risk of aircraft test, and provide technical support for aircraft design and flight.

10.5.4 Environmental test

Environment outside flight vehicle consists of the effects of physics, chemistry, and biology, such as weather and noise etc. Environmental tests are mainly composed of natural environmental tests, mechanical environmental tests, and electromagnetic environment tests, and so on. The purpose of environmental tests is to verify the adaptability of flight vehicles in the required environment.

10.5.4.1 Natural environmental tests

The natural environmental tests main include the high temperature and low temperature test, temperature shock test, humid-hot test, fungus test, salt-fog test, sand dust test, drip-proof test and so on. To validate the adaptability of flight vehicles and their components under the condition of the above environments.

10.5.4.2 Mechanical environmental test

The purpose of mechanical environmental test is to verify the adaptability of flight vehicles and their components in the required mechanical environments. The mechanical environmental tests mainly include impact test, collision test, free drop test, vibration test, acceleration test, vibration test, carrier test and etc.

10.5.4.3 Electromagnetic compatibility test

As known, there are many types of complex electronic devices and components on flight vehicles. Consequently, there are electromagnetic environment both inside and outside flight vehicles. Electromagnetic compatibility (EMC) refers to the ability of a device or system to operate within its electromagnetic environment, and without causing intolerable electromagnetic interference to any device in its environment. As a result, EMC test is necessary to validate whether flight vehicles satisfy the requirement of electromagnetic compatibility.

10.5.5 System performance test

In order to validate the cross-link performance of the whole flight system before flight test, system performance tests should be finished. It includes construction static test, construction modal test, guidance system test, propulsion system test, and so on. At the same time, the tactical and technical indexes of the entire flight vehicle system must be tested according to the requirements.

10.5.6 Reliability test

Reliability test is all kinds of tests to understand, evaluate, analyze and improve the reliability of the products. Flight vehicle reliability test means testing and verifying the reliability of the flight vehicle including its parts and components through research tests with limited samples, time and using conditions. And its principle is all the equipment satisfy the reliability indexes.

10.6 Summary

In short, since the successful flight of the Wright brothers' Flyer-1 in 1903 as the first powered controlled continuous flight of human beings, various types of aircraft have been developing rapidly. At the same time, with the rapid development of science and technology, aeronautical and astronautical science and technology have been developing continuously, to promote the design and development of flight vehicles.

In particular, for the research and development of the flight vehicle, it is very important for person to know much more about the flight vehicle system, the system compositions and system working principles. And the mutual relations and mutual reactions between the subsystems should be analyzed clearly. The flight vehicle should be studied by means of the system engineering views and system engineering methods. At the same time, the aircraft design and research and development are closely connected with the development of science and technology.

Questions

1. Describe the research and development procedure of a missile.
2. Give some basic concepts of the flight vehicle design.
3. Please give some ideas about how to design a flight vehicle by yourself.
4. Describe main steps of technology and management return to zero.
5. Describe the compositions of a ground-to-air missile system.
6. Describe components of design structure matrix (DSM).
7. Compare functions of each typical tests during flight vehicle system research and development process.
8. Describe the advantages and disadvantages of the free-flight test.

Words and phrases

RAND (Research and Development Corporation) （美）兰德公司

R&D (Research and Development) 研究与开发

validation 确认

iterate 重复, 迭代

Space systems-Closed Loop Problem Solving Management 航天质量问题归零管理

wind tunnel test　风洞试验
venturi　文氏管,文丘里管（一种流体流量测定装置,亦作 venturitube）
Bernoulli's theorem　伯努利定律
brown-out　部分停电
Langley Research Center　兰利研究中心
evacuated　抽空
DOF（Degree of Freedom）　自由度
usher　引领,开辟
philosophy　理念
maintainability　可维护性
integrate　集成

DARPA（Defense Advanced Research Projects Agency）（美）国防部高级研究计划局
model free-flight test　模型自由飞试验
free flight model　自由飞模型
humid-hot test　湿热试验
fungus test　霉菌试验
salt-fog test　盐雾试验
sand dust test　沙尘试验
drip-proof test　防滴试验
mechanical environmental test　力学环境试验
electromagnetic environment test　电磁环境试验

References

［1］段卓毅.X 系列飞行器概览［M］.北京:航空工业出版社,2017.
［2］冯志高,关成启,张洪文.高超声速飞行器概论［M］.北京:北京理工大学出版社,2016.
［3］理查德·布洛克利,史维.航空航天科技出版工程 8:系统工程［M］.唐胜景,马东立,林海,等译.北京:北京理工大学出版社,2016.
［4］果琳丽,杨宏,田林,等.载人深空探测技术导论(上册、下册)［M］.北京:北京理工大学出版社,2019.
［5］昂海松,余雄庆.飞行器先进设计技术［M］.2 版.北京:国防工业出版社,2014.
［6］宋振铎.反坦克制导兵器论证与试验［M］.北京:国防工业出版社,2013.
［7］金振中,李晓斌,等.战术导弹试验设计［M］.北京:国防工业出版社,2013.
［8］张守言.模型自由飞试验［M］.北京:国防工业出版社,2002.
［9］梁相文.未来试飞新技术挑战［M］.北京:航空工业出版社,2014.